COMPUTATIONAL MATHEMATICS SERIES

INSIDE the FFT BLACK BOX

Serial and Parallel Fast Fourier Transform Algorithms

COMPUTATIONAL MATHEMATICS SERIES

INSIDE the FFT BLACK BOX

Serial and Parallel Fast Fourier Transform Algorithms

Eleanor Chu

University of Guelph
Ontario, Canada

Alan George

University of Waterloo
Ontario, Canada

CRC Press

Boca Raton London New York Washington, D.C.

Library of Congress Cataloging-in-Publication Data

Catalog record is available from the Library of Congress.

No claim to original U.S. Government works
International Standard Book Number 0-8493-0270-6
Library of Congress Card Number 99-048017
Printed in the United States of America 1 2 3 4 5 6 7 8 9 0
Printed on acid-free paper

Preface

The fast Fourier transform (FFT) algorithm, together with its many successful applications, represents one of the most important advancements in scientific and engineering computing in this century. The wide usage of computers has been instrumental in driving the study of the FFT, and a very large number of articles have been written about the algorithm over the past thirty years. Some of these articles describe modifications of the basic algorithm to make it more efficient or more applicable in various circumstances. Other work has focused on implementation issues, in particular, the development of parallel computers has spawned numerous articles about implementation of the FFT on multiprocessors. However, to many computing and engineering professionals, the large collection of serial and parallel algorithms remain hidden inside the FFT black box because: (1) coverage of the FFT in computing and engineering textbooks is usually brief, typically only a few pages are spent on the algorithmic aspects of the FFT; (2) cryptic and highly variable mathematical and algorithmic notation; (3) limited length of journal articles; and (4) important ideas and techniques in designing efficient algorithms are sometimes buried in software or hardware-implemented FFT programs, and not published in the open literature.

This book is intended to help rectify this situation. Our objective is to bring these numerous and varied ideas together in a common notational framework, and make the study of FFT an inviting and relatively painless task. In particular, the book employs a unified and systematic approach in developing the multitude of ideas and computing techniques employed by the FFT, and in so doing, it closes the gap between the often brief introduction in textbooks and the equally often intimidating treatments in the FFT literature. The unified notation and approach also facilitates the development of new parallel FFT algorithms in the book.

This book is self-contained at several levels. First, because the fast Fourier transform (FFT) is a fast "algorithm" for computing the discrete Fourier transform (DFT), an "algorithmic approach" is adopted throughout the book. To make the material fully accessible to readers who are not familiar with the design and analysis of computer algorithms, two appendices are given to provide necessary background. Second, with the help of examples and diagrams, the algorithms are explained in full. By exercising the appropriate notation in a consistent manner, the algorithms are explicitly connected to the mathematics underlying the FFT—this is often the "missing link" in the literature. The algorithms are presented in pseudo-code and a complexity analysis of each is provided.

Features of the book

• *The book is written to bridge the gap between textbooks and literature.* We believe this book is unique in this respect. The majority of textbooks largely focus on the underlying mathematical transform (DFT) and its applications, and only a small part is devoted to the FFT, which is a fast algorithm for computing the DFT.

• *The book teaches up-to-date computational techniques relevant to the FFT.* The book systematically and thoroughly reviews, explains, and unifies FFT ideas from journals across the disciplines of engineering, mathematics, and computer science from 1960 to 1999. In addition, the book contains several parallel FFT algorithms that are believed to be new.

• *Only background found in standard undergraduate mathematical science, computer science, or engineering curricula is required.* The notations used in the book are fully explained and demonstrated by examples. As a consequence, this book should make FFT literature accessible to senior undergraduates, graduate students, and computing professionals. The book should serve as a self-teaching guide for learning about the FFT. Also, many of the ideas discussed are of general importance in algorithm design and analysis, efficient numerical computation, and scientific programming for both serial or parallel computers.

Use of the book

It is expected that this book will be of interest and of use to senior undergraduate students, graduate students, computer scientists, numerical analysts, engineering professionals, specialists in parallel and distributed computing, and researchers working in computational mathematics in general.

The book also has potential as a supplementary text for undergraduate and graduate courses offered in mathematical science, computer science, and engineering programs. Specifically, it could be used for courses in scientific computation, numerical analysis, digital signal processing, the design and analysis of computer algorithms, parallel algorithms and architectures, parallel and distributed computing, and engineering courses treating the discrete Fourier transform and its applications.

Scope of the book

The book is organized into 24 chapters and 2 appendices. It contains 97 figures and 38 tables, as well as 25 algorithms presented in pseudo-code, along with numerous code segments. The bibliography contains more than 100 references dated from 1960 to 1999. The chapters are organized into three parts.

I. Preliminaries Part I presents a brief introduction to the discrete Fourier transform through a simple example involving trigonometric interpolation. This part is included to make the book self-contained. Some details about floating point arithmetic as it relates to FFT computation is also included in Part I.

II. Sequential FFT Algorithms This part contains fourteen relatively short chapters (3 through 16). Although the FFT, like binary search and quicksort, is commonly used in textbooks to illustrate the divide and conquer paradigm and recursive algorithms, the FFT has a unique feature: the application of the basic FFT algorithm to "naturally ordered" input, if performed "in place," yields output in "bit-reversed" order. While this feature may be taken for granted by FFT insiders, it is often not addressed in detail in textbooks. Again, partly because of the lack of notation linking the underlying mathematics to the algorithm, and because it is understood by FFT professionals, this aspect of the FFT is either left unexplained or explained very briefly in the literature. This phenomenon, its consequences, and how to deal with it, is one of the topics of Part II.

Similarly, the basic FFT algorithm is generally introduced as most efficient when applied to vectors whose length N is a power of two, although it can be made even more efficient if N is a power of four, and even more so if it is a power of eight, and so on. These situations, as well as the case when N is arbitrary, are considered in Part II. Other special situations, such as when the input is real rather than complex, and various programming "tricks," are also considered in Part II, which concludes with a chapter on selected applications of FFT algorithms.

III. Parallel FFT Algorithms The last part deals with the many and varied issues that arise in implementing FFT algorithms on multiprocessor computers. Part III begins with a chapter that discusses the mapping of data to processors, because the designs of the parallel FFTs are mainly driven by data distribution, rather than by the way the processors are physically connected (through shared memory or by way of a communication network.) This is a feature not shared by parallel numerical algorithms in general.

Distributed-memory multiprocessors are discussed next, because implementing the algorithms on shared-memory architecture is straightforward. The hypercube multiprocessor architecture is particularly considered because it is so naturally compatible with the FFT algorithm. However, the material discussed later does not specifically depend on the hypercube architecture.

Following that, a series of chapters contains a large collection of parallel algorithms, including some that are believed to be new. All of the algorithms are described using a common notation that has been derived from one introduced in the literature. As in part II, dealing with the bit-reversal phenomenon is considered, along with balancing the computational load and avoiding communication congestion. The last two chapters deal with two-dimensional FFTs and the task of distributing the "twiddle factors" among the individual processors.

Appendix A contains basic information about efficient computation, together with some fundamentals on complexity notions and notation. Appendix B contains techniques that are helpful in solving recurrence equations. Since FFT algorithms are recursive, analysis of their complexity leads naturally to such equations.

Acknowledgments

This book resulted from our teaching and research activities at the University of Guelph and the University of Waterloo. We are grateful to both Universities for providing the environment in which to pursue these activities, and to the Natural Sciences and Engineering Research Council of Canada for our research support. At a personal level, Eleanor Chu owes a special debt of gratitude to her husband, Robert Hiscott, for his understanding, encouragement, and unwavering support.

We thank the reviewers of our book proposal and draft manuscript for their helpful suggestions and insightful comments which led to many improvements.

Our sincere thanks also go to Robert Stern (Publisher) and his staff at CRC Press for their enthusiastic support of this project.

Eleanor Chu
Guelph, Ontario

Alan George
Waterloo, Ontario

Contents

List of Figures

List of Tables

List of Algorithms

Part I

Preliminaries

Chapter 1

An Elementary Introduction to the Discrete Fourier Transform

This chapter is intended to provide a brief introduction to the discrete Fourier transform (DFT). It is not intended to be comprehensive; instead, through a simple example, it provides an illustration of how the computation that is the subject of this book arises, and how its results can be used. The DFT arises in a multitude of other contexts as well, and a dozen more DFT-related applications, together with information on a number of excellent references, are presented in Chapter 16 in Part II of this book. Readers familiar with the DFT may safely skip this chapter.

A major application of Fourier transforms is the analysis of a series of observations: $x_\ell, \ell = 0, \ldots, N-1$. Typically, N will be quite large: 10000 would not be unusual. The sources of such observations are many: ocean tidal records over many years, communication signals over many microseconds, stock prices over a few months, sonar signals over a few minutes, and so on. The assumption is that there are repeating patterns in the data that form *part* of the x_ℓ. However, usually there will be other phenomena which may not repeat, or repeat in a way that is not discernably cyclic. This is called "noise." The DFT helps to identify and quantify the cyclic phenomena. If a pattern repeats itself m times in the N observations, it is said to have *Fourier frequency m*.

To make this more specific, suppose one measures a signal from time $t = 0$ to $t = 680$ in steps of 2.5 seconds, giving 273 observations. The measurements might appear as shown in Figure 1.1. How does one make any sense out of it? As shown later, the DFT can help.

1.1 Complex Numbers

Effective computation of the DFT relies heavily on the use of complex numbers, so it is useful to review their basic properties. This material is elementary and probably well-known to most readers of this book, but it is included for completeness. Historically, complex numbers were introduced to deal with polynomial equations, such as $x^2 + 1 =$

3

Figure 1.1 Example of a noisy signal.

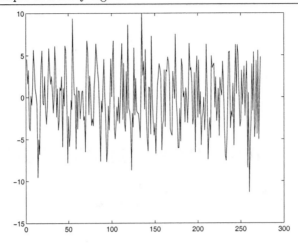

0, which have no real solutions. Informally, they can be defined as the set $\mathcal{C}$ of all "numbers" of the form $a + jb$ where a and b are real numbers and $j^2 = -1$.

Addition, subtraction, and multiplication are performed among complex numbers by treating them as binomials in the unknown j and using $j^2 = -1$ to simplify the result. Thus

$$(a + jb) + (c + jd) = (a + c) + j(b + d)$$

and

$$(a + jb) \times (c + jd) = (ac - bd) + j(ad + bc).$$

For the complex number $z = a + jb$, a is the *real part* of z and b is the *imaginary part* of z. The zero element of $\mathcal{C}$ is $0 + 0i$, and the additive inverse of $z = a + jb$ is $-a + i(-b)$. The multiplicative inverse z^{-1} is

$$z^{-1} = \frac{a - jb}{a^2 + b^2}.$$

The complex conjugate of $z = a + jb$ is denoted by $\bar{z}$ and is equal to $a - jb$. The *modulus* of z, denoted by $|z|$, is

$$\sqrt{z\bar{z}} = \sqrt{a^2 + b^2}.$$

Some additional facts that will be used later are

$$e^z = e^{(a+jb)} = e^a e^{jb} \quad \text{and} \quad e^{jb} = \cos b + j \sin b.$$

Thus, $Re(e^z) = e^a \cos b$ and $Im(z) = e^a \sin b$.

Just as a real number can be pictured as a point lying on a line, a complex number can be pictured as a point lying in a plane. With each complex number $a + jb$ one can associate a vector beginning at the origin and terminating at the point (a, b). These notions are depicted in Figure 1.2.

Figure 1.2 Visualizing complex numbers.

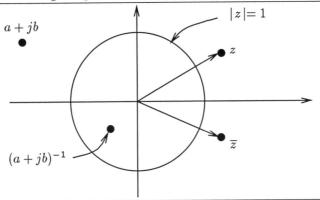

Instead of the pair (a, b), one can use the "length" (modulus) together with the angle the number makes with the real axis. Thus, $a + jb$ can be represented as $r \cos \theta + jr \sin \theta = re^{j\theta}$, where $r = |z| = \sqrt{a^2 + b^2}$ and $\theta = \arctan(b/a)$. This representation of a complex number is depicted in Figure 1.3.

Figure 1.3 Polar representation of a complex number.

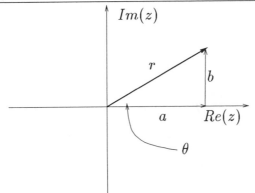

Multiplication of complex numbers in polar form is straightforward: if $z_1 = a + jb = r_1 e^{j\theta_1}$ and $z_2 = c + jd = r_2 e^{j\theta_2}$, then

$$z_1 z_2 = r_1 r_2 e^{j(\theta_1 + \theta_2)}.$$

The moduli are multiplied together, and the angles are added. Note that if $z = e^{j\theta}$, then $|z| = 1$ *for all values of* θ.

1.2 Trigonometric Interpolation

Suppose a function $f(\theta)$ is defined on the interval $(0, 2\pi)$, with f assumed to be periodic on the interval; thus, $f(\theta) = f(\theta \pm 2\pi)$.

Now consider constructing a trigonometric polynomial $p(\theta)$ to interpolate $f(\theta)$ of the form

(1.1)
$$p(\theta) = a_0 + \sum_{k=1}^{n} a_k \cos k\theta + b_k \sin k\theta.$$

This function has $2n + 1$ coefficients, so it should be possible to interpolate f at $2n + 1$ points. In the applications considered in this book, the points at which to interpolate are always *equally spaced* on the interval:

(1.2)
$$\theta_\ell = \frac{2\ell\pi}{2n + 1}, \quad \ell = 0, 1, \dots, 2n.$$

Let $x_\ell = f(\theta_\ell)$, and consider an example with $n = 2$. Then the interpolation conditions are $x_\ell = p(\theta_\ell)$, or

$$x_\ell = a_0 + a_1 \cos \theta_\ell + b_1 \sin \theta_\ell + a_2 \cos 2\theta_\ell + b_2 \sin 2\theta_\ell, \quad \ell = 0, 1, \dots, 4.$$

This leads to the system of equations

$$\begin{bmatrix} 1 & \cos \theta_0 & \sin \theta_0 & \cos 2\theta_0 & \sin 2\theta_0 \\ 1 & \cos \theta_1 & \sin \theta_1 & \cos 2\theta_1 & \sin 2\theta_1 \\ 1 & \cos \theta_2 & \sin \theta_2 & \cos 2\theta_2 & \sin 2\theta_2 \\ 1 & \cos \theta_3 & \sin \theta_3 & \cos 2\theta_3 & \sin 2\theta_3 \\ 1 & \cos \theta_4 & \sin \theta_4 & \cos 2\theta_4 & \sin 2\theta_4 \end{bmatrix} \begin{bmatrix} a_0 \\ a_1 \\ b_1 \\ a_2 \\ b_2 \end{bmatrix} = \begin{bmatrix} x_0 \\ x_1 \\ x_2 \\ x_3 \\ x_4 \end{bmatrix}.$$

Recall that $e^{j\theta} = \cos \theta + j \sin \theta$, which implies that

$$\cos \theta = \frac{e^{j\theta} + e^{-j\theta}}{2} \quad \text{and} \quad \sin \theta = \frac{e^{j\theta} - e^{-j\theta}}{2j}.$$

Using these in (1.1) with $n = 2$ yields

$$\begin{aligned} p(\theta) &= a_0 + \left(\frac{a_1}{2}\right) e^{j\theta} + \left(\frac{a_1}{2}\right) e^{-j\theta} + \left(\frac{b_1}{2j}\right) e^{j\theta} - \left(\frac{b_1}{2j}\right) e^{-j\theta} \\ &\quad + \left(\frac{a_2}{2}\right) e^{2j\theta} + \left(\frac{a_2}{2}\right) e^{-2j\theta} + \left(\frac{b_2}{2j}\right) e^{2j\theta} - \left(\frac{b_2}{2j}\right) e^{-2j\theta} \\ &= \left(\frac{a_2 + j\, b_2}{2}\right) e^{-2j\theta} + \left(\frac{a_1 + j\, b_1}{2}\right) e^{-j\theta} \\ &\quad + a_0 + \left(\frac{a_1 - j\, b_1}{2}\right) e^{j\theta} + \left(\frac{a_2 - j\, b_2}{2}\right) e^{2j\theta}. \end{aligned}$$

Giving the coefficients names corresponding to the powers of $e^{j\theta}$ yields

(1.3)
$$p(\theta) = X_{-2} e^{-2j\theta} + X_{-1} e^{-j\theta} + X_0 + X_1 e^{j\theta} + X_2 e^{2j\theta}.$$

Note that the coefficients appear in complex conjugate pairs. When the x_ℓ are real, it is straightforward to show that this is true in general. (See the next section.)

Recall (see (1.2)) that the points at which interpolation occurs are *evenly spaced*; that is, $\theta_\ell = \ell\theta_1$. Let $\omega = e^{j\theta_1} = e^{\frac{2j\pi}{2n+1}}$. Then all $e^{j\theta_\ell}$ can be expressed in terms of ω:

$$e^{j\theta_\ell} = e^{j\ell\theta_1} = \omega^\ell, \ell = 0, 1, \dots, 2n.$$

Also, note that $\omega^\ell = \omega^{\ell \pm (2n+1)}$ and $\omega^{-\ell} = \omega^{-\ell \pm (2n+1)}$. For the example with $n = 2$, $\omega = e^{\frac{2j\pi}{5}}$, and the interpolation condition at θ_ℓ in (1.3) is

$$f(\theta_\ell) = x_\ell = p(\theta_\ell) = X_{-2}\omega^{-2\ell} + X_{-1}\omega^{-\ell} + X_0\omega^0 + X_1\omega^\ell + X_2\omega^{2\ell}.$$

Using the fact that $\omega^{-\ell} = \omega^{(2n+1-\ell)}$, and renaming the coefficients similarly ($X_{-\ell} \rightarrow X_{2n+1-\ell}$), the interpolation condition at x_ℓ becomes

$$x_\ell = X_0 + X_1\omega^\ell + X_2\omega^{2\ell} + X_3\omega^{3\ell} + X_4\omega^{4\ell},$$

which has to be satisfied for $\ell = 0, 1, \ldots, 4$:

(1.4)
$$\begin{bmatrix} 1 & 1 & 1 & 1 & 1 \\ 1 & \omega & \omega^2 & \omega^3 & \omega^4 \\ 1 & \omega^2 & \omega^4 & \omega^6 & \omega^8 \\ 1 & \omega^3 & \omega^6 & \omega^9 & \omega^{12} \\ 1 & \omega^4 & \omega^8 & \omega^{12} & \omega^{16} \end{bmatrix} \begin{bmatrix} X_0 \\ X_1 \\ X_2 \\ X_3 \\ X_4 \end{bmatrix} = \begin{bmatrix} x_0 \\ x_1 \\ x_2 \\ x_3 \\ x_4 \end{bmatrix}.$$

This can be written as a matrix equation

$$MX = x.$$

It will be useful to have some additional properties of ω. First note that

$$1 + \omega + \omega^2 + \ldots + \omega^{2n} = 0.$$

This can be established by observing that the expression on the left side is a geometric sum equal to

$$\frac{1 - \omega^{2n+1}}{1 - \omega},$$

and this quantity is zero because $\omega^{2n+1} = 1$. For integers r and s one can show in a similar way that

(1.5)
$$\sum_{k=0}^{2n} \omega^{(kr-ks)} = \begin{cases} 0 & \text{if } r \neq s \\ 2n+1 & \text{if } r = s. \end{cases}$$

These simple results make solving $MX = x$ easy. To begin, let

$$\overline{M} = \begin{bmatrix} 1 & 1 & 1 & 1 & 1 \\ 1 & \bar{\omega} & \bar{\omega}^2 & \bar{\omega}^3 & \bar{\omega}^4 \\ 1 & \bar{\omega}^2 & \bar{\omega}^4 & \bar{\omega}^6 & \bar{\omega}^8 \\ 1 & \bar{\omega}^3 & \bar{\omega}^6 & \bar{\omega}^9 & \bar{\omega}^{12} \\ 1 & \bar{\omega}^4 & \bar{\omega}^8 & \bar{\omega}^{12} & \bar{\omega}^{16} \end{bmatrix}.$$

Then using (1.5) above, together with the fact that $\bar{\omega}^\ell = \omega^{-\ell}$, shows that $\overline{M}M$ is

$$\begin{bmatrix} 5 & 0 & 0 & 0 & 0 \\ 0 & 5 & 0 & 0 & 0 \\ 0 & 0 & 5 & 0 & 0 \\ 0 & 0 & 0 & 5 & 0 \\ 0 & 0 & 0 & 0 & 5 \end{bmatrix}.$$

Thus, $M^{-1} = \frac{1}{5}\overline{M}$, and

(1.6)
$$
\begin{bmatrix} X_0 \\ X_1 \\ X_2 \\ X_3 \\ X_4 \end{bmatrix}
= \frac{1}{5}
\begin{bmatrix}
1 & 1 & 1 & 1 & 1 \\
1 & \omega^{-1} & \omega^{-2} & \omega^{-3} & \omega^{-4} \\
1 & \omega^{-2} & \omega^{-4} & \omega^{-6} & \omega^{-8} \\
1 & \omega^{-3} & \omega^{-6} & \omega^{-9} & \omega^{-12} \\
1 & \omega^{-4} & \omega^{-8} & \omega^{-12} & \omega^{-16}
\end{bmatrix}
\begin{bmatrix} x_0 \\ x_1 \\ x_2 \\ x_3 \\ x_4 \end{bmatrix}.
$$

It is a simple exercise to carry out this development for general n, yielding the following formula for the DFT:

(1.7)
$$
X_r = \frac{1}{2n+1}\sum_{\ell=0}^{2n} x_\ell \omega^{-r\ell}, \quad r = 0, 1, \ldots, 2n.
$$

Similarly, the *inverse* DFT (IDFT) has the form

(1.8)
$$
x_\ell = \sum_{r=0}^{2n} X_r \omega^{r\ell}, \quad \ell = 0, 1, \ldots, 2n.
$$

1.3   Analyzing the Series

What information can the X_r provide? As noted earlier for the example with $n = 2$, when the given data x are real, the X_r appear in complex conjugate pairs. To establish this, note that

(1.9)
$$
X_{2n+1-r} = \frac{1}{2n+1}\sum_{\ell=0}^{2n} x_\ell \omega^{-(2n+1-r)\ell} = \frac{1}{2n+1}\sum_{\ell=0}^{2n} x_\ell \omega^{r\ell} = \frac{1}{2n+1}\sum_{\ell=0}^{2n} \bar{x}_\ell \bar{\omega}^{-r\ell} = \bar{X}_r.
$$

Writing (1.8) as

$$
x_\ell = X_0 + \sum_{r=1}^{n}\left(X_r \omega^{r\ell} + X_{2n+1-r}\omega^{\ell(2n+1-r)} \right)
$$

and using the fact that $\omega^{\ell(2n+1-r)} = \omega^{-r\ell} = \bar{\omega}^{r\ell}$ and $X_{2n+1-r} = \bar{X}_r$, x_ℓ can be expressed as

(1.10)
$$
x_\ell = X_0 + \sum_{r=1}^{n}\left(X_r \omega^{r\ell} + \bar{X}_r \bar{\omega}^{r\ell} \right).
$$

Recall that if a and b are complex numbers, then

$$
a + \bar{a} = 2Re(a), \quad ab = |a||b|e^{j(\theta_a + \theta_b)} \quad \text{and} \quad Re(ab) = |a||b|\cos(\theta_a + \theta_b).
$$

Using these, (1.10) can be written as

$$
x_\ell = X_0 + 2\sum_{r=1}^{n} Re(X_r \omega^{r\ell}) \quad = \quad X_0 + 2\sum_{r=1}^{n} |X_r|\cos\left(\frac{2r\ell\pi}{2n+1} + \phi_r \right),
$$

where $\frac{2r\ell\pi}{2n+1}$ is the phase angle of $\omega^{r\ell}$ and ϕ_r is the phase angle of X_r. Thus, after computing the coefficients $X_r, r = 0, 1, \ldots, 2n$, the interpolating function can be evaluated at any point in the interval $[0, 2\pi]$ using the formula

$$p(\theta) = X_0 + 2 \sum_{r=1}^{n} |X_r| \cos(r\theta + \phi_r).$$

In many applications, it is the *amplitudes* (the size of $2\,|X_r|$) that are of interest. They indicate the strength of each frequency in the signal.

To make this discussion concrete, consider the signal shown in Figure 1.1, where the 273 measurements are plotted. Using Matlab,[1] one can compute and plot $|X|$, as shown on the left in Figure 1.4. Note that apart from the left endpoint (corresponding

---

Figure 1.4 Plot of $|X|$ for the example in Figure 1.1.

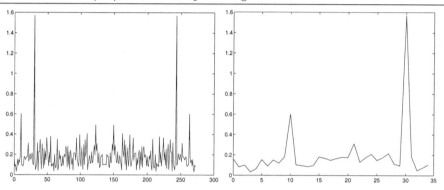

---

to $|X_0|$), the plot of the entire $|X|$ is symmetric, as expected; as shown above, the X_r occur in complex conjugate pairs, with $X_r = \bar{X}_{2n+1-r}$. The plot on the right in Figure 1.4 contains the first 30 components of $|X|$ so that more detail can be seen. It suggests that the signal has two dominant Fourier frequencies: 10 and 30.

1.4 Fourier Frequency Versus Time Frequency

As θ goes from 0 to 2π, $\cos(r\theta + \phi_r)$ makes r cycles. Suppose the x_ℓ are collected over an interval of T seconds. As θ goes from 0 to 2π, t goes from 0 to T. Thus, $\cos(r\theta + \phi_r)$ oscillates at r/T cycles per second. Making a change of variable yields

(1.11)
$$p(t) = X_0 + 2 \sum_{r=1}^{n} |X_r| \cos\left(2\pi \left(\frac{r}{T}\right) t + \phi_r\right).$$

The usual practice is to express the phase shift ϕ_r in terms of a negative shift in time t. Thus, (1.11) is often written in the form

(1.12)
$$p(t) = X_0 + 2 \sum_{r=1}^{n} |X_r| \cos\left(2\pi \left(\frac{r}{T}\right)(t - t_r)\right)$$

---

[1]Matlab is a proprietary commercial product of The MathWorks, Inc., Natick, MA. Web URL: http://www.mathworks.com

where

$$t_r = \frac{-\phi_r}{2\pi \left(\frac{r}{T}\right)}.$$

Returning to the signal shown in Figure 1.1, recall that the 273 data elements were collected at intervals of 2.5 seconds over a period of $T = 680$ seconds. Thus, since the dominant Fourier frequencies in the signal appear to be 10 and 30, the dominant frequencies in cycles per second would be 0.014706 and 0.04418 cycles per second. Figure 1.5 contains a plot of the first 40 amplitudes ($2 \mid X_r \mid$) against cycles per second.

Figure 1.5 Plot of amplitudes against cycles per second for the example in Figure 1.1.

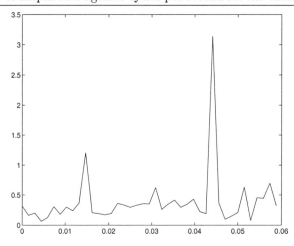

1.5 Filtering a Signal

Suppose $\mid X_d \mid$ is much larger than the other coefficients. If one assumes the other frequencies are the result of noise, one can "clean up" the signal by setting all but X_d to zero. Thus, (1.11) might be replaced by the *filtered signal*

$$p_{clean}(t) = X_0 + 2 \mid X_d \mid \cos\left(2\pi \left(\frac{d}{T}\right) t + \phi_d\right).$$

Of course there may be several apparently dominant frequencies, in which case more than one of the elements of X would be retained. As an illustration, again consider the example of Figure 1.1. The dominant signals appear to be Fourier frequency 10 and 30. Discarding all elements of X for which $\mid X_r \mid < 0.6$ yields a "cleaned up" signal. Evaluating (1.12) from $t = 0$ to $t = 250$ yields the signal shown on the right in Figure 1.6. The plot on the left is the corresponding part of the original signal shown in Figure 1.1.

There is vast literature on digital filtering, and the strategy described here is intended only to illustrate the basic idea. For a comprehensive introduction to the topic, see Terrell [103].

Figure 1.6 Plot of part of the original and clean signals for the example in Figure 1.1.

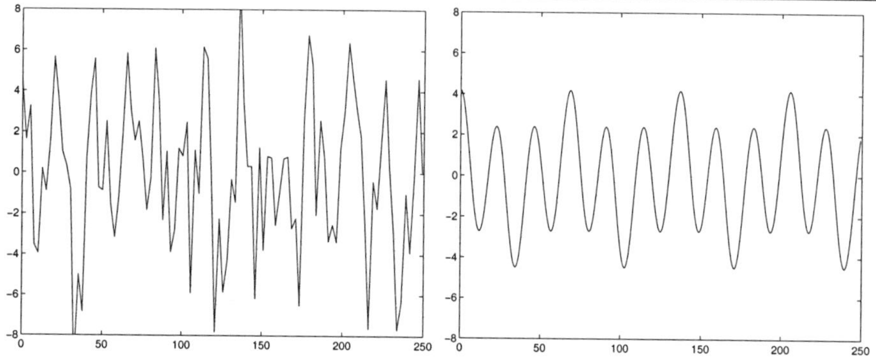

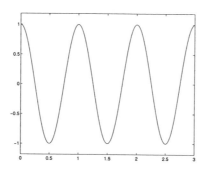

1.6 How Often Does One Sample?

In performing the analysis of a time series, one has the values of a certain (unknown) function $f(t)$ at equally spaced intervals of time. Let δT be the time interval between successive observations in seconds. Then $1/\delta T$ is called the *sampling rate*. This is the number of observations taken each second. If the sampling rate is $1/\delta T$, what frequencies can the Fourier transform reliably detect?

An intuitive argument is as follows. Consider a pure cosine signal with frequency 1, sampled over $T = 3$ seconds as shown in Figure 1.7. In the representation

Figure 1.7 A pure cosine signal.

$$p(t) = X_0 + 2 \sum_{r=1}^{n} |X_r| \cos \left(2\pi \left(\frac{r}{T} \right) t + \phi_r \right),$$

it is evident that in order to be able to represent this signal, the $\cos(2\pi \left(\frac{3}{T} \right) t + \phi_3)$ term must be present. This implies that one needs $n \geq 3$ or $2n + 1 \geq 7$. That is, more than 2 samples per second, or at least 7 sample points.

Another way to look at it is as follows. One needs to sample often enough to detect *all* the oscillations if the true signal is to be detected. In order to detect oscillations up to s cycles per second, one must sample at a rate at least $2s$ times per second. In most practical situations, there is a range of frequencies of interest, and the sampling rate is chosen accordingly.

1.7   Notes and References

The trigonometric polynomial

$$p(\theta) = a_0 + \sum_{k=1}^{n} a_k \cos k\theta + b_k \sin k\theta$$

was used in Section 1.2 to interpolate a periodic function $f(\theta)$. This polynomial has $2n + 1$ coefficients, and by interpolating $f(\theta)$ at the $2n + 1$ primitive roots of unity, a DFT of length $N = 2n + 1$ results.

DFTs of length $N = 2^s$ are the most convenient and most efficient to compute. Such a DFT can be obtained by using a a trigonometric polynomial having a slightly different form than the one above, namely

$$(1.13) \qquad p(\theta) = \frac{a_0}{2} + \frac{a_{n+1}}{2} \cos (n+1)\theta + \sum_{k=1}^{n} a_k \cos k\theta + b_k \sin k\theta.$$

This polynomial has $2n + 2$ coefficients; thus, n can be chosen so that $N = 2n + 2$ is a power of two. The derivation of the DFT using (1.13) is similar to the derivation done for the case where N is odd, and is left as an exercise.

Chapter 2

Some Mathematical and Computational Preliminaries

The development in Chapter 1 showed that the computation of the DFT involves the multiplication of a matrix $\overline{M}$ by a vector x, where the matrix has very special structure. In particular, it is symmetric, and each of its elements is a power of a single number ω, where ω depends on the order $N = 2n+1$ of the matrix[1]. These numbers are called the *twiddle factors*. Moreover, since $\omega^{-\ell} = \omega^{-\ell \pm N}$, only N of the N^2 entries in the matrix are actually different. Finally, since ω depends on N, in many contexts it will be necessary to distinguish ω's corresponding to different values of N; to do this, the notation ω_N^ℓ will be used.

For example, the matrix $\overline{M}$ in (1.6) satisfies

$$
\begin{bmatrix}
1 & 1 & 1 & 1 & 1 \\
1 & \omega^{-1} & \omega^{-2} & \omega^{-3} & \omega^{-4} \\
1 & \omega^{-2} & \omega^{-4} & \omega^{-6} & \omega^{-8} \\
1 & \omega^{-3} & \omega^{-6} & \omega^{-9} & \omega^{-12} \\
1 & \omega^{-4} & \omega^{-8} & \omega^{-12} & \omega^{-16}
\end{bmatrix}
=
\begin{bmatrix}
1 & 1 & 1 & 1 & 1 \\
1 & \omega^{-1} & \omega^{-2} & \omega^{-3} & \omega^{-4} \\
1 & \omega^{-2} & \omega^{-4} & \omega^{-1} & \omega^{-3} \\
1 & \omega^{-3} & \omega^{-1} & \omega^{-4} & \omega^{-2} \\
1 & \omega^{-4} & \omega^{-3} & \omega^{-2} & \omega^{-1}
\end{bmatrix}.
$$

Given these features, it is not surprising that when N is a power of two, the structure of the matrix can be exploited to reduce the cost[2] of computing X from $\Theta\left(N^2\right)$ resulting from a straightforward matrix times vector computation to $\Theta\left(N \log_2 N\right)$. Indeed, exploring the numerous variants of the fast Fourier transform (FFT) algorithm which exploit this structure is a main topic of this book. However, the price of this reduction is the use of complex arithmetic. This chapter deals with various aspects of complex arithmetic, together with the efficient computation of the twiddle factors. Usually the twiddle factors are computed in advance of the rest of the computation, although in some contexts they may be computed "on the fly." This issue is explored more fully in Chapter 4 for implementing sequential FFTs and in Chapter 24 for implementing parallel FFTs.

---

[1] In the remainder of this book, the order of the matrices M and $\overline{M}$ will be denoted by N. Moreover, as will be apparent in the development in the next chapter, N will often be a power of 2.

[2] The Θ-notation defined in Appendix A is used to denote the arithmetic cost of each algorithm.

2.1   Computing the Twiddle Factors

Let N be a power of 2, and recall that

$$\omega_N^r = e^{jr\theta} = \cos(r\theta) + j\sin(r\theta), \quad \theta = \frac{2\pi}{N}, \quad j = \sqrt{-1}\,.$$

Since the ω_N^r's are the complex roots of unity, they are symmetric and equally spaced on the unit circle. Thus, if $a + jb$ is a twiddle factor, then so are $\pm a \pm jb$ and $\pm b \pm ja$. By exploiting this property, one needs only to compute the first $N/8 - 1$ values, namely ω_N^r for $r = 1, \ldots, N/8 - 1$. (Note that $\omega_N^r = 1$ for $r = 0$.)

The most straightforward approach is to use the standard trigonometric function library procedures for each of the $N/8 - 1$ values of $\cos(r\theta)$ and $\sin(r\theta)$. The cost will be $N/4 - 2$ trigonometric function calls, and each call will require several floating-point arithmetic operations.

To avoid the relatively expensive trigonometric function calls, one can use the following algorithm proposed by Singleton [83]. It makes use of the trigonometric identities.

$$\cos(a + b) = \cos a \cos b - \sin a \sin b$$

$$\cos a = \cos\left(2 \times \frac{a}{2}\right) = 1 - 2\sin^2\left(\frac{a}{2}\right)$$

$$\sin(a + b) = \sin a \cos b + \sin b \cos a$$

Letting $\theta = 2\pi/N$ as above, $\cos((r + 1)\theta)$ can be computed in terms of $\cos(r\theta)$ and $\sin(r\theta)$ according to the formula derived below.

$$\begin{aligned}
\cos((r + 1)\theta) &= \cos(r\theta + \theta) \\
&= \cos(r\theta)\cos(\theta) - \sin(r\theta)\sin(\theta) \\
&= \cos(r\theta)\left(1 - 2\sin^2(\theta/2)\right) - \sin(r\theta)\sin(\theta) \\
&= C \times \cos(r\theta) - S \times \sin(r\theta)\,,
\end{aligned}$$

where the constants C and S are

$$C = 1 - 2\sin^2(\theta/2) \quad \text{and} \quad S = \sin(\theta).$$

When $r = 0$, the initial values are $\cos(0) = 1$ and $\sin(0) = 0$. Using these same constants C and S, $\sin((r + 1)\theta)$ can be computed in terms of $\cos(r\theta)$ and $\sin(r\theta)$ in a similar way:

$$\begin{aligned}
\sin((r + 1)\theta) &= \sin(r\theta + \theta) \\
&= \sin(r\theta)\cos(\theta) + \cos(r\theta)\sin(\theta) \\
&= \sin(r\theta)\left(1 - 2\sin^2(\theta/2)\right) + \cos(r\theta)\sin(\theta) \\
&= S \times \cos(r\theta) + C \times \sin(r\theta)\,.
\end{aligned}$$

A pseudo-code program for computing the $N/2$ twiddle factors is given as Algorithm 2.1 below. Note that the array *wcos* stores the real part of the twiddle factors and the array *wsin* stores the imaginary part of the twiddle factors.

Algorithm 2.1 Singleton's method for computing the $N/2$ twiddle factors.

---

begin

   $\theta := 2\pi/N$

   $S := \sin(\theta); \ C := 1 - 2 * \sin^2(\theta/2)$ Call library function sin to compute S and C

   $wcos[0] := 1; \ wsin[0] := 0$ Initialize $w_N^0 = \cos(0) + j\sin(0)$

   for $K := 0$ **to** $N/8 - 2$ **do**   Compute the first $N/8$ factors

      $wcos[K + 1] := C * wcos[K] - S * wsin[K]$

      $wsin[K + 1] := S * wcos[K] + C * wsin[K]$

   end for

   $L := N/8$ Store the next $N/8$ factors

   $wcos[L] := \sqrt{2}/2; \ wsin[L] := \sqrt{2}/2$ $w_N^{N/8} = \cos(\pi/4) + j\sin(\pi/4)$

   for $K := 1$ **to** $N/8 - 1$ **do**

      $wcos[L + K] := wsin[L - K]$

      $wsin[L + K] := wcos[L - K]$

   end for

   $L := N/4$ Store the next $N/4$ factors

   $wcos[L] := 0; \ wsin[L] := 1$ $w_N^{N/4} = \cos(\pi/2) + j\sin(\pi/2)$

   for $K := 1$ **to** $N/4 - 1$ **do**

      $wcos[L + K] := -wcos[L - K]$

      $wsin[L + K] := wsin[L - K]$

   end for

end

2.2   Multiplying Two Complex Numbers

Recall from Chapter 1 that given complex numbers $z_1 = a + jb$ and $z_2 = c + jd$,

(2.1) $z_1 + z_2 = (a + jb) + (c + jd) = (a + c) + j(b + d) \,,$

(2.2) $z_1 \times z_2 = (a + jb) \times (c + jd) = (a \times c - b \times d) + j(a \times d + b \times c) \,.$

Note that the relation $j^2 = -1$ has been used in obtaining (2.2).

2.2.1   Real floating-point operation (FLOP) count

Since real floating-point binary operations are usually the only ones implemented by the computer hardware, the operation count of a computer algorithm is almost universally expressed in terms of "flops"; that is, real floating-point operations. According to rule (2.1), adding two complex numbers requires two real additions; according to rule (2.2), multiplying two complex numbers requires four real multiplications and two real additions/subtractions.

   It is well-known that multiplying two complex numbers can also be done using three multiplications and five additions. Letting $\lambda = z_1 \times z_2 = (a + jb) \times (c + jd)$, the real part of λ, denoted by $\text{Re}(\lambda)$, and the imaginary part of λ, denoted by $\text{Im}(\lambda)$, can be obtained as shown in (2.2) below.

$$m_1 = (a + b) \times c$$
$$m_2 = (d + c) \times b$$
(2.3)
$$m_3 = (d - c) \times a$$
$$\text{Re}(\lambda) = m_1 - m_2$$
$$\text{Im}(\lambda) = m_1 + m_3$$

Compared to directly evaluating the right-hand side of (2.2), the method described in (2.3) saves one real multiplication at the expense of three additional real additions/subtractions. Consequently, this method is more economical only when a multiplication takes significantly longer than an addition/subtraction, which may have been true for some ancient computers, but is almost certainly not true today. Indeed, such operations usually take equal time. Note also that the total flop count is increased from *six* in (2.2) to *eight* in (2.3).

2.2.2   Special considerations in computing the FFT

Since a pre-computed twiddle factor is always one of the two operands involved in each complex multiplication in the FFT, any intermediate results involving only the real and imaginary parts of a twiddle factor can be pre-computed and stored for later use. For example, if $\omega_N^\ell = c + jd$, one may pre-compute and store $\delta = d + c$ and $\gamma = d - c$, which are the intermediate results used by (2.3). With δ, γ, and the real part c stored, each complex multiplication in the FFT involving $x_\ell = a + jb$ and $\omega_N^\ell = c + jd$ can be computed using three real multiplications and three real additions/subtractions following (2.4) below.

$$m_1 = (a + b) \times c$$
$$m_2 = \delta \times b$$
(2.4)
$$m_3 = \gamma \times a$$
$$\text{Re}(\lambda) = m_1 - m_2$$
$$\text{Im}(\lambda) = m_1 + m_3$$

Compared to using formula (2.2), the total flop count in (2.4) remains six, but one multiplication has been exchanged for one addition/subtraction. This will result in lower execution time only if an addition/subtraction costs less than one multiplication. As noted earlier, this is usually not the case on modern computers. The disadvantage is that 50% more space is needed to store the pre-computed intermediate results involving the twiddle factors; c, γ and δ must be stored, rather than c and d.

The paragraph above, together with (2.4), explains the common practice by researchers to count a total of six flops for each complex multiplication in evaluating the various FFT algorithms. Following this practice, all complexity results provided in this book are obtained using six flops as the cost of a complex multiplication.

2.3 Expressing Complex Multiply-Adds in Terms of Real Multiply-Adds

As noted earlier, most high-performance workstations can perform multiplications as fast as additions. Moreover, many of them can do a multiplication and an addition *simultaneously*. The latter is accomplished on some machines by a single multiply-add instruction. Naturally, one would like to exploit this capability [48]. To make use of such a multiply-add instruction, the computation of complex $z = z_1 + w \times z_2$ may be formulated as shown below, where $z_1 = a + jb$, $w = c + js$, and $z_2 = d + je$.

$$\delta = s/c$$
$$m_1 = d - \delta \times e$$
(2.5)
$$m_2 = e + \delta \times d$$
$$\mathrm{Re}(z) = a + c \times m_1$$
$$\mathrm{Im}(z) = b + c \times m_2.$$

Thus, in total, one division and four multiply-adds are required to compute z. Formula (2.5) is derived below.

$$
\begin{aligned}
z &= z_1 + w \times z_2 \\
&= (a + jb) + (c + js) \times (d + je) \\
&= (a + jb) + (c \times d - s \times e) + j(s \times d + c \times e) \\
&= a + (c \times d - s \times e) + j\left(b + (c \times e + s \times d)\right) \\
&= a + c \times \left(d - \left(\frac{s}{c}\right) \times e\right) + j\left(b + c \times \left(e + \left(\frac{s}{c}\right) \times d\right)\right) \\
&= a + c \times (d - \delta \times e) + j(b + c \times (e + \delta \times d)) \\
&= a + c \times m_1 + j(b + c \times m_2).
\end{aligned}
$$

As will be apparent in the following chapters, the FFT computation is dominated by complex multiply-adds. In [48], the idea of pairing up multiplications and additions is exploited fully in the implementation of the radix 2, 3, 4, and 5 FFT kernels. However, the success of this strategy depends on whether a compiler can generate efficient machine code for the new FFT kernels as well as on other idiosyncrasies of different machines. The actual execution time may not be improved. See [48] for details on timing and accuracy issues associated with this strategy.

2.4 Solving Recurrences to Determine an Unknown Function

In various parts of subsequent chapters, it will be necessary to determine the computational cost of various algorithms. These algorithms are recursive; the solution of a problem of size N is the result of (recursively) solving two problems of size $N/2$, together with combining their solutions to obtain the solution to the original problem of size N. Thus, the task of determining the arithmetic cost of such algorithms is to determine a function $T(N)$, where

$$(2.6) \qquad T(N) = \begin{cases} 2T\left(\frac{N}{2}\right) + \beta N & \text{if } N = 2^k > 1, \\ \gamma & \text{if } N = 1. \end{cases}$$

Here βN is the cost of combining the solutions to the two half-size problems. The solution to this problem,

$$(2.7) \qquad T(N) = \beta N \log_2 N + \Theta(N),$$

is derived in Appendix B; the Θ-notation is defined in Appendix A. A slightly more complicated recurrence arises in the analysis of some generalized FFT algorithms that are considered in subsequent chapters. For example, some algorithms provide a solution to the original problem of size N by (recursively) solving α problems of size N/α and then combining their solutions to obtain the solution to the original problem of size N. The appropriate recurrence is shown below, where now βN is the cost of combining the solutions to the α problems of size N/α.

$$(2.8) \qquad T(N) = \begin{cases} \alpha T\left(\frac{N}{\alpha}\right) + \beta N & \text{if } N = \alpha^k > 1, \\ \gamma & \text{if } N = 1. \end{cases}$$

The solution to this problem is

$$(2.9) \qquad T(N) = \beta N \log_\alpha N + \Theta(N).$$

These results, their derivation, together with a number of generalizations, can be found in Appendix B. Some basic information about efficient computation together with some fundamentals on complexity notions and notation, such as the "big-Oh" notation, the "big-Omega" notation, and the Θ-notation, are contained in Appendix A.

Part II

Sequential FFT Algorithms

Chapter 3

The Divide-and-Conquer Paradigm and Two Basic FFT Algorithms

As noted earlier, the computation of the DFT involves the multiplication of a matrix $\overline{M}$ by a vector $\boldsymbol{x}$, where the matrix has very special structure. FFT algorithms exploit that structure by employing a a *divide-and-conquer paradigm*. Developments over the past 30 years have led to a host of variations of the basic algorithm; these are the topics of subsequent chapters. In addition, the development of multiprocessor computers has spurred the development of FFT algorithms specifically tailored to run well on such machines. These too are considered in subsequent chapters of this book.

The purpose of this chapter is to introduce the main ideas of FFT algorithms. This will serve as a basis and motivation for the material presented in subsequent chapters, where numerous issues related to its efficient implementation are considered.

The three major steps of the divide-and-conquer paradigm are

Step 1. Divide the problem into two or more subproblems of smaller size.

Step 2. Solve each subproblem *recursively* by the same algorithm. Apply the boundary condition to terminate the recursion when the sizes of the subproblems are small enough.

Step 3. Obtain the solution for the original problem by combining the solutions to the subproblems.

The radix-2 FFT is a recursive algorithm obtained from dividing the given problem (and each subproblem) into two subproblems of half the size. Within this framework, there are two commonly-used FFT variants which differ in the way the two half-size subproblems are defined. They are referred to as the DIT (*decimation in time*) FFT and the DIF (*decimation in frequency*) FFT, and are derived below.

It is intuitively apparent that a divide-and-conquer strategy will work best when N is a power of two, since subdivision of the problems into successively smaller ones can proceed until their size is one. Of course, there are many circumstances when it is not possible to arrange that N is a power of two, and so the algorithms must be modified accordingly. Such modifications are dealt with in detail in later chapters. However, in this chapter it is assumed that $N = 2^n$. Also, since the algorithm involves solving problems of different sizes, it is necessary to distinguish among their respective ω's; ω_q will refer to the ω corresponding to a problem of size q.

In addition, to simplify the presentation in the remainder of the book, and to avoid notational clutter, two adjustments have been made in the notation in the sequel. First, the factor $\frac{1}{N}$ has been omitted from consideration in the computations. This obviously does not materially change the computation. Second, ω has been implicitly redefined as $\bar{\omega}$, so the negative superscript in (1.7) can be omitted. Again, this does not change things in any material way, but does make the presentation somewhat cleaner. Thus, the equation actually studied in the remainder of this book is

$$(3.1) \qquad X_r = \sum_{\ell=0}^{N-1} x_\ell \omega^{r\ell}, r = 0, 1, \dots, N-1.$$

The following identities involving the twiddle factors are used repeatedly in what follows.

$$(3.2) \qquad (\omega_N)^{\frac{N}{2}} = -1, \quad \omega_{\frac{N}{2}} = \omega_N^2, \quad \omega_N^N = 1.$$

3.1   Radix-2 Decimation-In-Time (DIT) FFT

The radix-2 DIT FFT is derived by first rewriting equation (3.1) as

$$(3.3) \qquad X_r = \sum_{k=0}^{\frac{N}{2}-1} x_{2k} \omega_N^{r(2k)} + \omega_N^r \sum_{k=0}^{\frac{N}{2}-1} x_{2k+1} \omega_N^{r(2k)}, \quad r = 0, 1, \dots, N-1.$$

Using the identity $\omega_{\frac{N}{2}} = \omega_N^2$ from (3.2), (3.3) can be written as

$$(3.4) \qquad X_r = \sum_{k=0}^{\frac{N}{2}-1} x_{2k} \omega_{\frac{N}{2}}^{rk} + \omega_N^r \sum_{k=0}^{\frac{N}{2}-1} x_{2k+1} \omega_{\frac{N}{2}}^{rk}, \quad r = 0, 1, \dots, N-1.$$

Since $\omega_{\frac{N}{2}}^{rk} = \omega_{\frac{N}{2}}^{(r+\frac{N}{2})k}$, it is necessary to compute only the sums for $r = 0, 1, \dots, \frac{N}{2} - 1$. Thus, each summation in (3.4) can be interpreted as a DFT of size $N/2$, the first involving the even-indexed set $\{x_{2k} | k = 0, \dots, N/2 - 1\}$, and the second involving the odd-indexed set $\{x_{2k+1} | k = 0, \dots, N/2 - 1\}$. (Hence the use of the term *decimation in time*.) Defining $y_k = x_{2k}$ and $z_k = x_{2k+1}$ in (3.4) yields the two subproblems below, each having a form identical to (3.1) with N replaced by $N/2$:

$$(3.5) \qquad Y_r = \sum_{k=0}^{\frac{N}{2}-1} y_k \omega_{\frac{N}{2}}^{rk}, \quad r = 0, 1, \dots, N/2 - 1,$$

and

$$(3.6) \qquad Z_r = \sum_{k=0}^{\frac{N}{2}-1} z_k \omega_{\frac{N}{2}}^{rk}, \quad r = 0, 1, \dots, N/2 - 1.$$

After these two subproblems are each (recursively) solved, the solution to the original problem of size N is obtained using (3.4). The first $N/2$ terms are given by

$$(3.7) \qquad X_r = Y_r + \omega_N^r Z_r, \quad r = 0, 1, \dots, N/2 - 1,$$

and using the fact that $\omega_N^{\frac{N}{2}+r} = -\omega_N^r$ and $\omega_N^{\frac{N}{2}} = 1$, the remaining terms are given by

$$
\begin{aligned}
X_{r+\frac{N}{2}} &= \sum_{k=0}^{\frac{N}{2}-1} y_k \omega_{\frac{N}{2}}^{(r+\frac{N}{2})k} + \omega_N^{r+\frac{N}{2}} \sum_{k=0}^{\frac{N}{2}-1} z_k \omega_{\frac{N}{2}}^{(r+\frac{N}{2})k} \\
(3.8) \qquad &= \sum_{k=0}^{\frac{N}{2}-1} y_k \omega_{\frac{N}{2}}^{rk} - \omega_N^r \sum_{k=0}^{\frac{N}{2}-1} z_k \omega_{\frac{N}{2}}^{rk} \\
&= Y_r - \omega_N^r Z_r, \quad r = 0, 1, \dots, N/2 - 1.
\end{aligned}
$$

Note that equations (3.7) and (3.8) can be applied to any problem of even size. Therefore, while they appear to represent only the last (combination) step, it is understood that when the problem size is a power of 2, the two subproblems defined by equations (3.5) and (3.6) (as well the subsequent subproblems of size ≥ 2) would have each been solved in exactly the same manner.

The computation represented by equations (3.7) and (3.8) is commonly referred to as a Cooley-Tukey butterfly in the literature, and is depicted by the annotated butterfly symbol in Figure 3.1 below.

Figure 3.1 The Cooley-Tukey butterfly.

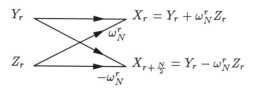

$$Y_r \qquad\qquad X_r = Y_r + \omega_N^r Z_r$$
$$\omega_N^r$$
$$Z_r \qquad\qquad X_{r+\frac{N}{2}} = Y_r - \omega_N^r Z_r$$
$$-\omega_N^r$$

3.1.1 Analyzing the arithmetic cost

Let $T(N)$ be the arithmetic cost of computing the radix-2 DIT FFT of size N, which implies that computing a half-size transform using the same algorithm costs $T\left(\frac{N}{2}\right)$. In order to set up the recurrence equation, one needs to relate $T(N)$ to $T\left(\frac{N}{2}\right)$. According to (3.7) and (3.8), N complex additions and $\frac{N}{2}$ complex multiplications are needed to complete the transform, assuming that the twiddle factors are pre-computed as suggested in Section 2.1.

Recall that that one complex addition incurs two real additions according to (2.1), and one complex multiplication (with pre-computed intermediate results involving the real and imaginary parts of a twiddle factor) incurs three real multiplications and three real additions according to (2.4).

Therefore, counting a floating-point addition or multiplication as one flop, $2N$ flops are incurred by the N complex additions, and $3N$ flops are incurred by the $\frac{N}{2}$ complex multiplications. In total, $5N$ flops are needed to complete the transform after the two half-size subproblems are each solved at the cost of $T\left(\frac{N}{2}\right)$. Accordingly, the arithmetic cost $T(N)$ is represented by the following recurrence.

$$(3.9) \qquad T(N) = \begin{cases} 2T\left(\frac{N}{2}\right) + 5N & \text{if } N = 2^n \geq 2, \\ 0 & \text{if } N = 1. \end{cases}$$

Comparing (3.9) with (2.6) and using (2.7) leads to the following expression for the arithmetic cost:

$$(3.10) \qquad T(N) = 5N \log_2 N .$$

3.2 Radix-2 Decimation-In-Frequency (DIF) FFT

As its name implies, the radix-2 DIF FFT algorithm is obtained by *decimating* the output frequency series into an even-indexed set $\{X_{2k} \mid k = 0, \ldots, N/2 - 1\}$ and an odd-indexed set $\{X_{2k+1} \mid k = 0, \ldots, N/2 - 1\}$. To define the two half-size subproblems, equation (3.1) is rewritten as

$$
X_r = \sum_{\ell=0}^{\frac{N}{2}-1} x_\ell \omega_N^{r\ell} + \sum_{\ell=\frac{N}{2}}^{N-1} x_\ell \omega_N^{r\ell} = \sum_{\ell=0}^{\frac{N}{2}-1} x_\ell \omega_N^{r\ell} + \sum_{\ell=0}^{\frac{N}{2}-1} x_{\ell+\frac{N}{2}} \omega_N^{r\left(\ell+\frac{N}{2}\right)}
$$

$$(3.11)$$

$$
= \sum_{\ell=0}^{\frac{N}{2}-1} \left(x_\ell + x_{\ell+\frac{N}{2}} \omega_N^{r\frac{N}{2}} \right) \omega_N^{r\ell}, \quad r = 0, 1, \ldots, N-1.
$$

For r even, using (3.2) in (3.11) yields

$$
X_{2k} = \sum_{\ell=0}^{\frac{N}{2}-1} \left(x_\ell + x_{\ell+\frac{N}{2}} \omega_N^{kN} \right) \omega_N^{2k\ell}
$$

$$(3.12)$$

$$
= \sum_{\ell=0}^{\frac{N}{2}-1} \left(x_\ell + x_{\ell+\frac{N}{2}} \right) \omega_{\frac{N}{2}}^{k\ell}, \quad k = 0, 1, \ldots, N/2 - 1.
$$

Defining $Y_k = X_{2k}$ and $y_\ell = x_\ell + x_{\ell+\frac{N}{2}}$ yields the half-size subproblem

$$(3.13) \qquad Y_k = \sum_{\ell=0}^{\frac{N}{2}-1} y_\ell \omega_{\frac{N}{2}}^{k\ell}, \quad k = 0, 1, \ldots, N/2 - 1.$$

Similarly, for r odd, using (3.2) in (3.11) yields

$$
X_{2k+1} = \sum_{\ell=0}^{\frac{N}{2}-1} \left(x_\ell + x_{\ell+\frac{N}{2}} \omega_N^{(2k+1)\frac{N}{2}} \right) \omega_N^{(2k+1)\ell}
$$

$$(3.14)$$

$$
= \sum_{\ell=0}^{\frac{N}{2}-1} \left(\left(x_\ell - x_{\ell+\frac{N}{2}} \right) \omega_N^\ell \right) \omega_{\frac{N}{2}}^{k\ell}, \quad k = 0, 1, \ldots, N/2 - 1.
$$

Defining $Z_k = X_{2k+1}$ and $z_\ell = \left(x_\ell - x_{\ell+\frac{N}{2}} \right) \omega_N^\ell$ yields the second half-size problem

$$(3.15) \qquad Z_k = \sum_{\ell=0}^{\frac{N}{2}-1} z_\ell \omega_{\frac{N}{2}}^{k\ell}, \quad k = 0, 1, \ldots, N/2 - 1.$$

Note that because $X_{2k} = Y_k$ in (3.13) and $X_{2k+1} = Z_k$ in (3.15), no more computation is needed to obtain the solution for the original problems after the two subproblems are solved. Therefore, in the implementation of the DIF FFT, the bulk of the work is done during the *subdivision* step, i.e., the set-up of appropriate subproblems, and there is no combination step. Consequently, the computation of $y_\ell = x_\ell + x_{\ell + \frac{N}{2}}$ and $z_\ell = (x_\ell - x_{\ell + \frac{N}{2}})\omega_N^\ell$ completes the first (subdivision) step.

The computation of y_ℓ and z_ℓ in the subdivision step as defined above is referred to as the Gentleman-Sande butterfly in the literature, and is depicted by the annotated butterfly symbol in Figure 3.2.

Figure 3.2 The Gentleman-Sande butterfly.

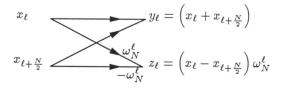

3.2.1 Analyzing the arithmetic cost

Observe that the computation of y_ℓ and z_ℓ in the subdivision step requires N complex additions and $\frac{N}{2}$ complex multiplications, which amount to the same cost as the combination step in the radix-2 DIT FFT algorithm discussed earlier, and they are the only cost in addition to solving the two half-size subproblems at the cost of $T\left(\frac{N}{2}\right)$ each. Accordingly, the total arithmetic cost of the radix-2 DIF FFT is also represented by the recurrence equation (3.9), and $T(N) = 5N \log_2 N$ from (3.10).

3.3 Notes and References

The basic form of the DIT (*decimation in time*) FFT presented in Section 3.1 was used by Cooley and Tukey [33]; the basic form of the DIF (*decimation in frequency*) FFT presented in Section 3.2 was found independently by Gentleman and Sande [47], and Cooley and Stockham according to [30].

An interesting account of the history of the fast Fourier transform may be found in the article by Cooley, Lewis, and Welch [32]. An account of Gauss and the history of the FFT is contained in a more recent article by Heideman, Johnson, and Burrus [52]. A bibliography of more than 3500 titles on the fast Fourier transform and convolution algorithms was published in 1995 [85].

Chapter 4

Deciphering the Scrambled Output from In-Place FFT Computation

In practice, FFT computations are normally performed *in place* in a one-dimensional array, with new values overwriting old values as implied by the butterflies introduced in the previous chapter. For example, Figure 4.2 implies that y_ℓ overwrites x_ℓ and z_ℓ overwrites $x_{\ell+\frac{N}{2}}$. A consequence of this, although the details may not yet be clear, is that the output is "scrambled"; the order of the elements of the vector X in the array will not generally correspond to that of the input x. For example, applying the DIF FFT to the data x stored in the array a will result in X "scrambled" in a when the computation is complete, as shown in Figure 4.1. One of the main objectives of this chapter is to develop machinery to facilitate a clear understanding of how this scrambling occurs. Some notation that will be useful in the remainder of the book will also be introduced. The DIF FFT algorithm will be used as the vehicle with which to carry out these developments.

Figure 4.1 The input x in array a is overwritten by scrambled output X.

Input:

x_0	x_1	x_2	x_3	x_4	x_5	x_6	x_7
$a[0]$	$a[1]$	$a[2]$	$a[3]$	$a[4]$	$a[5]$	$a[6]$	$a[7]$

Output:

X_0	X_4	X_2	X_6	X_1	X_5	X_3	X_7
$a[0]$	$a[1]$	$a[2]$	$a[3]$	$a[4]$	$a[5]$	$a[6]$	$a[7]$

Consider the first subdivision step of the radix-2 DIF FFT, which is depicted by the Gentleman-Sande butterfly in Figure 4.2. Recall that by defining $y_\ell = x_\ell + x_{\ell+\frac{N}{2}}$

Figure 4.2 The Gentleman-Sande butterfly.

$$x_\ell \qquad\qquad\qquad y_\ell = \left(x_\ell + x_{\ell+\frac{N}{2}}\right)$$

$$x_{\ell+\frac{N}{2}} \qquad \begin{array}{c}\omega_N^\ell\\ -\omega_N^\ell\end{array} \qquad z_\ell = \left(x_\ell - x_{\ell+\frac{N}{2}}\right)\omega_N^\ell$$

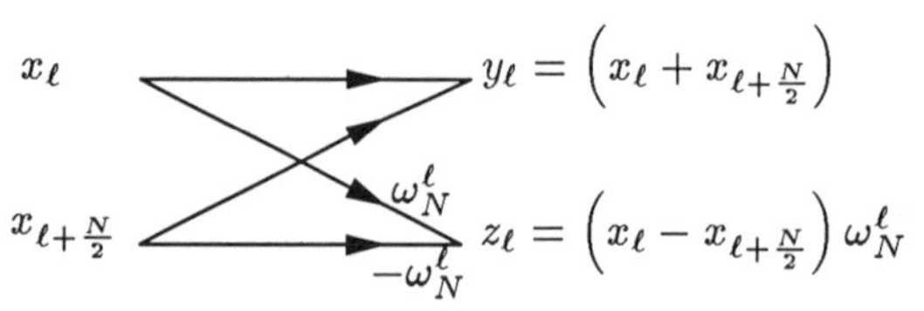

and $z_\ell = \left(x_\ell - x_{\ell+\frac{N}{2}}\right)\omega_N^\ell$ the two half-size subproblems to be solved are

(4.1) $$Y_k = X_{2k} = \sum_{\ell=0}^{\frac{N}{2}-1} y_\ell \omega_{\frac{N}{2}}^{k\ell}, \quad k = 0, 1, \ldots, N/2 - 1,$$

and

(4.2) $$Z_k = X_{2k+1} = \sum_{\ell=0}^{\frac{N}{2}-1} z_\ell \omega_{\frac{N}{2}}^{k\ell}, \quad k = 0, 1, \ldots, N/2 - 1.$$

Definition 4.1 The input data $x_0, x_1, \cdots, x_{N-1}$ are said to be in "natural order" if x_i and x_{i+1} are stored in consecutive locations in a for all $0 \le i \le N - 2$. Similarly, the output data $X_0, X_1, \cdots, X_{N-1}$ are said to be in natural order if X_i and X_{i+1} are stored in consecutive locations in a for all $0 \le i \le N - 2$.

For example, the eight input elements $x_0, x_1, \ldots, x_7$ in Figure 4.1 are in natural order but the output elements $X_0, X_1, \ldots, X_7$ are not. Throughout this chapter it is assumed that the input data $x_i, 0 \le i < N$ are stored in the one-dimensional array a in natural order.

Since the FFT computations are performed in place repeatedly on subproblems of various sizes, determining the location of X_r in the array at the end of the computation is not immediately obvious. The butterfly notation does not explicitly relate the locations of the input data to the locations of the output data – as noted earlier, the Gentleman-Sande butterfly in Figure 4.2 displays only the first subdivision step of the recursive DIF FFT algorithm. Thus, there is a gap between the elegant (but somewhat implicit) butterfly notation and the detailed specification of the positions of the outputs X_r contained in the repeatedly modified array.

The purpose of the following sections is to close this gap by developing an iterative form of the radix-2 DIF FFT algorithm and using a simple notation to assist in its specification. The notation adopted will also facilitate the adaptation of these FFT algorithms to parallel processing, which is the focus of Part III of this book.

4.1   Iterative Form of the Radix-2 DIF FFT

Recall that each subdivision step is defined recursively by the Gentleman-Sande butterfly. Since there is no combination step, it is straightforward to derive an iterative algorithm by simply subdividing each resulting subproblem iteratively until the problem size is one. This iterative subdividing process is presented in Algorithm 4.1.

Algorithm 4.1 The skeleton of an iterative radix-2 DIF FFT algorithm.

begin

    NumOfProblems := 1   Initially: One problem of size N

    ProblemSize := *N*

    while *ProblemSize* > 1 **do**   Halve each problem

        for *K* := 1 **to** *NumOfProblems* **do**

            Compute the Gentleman-Sande butterfly to divide

            the *K*th problem into two halves.

        end for

        NumOfProblems := *NumOfProblems* * 2

        ProblemSize := *ProblemSize*/2

    end while

end

As discussed in Section 2.1, the twiddle factors are assumed to have been pre-computed and stored in an array w, with $w[\ell] = \omega_N^\ell, 0 \leq \ell < N/2$. To halve the Kth subproblem, the pseudo-code shown in Figure 4.3 can now be inserted inside the for-loop in Algorithm 4.1 The result is Algorithm 4.2.

Figure 4.3 The pseudo-code implementing the Gentleman-Sande butterfly.

W := *An appropriate twiddle factor from* w

Temp := $a[J]$

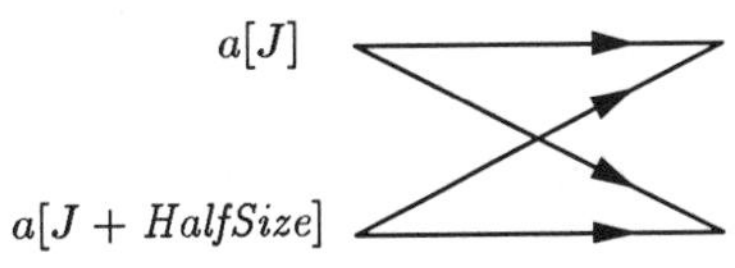

$a[J]$:= *Temp* + $a[J + HalfSize]$

$a[J + HalftSize]$:= $W * (Temp - a[J + HalfSize])$

Algorithm 4.2 The iterative radix-2 DIF FFT algorithm in pseudo-code.

---

begin
    NumOfProblems := 1   Initially: One problem of size N
    ProblemSize := N
    while *ProblemSize* > 1 **do**   Halve each problem
        HalfSize := *ProblemSize*/2
        for *K* := 0 **to** *NumOfProblems* − 1 **do**
            JFirst := *K* ∗ *ProblemSize*
            JLast := *JFirst* + *HalfSize* − 1
            Jtwiddle := 0
            for *J* := *JFirst* **to** *JLast* **do**
                W := *w*[*Jtwiddle*]   Access pre-stored $w[\ell] = \omega_N^{\ell}$
                Temp := *a*[*J*]
                a[*J*] := *Temp* + *a*[*J* + *HalfSize*]
                a[*J* + *HalfSize*] := *W* ∗ (*Temp* − *a*[*J* + *HalfSize*])
                Jtwiddle := *Jtwiddle* + *NumOfProblems*
            end for
        end for
        NumOfProblems := 2 ∗ *NumOfProblems*
        ProblemSize := *HalfSize*
    end while
end

4.2 Applying the Iterative DIF FFT to a $N = 32$ Example.

To have a concrete example for use throughout this book, the iterative DIF FFT Algorithm 4.2 is applied to an input sequence of size $N = 32 = 2^5$. Figure 4.4 displays the initial contents of a, the five stages of butterfly computation, and the resulting scrambled sequence (which will be deciphered in Section 4.4.)

To help identify the pairs of subproblems resulting from every stage of butterfly computation, the subproblems involving $a[0]$, which initially contains the input element x_0, are highlighted in Figure 4.4.

Note that, in general, for an input sequence of $N = 2^n$ elements, there are exactly $N/2$ butterflies in each of the $\log_2 N$ stages of butterfly computations.

Figure 4.4 The DIF FFT with naturally ordered input and bit-reversed output.

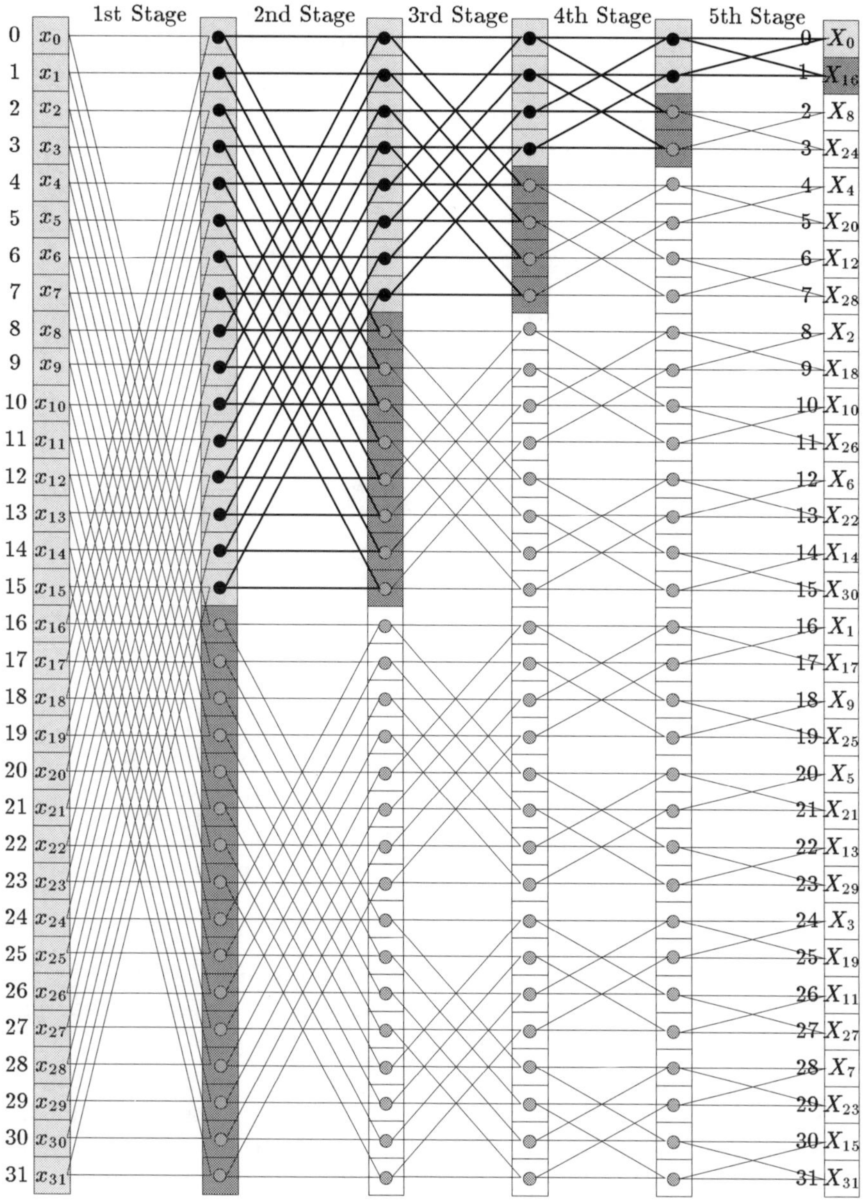

Adapted from E. Chu and A. George [28], *Linear Algebra and its Applications*, 284:95–124, 1998. With permission.

4.3 Storing and Accessing Pre-computed Twiddle Factors

The actual performance of an algorithm often depends on whether the algorithm accesses data in the computer memory efficiently. This is particularly true on today's high-performance machines, where the floating-point arithmetic is so much faster than the time required to transfer data in and out of memory. Using the example in Figure 4.4, it is a simple task to verify that after the first stage of Algorithm 4.2 is completed, only a subset of the $N/2$ pre-computed ω_N^{ℓ} elements are accessed in each of the subsequent stages. This subset becomes smaller and smaller as depicted in Figure 4.5 below for $N/2 = 16$.

Figure 4.5 Accessing the *short* w vector by the DIF FFT.

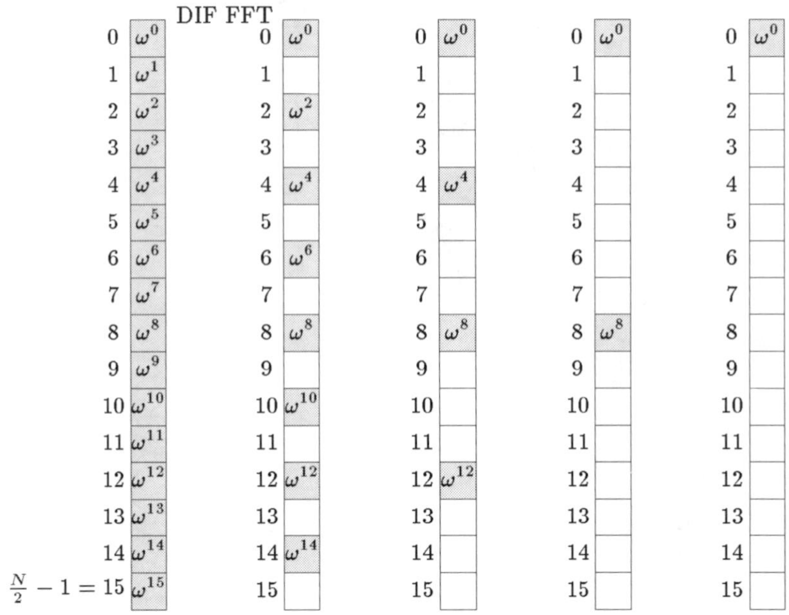

DIF FFT

Index	Stage 1	Stage 2	Stage 3	Stage 4	Stage 5
0	ω^0	ω^0	ω^0	ω^0	ω^0
1	ω^1				
2	ω^2	ω^2			
3	ω^3				
4	ω^4	ω^4	ω^4		
5	ω^5				
6	ω^6	ω^6			
7	ω^7				
8	ω^8	ω^8	ω^8	ω^8	
9	ω^9				
10	ω^{10}	ω^{10}			
11	ω^{11}				
12	ω^{12}	ω^{12}	ω^{12}		
13	ω^{13}				
14	ω^{14}	ω^{14}			
$\frac{N}{2}-1=15$	ω^{15}				

Figure 4.5 shows that the individual factors are accessed in array locations apart by a stride equal to a power of two. This causes the so-called "power-of-2" problem on machines with a hierarchy of memory banks. To avoid this problem, it is common to store the individual twiddle factors required in each stage in consecutive locations of a *long* w vector of size $N - 1$ as shown in Figure 4.6. It is a simple task to modify the pseudo-code program of the FFT Algorithm 4.2 to employ the long w vector, and it is left as an exercise.

Figure 4.6 Accessing the *long* w vector by the DIF FFT.

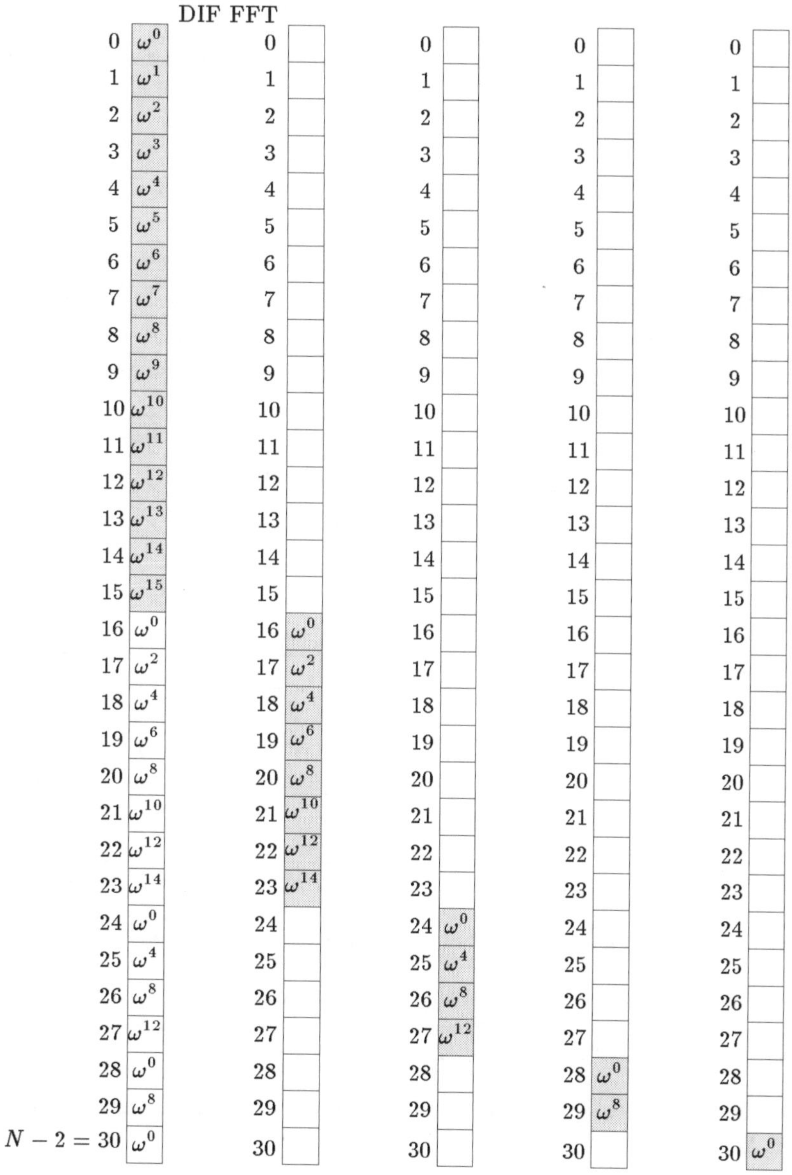

DIF FFT

#	col 1	col 2	col 3	col 4	col 5
0	ω^0				
1	ω^1				
2	ω^2				
3	ω^3				
4	ω^4				
5	ω^5				
6	ω^6				
7	ω^7				
8	ω^8				
9	ω^9				
10	ω^{10}				
11	ω^{11}				
12	ω^{12}				
13	ω^{13}				
14	ω^{14}				
15	ω^{15}				
16	ω^0	ω^0			
17	ω^2	ω^2			
18	ω^4	ω^4			
19	ω^6	ω^6			
20	ω^8	ω^8			
21	ω^{10}	ω^{10}			
22	ω^{12}	ω^{12}			
23	ω^{14}	ω^{14}			
24	ω^0		ω^0		
25	ω^4		ω^4		
26	ω^8		ω^8		
27	ω^{12}		ω^{12}		
28	ω^0			ω^0	
29	ω^8			ω^8	
$N-2=30$	ω^0				ω^0

4.4   A Binary Address Based Notation and the Bit-Reversed Output

To provide background, and to make this section self-contained, a brief review of the binary number system is presented.

4.4.1   Binary representation of positive decimal integers

Definition 4.2 The n-bit sequence $i_{n-1} \cdots i_1 i_0$, where $i_k = 0$ or 1 for $0 \leq k \leq n-1$, represents positive integer value L in the range 0 to $2^n - 1$, where

$$L = i_{n-1} \times 2^{n-1} + \cdots + i_1 \times 2^1 + i_0 \times 2^0.$$

Given below is an example for $0 \leq L \leq 2^n - 1 = 7$.

Table 4.1 Binary representation of integers $0 \leq L \leq 2^n - 1 = 7$.

Decimal L	Binary $i_2 i_1 i_0$	Verification: $i_2 \times 2^2 + i_1 \times 2^1 + i_0 \times 2^0 = L$
0	000	$0 \times 2^2 + 0 \times 2^1 + 0 \times 2^0 = 0$
1	001	$0 \times 2^2 + 0 \times 2^1 + 1 \times 2^0 = 1$
2	010	$0 \times 2^2 + 1 \times 2^1 + 0 \times 2^0 = 2$
3	011	$0 \times 2^2 + 1 \times 2^1 + 1 \times 2^0 = 3$
4	100	$1 \times 2^2 + 0 \times 2^1 + 0 \times 2^0 = 4$
5	101	$1 \times 2^2 + 0 \times 2^1 + 1 \times 2^0 = 5$
6	110	$1 \times 2^2 + 1 \times 2^1 + 0 \times 2^0 = 6$
7	111	$1 \times 2^2 + 1 \times 2^1 + 1 \times 2^0 = 7$

Consider $N = 2^n$ and $0 \leq L \leq N-1$. Given below are some useful properties of L's n-bit binary representation $i_{n-1} \cdots i_1 i_0$. These properties follow from Definition 4.2 immediately, and they can be easily verified using the example in Table 4.1.

◆ **Property 1.** L is an *even* number if and only if the right-most i_0 bit is 0.

◆ **Property 2.** L is an *odd* number if and only if the right-most i_0 bit is 1.

◆ **Property 3.** $L < N/2$ if and only if the left-most i_{n-1} bit is 0.

◆ **Property 4.** $L \geq N/2$ if and only if the left-most i_{n-1} bit is 1.

◆ **Property 5.** For $0 \leq L < M \leq N - 1$, L and M differ in the binary i_k bit if and only if $M - L = 2^k$.

4.4.2   Deciphering the scrambled output

By making use of the binary number properties, the scrambled FFT output can now be easily deciphered if a simple binary address based notation is used to assist in the specification of Algorithm 4.2. That is, instead of referring to the decimal value of J

in $a[J]$, one uses the binary representation of J and does all the "thinking" in terms of binary numbers.

Observe that in Algorithm 4.2, the initial $ProblemSize = N = 2^n$, and the variable $HalfSize$ takes on the values 2^{n-1}, 2^{n-2}, $\cdots$, 2, 1. Since the distance between $a[J]$ and $a[J + HalfSize]$ is a power of 2, the binary number property 5 applies here. That is, the addresses $a[J]$ and $a[J + HalfSize]$, denoted as *binary numbers* $i_{n-1} \cdots i_1 i_0$, differ only in bit i_k when $HalfSize = 2^k$ and the algorithm can be expressed in terms of binary addresses as shown below.

Algorithm 4.3 The radix-2 DIF FFT algorithm in terms of binary addresses:

---

begin

    $k := n - 1$ Initial problem size $N = 2^n$

    while $k \geq 0$ **do**   Halve each problem

            Apply Gentleman-Sande butterfly computation

            to all pairs of array elements whose

            binary addresses differ in bit i_k

            $k := k - 1$

    end while

end

For $N = 8 = 2^3$, the DIF FFT algorithm as described above consists of three stages involving the sequence $k = 2, 1, 0$, with butterflies being applied to all pairs of array elements whose binary addresses differ in bit i_k. In Table 4.2, the binary addresses of all four pairs of $a[J]$ and $a[J + HalfSize]$ are given for all three stages of the butterfly computation.

Table 4.2 Decimal and binary addresses of $a[J]$ and $a[J + HalfSize]$ for $N = 8$, $k = 2, 1, 0$.

Butterfly Computation	Current HalfSize	Decimal $a[J]$	Binary J	Decimal $a[J + HalfSize]$	Binary $J + HalfSize$
Stage 1	$N/2 = 2^2 = 4$	$a[0]$	000	$a[4]$	100
($k = 2$)		$a[1]$	001	$a[5]$	101
		$a[2]$	010	$a[6]$	110
		$a[3]$	011	$a[7]$	111
Stage 2	$N/2^2 = 2^1 = 2$	$a[0]$	000	$a[2]$	010
($k = 1$)		$a[1]$	001	$a[3]$	011
	$N/2^2 = 2^1 = 2$	$a[4]$	100	$a[6]$	110
		$a[5]$	101	$a[7]$	111
Stage 3	$N/2^3 = 2^0 = 1$	$a[0]$	000	$a[1]$	001
($k = 0$)	$N/2^3 = 2^0 = 1$	$a[2]$	010	$a[3]$	011
	$N/2^3 = 2^0 = 1$	$a[4]$	100	$a[5]$	101
	$N/2^3 = 2^0 = 1$	$a[6]$	110	$a[7]$	111

Since the algorithm can be expressed in terms of binary addresses, it is convenient to adopt binary notation for the indices of a and the subscripts of x. Thus, in what follows the notation $x_{i_2 i_1 i_0}$ will mean x_r, where $i_2 i_1 i_0$ is the binary representation of r. Similarly, $a[i_2 i_1 i_0]$ refers to the element $a[m]$, where the binary representation of m is $i_2 i_1 i_0$.

Then, the entire Table 4.2 can be replaced by a one-line shorthand notation, which describes the three-stage DIF FFT process being applied to $a[m]$ as the sequence

$$\boxed{\quad \overset{\blacktriangledown}{i_2} i_1 i_0 \qquad \tau_2 \overset{\blacktriangledown}{i_1} i_0 \qquad \tau_2 \tau_1 \overset{\blacktriangledown}{i_0} \quad}$$

where $\overset{\blacktriangledown}{i_k}$ indicates that the butterfly computation involving the pairs in a different in bit i_k is being performed in the current stage, and τ_k indicates that the corresponding butterfly operations were performed in a previous stage. The transformation is completed when butterflies in all stages have been performed.

Which element of output X will be found in $a[i_2 i_1 i_0]$? If equations (4.1) and (4.2) are interpreted as implied by the butterfly in Figure 4.2, the first half of a will contain the even numbered X's after all stages are completed, and the second half will contain the odd-numbered X's. This in turn means that after the first butterfly step identified by operation $\overset{\blacktriangledown}{i_2} i_1 i_0$ is completed, it is known that $a[0 i_1 i_0]$ will ultimately contain output X_r, where r must be even, so the right-most bit of r is $i_2 = 0$. Similarly, it is also known that $a[1 i_1 i_0]$ will ultimately contain output X_r, where r must be odd, so the right-most bit of r is $i_2 = 1$. That is, the right-most bit of r for each X_r in array location $a[i_2 i_1 i_0]$ is now known to be i_2.

To continue, observe that the updated data in $a[0 i_1 i_0]$ define one subproblem to be further subdivided, and the updated data in $a[1 i_1 i_0]$ define the other subproblem to be further subdivided. By exactly the same argument applied to the binary addresses $i_1 i_0$, one can conclude that the second from the right-most bit of r for each X_r in $a[i_2 i_1 i_0]$ will be i_1.

Repeating the same argument one more time on successively halved portions of the array a yields the conclusion that $a[i_2 i_1 i_0]$ will finally contain $X_{i_0 i_1 i_2}$. In other words, $X_{i_0 i_1 i_2}$ will occupy the position originally occupied by $x_{i_2 i_1 i_0}$. That is, the output X is in *bit-reversed* order with respect to the subscript of the input element which it overwrites. For example, as shown in Figure 4.7, x_{001} is overwritten by X_{100}, x_{110} is overwritten by X_{011}, and so on. It is easy to verify that the bit-reversed output in Figure 4.7 represents indeed the scrambled output $\{X_0, X_4, X_2, X_6, X_1, X_5, X_3, X_7\}$ previously shown in Figure 4.1.

Figure 4.7 The input x in array a is overwritten by bit-reversed output X.

Input a: | x_{000} | x_{001} | x_{010} | x_{011} | x_{100} | x_{101} | x_{110} | x_{111} |

Address: 000   001   010   011   100   101   110   111

Output a: | X_{000} | X_{100} | X_{010} | X_{110} | X_{001} | X_{101} | X_{011} | X_{111} |

Address: 000   001   010   011   100   101   110   111

The conclusion can now be immediately extended to the example for $N = 32$. The entire Figure 4.4 can be replaced by the sequence

$$\boxed{\overset{\triangledown}{i_4 i_3 i_2 i_1 i_0} \quad \overset{\triangledown}{\tau_4 i_3 i_2 i_1 i_0} \quad \overset{\triangledown}{\tau_4 \tau_3 i_2 i_1 i_0} \quad \overset{\triangledown}{\tau_4 \tau_3 \tau_2 i_1 i_0} \quad \overset{\triangledown}{\tau_4 \tau_3 \tau_2 \tau_1 i_0}}$$

with the understanding that on input, $a[i_4 i_3 i_2 i_1 i_0]$ contains $x_{i_4 i_3 i_2 i_1 i_0}$, and on output, $a[i_4 i_3 i_2 i_1 i_0]$ contains the bit-reversed $X_{i_0 i_1 i_2 i_3 i_4}$. Refer to Figure 4.4 for the decimal subscripts of all 32 bit-reversed output elements $X_{i_0 i_1 i_2 i_3 i_4}$.

4.5 Shorthand Notation for the Twiddle Factors

The twiddle factors corresponding to the three-stage DIF FFT may be specified by the binary representation of the address *Jtwiddle* in Algorithm 4.2 as shown in Table 4.3.

Table 4.3 Shorthand notations for twiddle factors ($N = 8$).

Butterfly Computation	Current *NumberOfProblems*	Decimal $w[Jtwiddle]$	Binary *Jtwiddle* w.r.t. modified $x_{i_2 i_1 i_0}$	Shorthand notation w.r.t. modified $x_{i_2 i_1 i_0}$
$\overset{\triangledown}{i_2 i_1 i_0}$	$NumberOfProblems = 1$ $(HalfSize = 4)$	$w[0] = \omega_N^0$ $w[1] = \omega_N^1$ $w[2] = \omega_N^2$ $w[3] = \omega_N^3$	$i_1 i_0 = 00$ $i_1 i_0 = 01$ $i_1 i_0 = 10$ $i_1 i_0 = 11$	$\omega_N^{i_1 i_0}$
$\overset{\triangledown}{\tau_2 i_1 i_0}$	$NumberOfProblems = 2$ $(HalfSize = 2)$	$w[0] = \omega_N^0$ $w[2] = \omega_N^2$	$i_0 0 = 00$ $i_0 0 = 10$	$\omega_N^{i_0 0}$
$\overset{\triangledown}{\tau_2 \tau_1 i_0}$	$NumberOfProblems = 4$ $(HalfSize = 1)$	$w[0] = \omega_N^0$	00	$\omega_N^0 = 1$

Observe that during the $\overset{\triangledown}{i_2 i_1 i_0}$ stage, the twiddle factor $w[i_1 i_0] = \omega_N^{i_1 i_0}$ is used to update $a[1 i_1 i_0]$ and during the $\overset{\triangledown}{\tau_2 i_1 i_0}$ stage that follows, the updating of $a[\tau_2 1 i_0]$ involves $w[i_0 0] = \omega_N^{i_0 0}$. Finally, the updating of $a[\tau_2 \tau_1 1]$ during the last $\overset{\triangledown}{\tau_2 \tau_1 i_0}$ stage all involves $w[00] = \omega_N^0 = 1$.

For $N = 32$, the twiddle factors corresponding to the five stages of butterfly computations are

$$\omega_N^{i_3 i_2 i_1 i_0}, \ \omega_N^{i_2 i_1 i_0 0}, \ \omega_N^{i_1 i_0 0 0}, \ \omega_N^{i_0 0 0 0}, \ \omega_N^0 = 1.$$

Chapter 5

Bit-Reversed Input to the Radix-2 DIF FFT

Technically speaking, the correctness of Algorithm 4.2 depends on the fact that x_m is initially contained in $a[m]$. For easy reference, the contents of $a[J]$ and $a[J + HalfSize]$ have been added to Table 4.2 to obtain Table 5.1. The notation $x_m^{(\ell)}$ is used to denote x_m's derivative, which overwrites x_m after an in-place butterfly computation in Stage ℓ. Accordingly, after Stage 3 is completed, $x_{i_2 i_1 i_0}^{(3)} = X_{i_0 i_1 i_2}$ as explained in Chapter 4.

Table 5.1 The contents of $a[J]$ and $a[J + HalfSize]$ in Algorithm 4.2.

Butterfly Computation	Current HalfSize	Decimal $a[J]$	Binary J	Decimal $a[J + HalfSize]$	Binary $J + HalfSize$
Stage 1	$N/2 = 2^2 = 4$	$a[0] = x_0$	000	$a[4] = x_4$	100
		$a[1] = x_1$	001	$a[5] = x_5$	101
		$a[2] = x_2$	010	$a[6] = x_6$	110
		$a[3] = x_3$	011	$a[7] = x_7$	111
Stage 2	$N/2^2 = 2^1 = 2$	$a[0] = x_0^{(1)}$	000	$a[2] = x_2^{(1)}$	010
		$a[1] = x_1^{(1)}$	001	$a[3] = x_3^{(1)}$	011
	$N/2^2 = 2^1 = 2$	$a[4] = x_4^{(1)}$	100	$a[6] = x_6^{(1)}$	110
		$a[5] = x_5^{(1)}$	101	$a[7] = x_7^{(1)}$	111
Stage 3	$N/2^3 = 2^0 = 1$	$a[0] = x_0^{(2)}$	000	$a[1] = x_1^{(2)}$	001
	$N/2^3 = 2^0 = 1$	$a[2] = x_2^{(2)}$	010	$a[3] = x_3^{(2)}$	011
	$N/2^3 = 2^0 = 1$	$a[4] = x_4^{(2)}$	100	$a[5] = x_5^{(2)}$	101
	$N/2^3 = 2^0 = 1$	$a[6] = x_6^{(2)}$	110	$a[7] = x_7^{(2)}$	111

However, the following program is correct, *regardless of where* $x_{i_2 i_1 i_0}^{(1)}$, $x_{i_2 i_1 i_0}^{(2)}$, *and* $x_{i_2 i_1 i_0}^{(3)}$ *are found in* a.

Algorithm 5.1 The iterative DIF FFT algorithm applied to the x elements.

---
begin
    $k := n - 1$ Initial problem size $N = 2^n$
    while $k \geq 0$ **do**   Halve each problem
        Apply Gentleman-Sande butterfly computation
        to all pairs of elements of x whose
        (binary) subscripts differ in bit i_k
        $k := k - 1$
    end while
end

Thus, the input data could be permuted arbitrarily; if the butterflies are applied correctly to the data, the correct answers would be obtained. Moreover, the element of a that initially contained $x_{i_2 i_1 i_0}$ would contain $x^{(3)}_{i_2 i_1 i_0} = X_{i_0 i_1 i_2}$ at the end of the computation.

5.1   The Effect of Bit-Reversed Input

Suppose the objective is to find an initial ordering of the input so that the resulting output is in natural order. The observation above implies that initially, $a[i_0 i_1 i_2]$ should contain $x_{i_2 i_1 i_0}$, since at the end of the computation one wants $a[i_0 i_1 i_2]$ to contain $x^{(3)}_{i_2 i_1 i_0} = X_{i_0 i_1 i_2}$. That is, the input should be placed in bit-reversed order before the computation begins as shown in Figure 5.1.

Figure 5.1 Bit-reversed input to the FFT and the naturally ordered output.

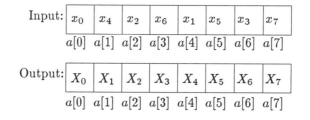

A separate program is needed to handle the bit-reversed input data. Algorithm 5.2 contains the pseudo-code. The twiddle factors are assumed to have been pre-computed and stored in array w *in bit-reversed order* as shown in Figure 5.2. This allows one to relate the binary addresses in w to the shorthand notations for the twiddle factors using $w[i_0 i_1] = w_N^{i_1 i_0}$, $w[0 i_0] = w_N^{i_0 0}$, and $w[00] = w_N^0 = 1$.

Figure 5.2 Store the $N/2 = 4$ pre-computed twiddle factors in bit-reversed order.

$$w \quad \boxed{\omega_N^0 \mid \omega_N^2 \mid \omega_N^1 \mid \omega_N^3}$$

$$w[0] \quad w[1] \quad w[2] \quad w[3]$$

Algorithm 5.2 The radix-2 DIF FFT algorithm for bit-reversed input.

begin
    $NumOfProblems := 1$ Initially: One problem of size N
    $ProblemSize := N$ $HalfSize = ProblemSize/2$
    $Distance := 1$
    while $ProblemSize > 1$ **do**   Halve each problem
            for $JFirst := 0$ **to** $NumOfProblems - 1$ **do**
                $J := JFirst;\; Jtwiddle := 0$
                while $J < N - 1$ **do**
                      $W := w[Jtwiddle]$ Assume $w[m] = \omega_N^\ell$, m bit-reversed ℓ
                      $Temp := a[J]$
                      $a[J] := Temp + a[J + Distance]$
                      $a[J + Distance] := (Temp - a[J + Distance]) * W$
                      $Jtwiddle := Jtwiddle + 1$ Access consecutive $w[m]$
                      $J := J + 2 * NumOfProblems$
                end while
            end for
        $NumOfProblems := NumOfProblems * 2$
        $ProblemSize := ProblemSize/2$
        $Distance := Distance * 2$
    end while
end

Applying Algorithm 5.2 to the bit-reversed data in the example with $N = 8$ in Figure 5.1, all pairs of $a[J]$ and $a[J + Distance]$ are identified in Table 5.2 and they show that each $x_m^{(\ell)}$ is paired up with the same partner as previously shown in Table 5.1.

Table 5.2 The contents of $a[J]$ and $a[J + Distance]$ in Algorithm 5.2.

Butterfly Computation	Current $Distance$ & $HalfSize$	Decimal $a[J]$	Binary J	Decimal $a[J + Distance]$	Binary $J + Distance$
Stage 1	$Distance = 1$	$a[0] = x_0$	000	$a[1] = x_4$	001
	($HalfSize = 4$)	$a[2] = x_2$	010	$a[3] = x_6$	011
		$a[4] = x_1$	100	$a[5] = x_5$	101
		$a[6] = x_3$	110	$a[7] = x_7$	111
Stage 2	$Distance = 2$	$a[0] = x_0^{(1)}$	000	$a[2] = x_2^{(1)}$	010
	($HalfSize = 2$)	$a[4] = x_1^{(1)}$	100	$a[6] = x_3^{(1)}$	011
		$a[1] = x_4^{(1)}$	001	$a[3] = x_6^{(1)}$	011
		$a[5] = x_5^{(1)}$	101	$a[7] = x_7^{(1)}$	111
Stage 3	$Distance = 4$	$a[0] = x_0^{(2)}$	000	$a[4] = x_1^{(2)}$	100
	($HalfSize = 1$)	$a[1] = x_4^{(2)}$	001	$a[5] = x_5^{(2)}$	101
		$a[2] = x_2^{(2)}$	010	$a[6] = x_3^{(2)}$	110
		$a[3] = x_6^{(2)}$	011	$a[7] = x_7^{(2)}$	111

5.2   A Taxonomy for Radix-2 FFT Algorithms

In this and the previous chapter, two similar but not identical DIF FFT algorithms were developed. The first accepts its input in natural order, and produces its output in bit-reversed order, while the second accepts its input in bit-reversed order, and produces its output in natural order. In later chapters, similar variations will be developed for the DIT FFT algorithm. In addition, versions of both DIT and DIF FFT algorithms will be developed that accept naturally ordered input and produce naturally ordered output. In order to be able to refer to these six variations in a succinct and suggestive way, a two-letter convention will be used in the sequel: NR will mean "naturally ordered input and bit-reversed output," with RN and NN denoting the obvious other possibilities mentioned above. Thus, the DIT$_{RN}$ algorithm would refer to the version of the DIT FFT algorithm that accepts its input in bit-reversed order and produces its output in natural order. If it makes no difference whether the DIT$_{NR}$ or DIF$_{NR}$ algorithms are intended, the term "an NR algorithm" will be used. It is common in the literature to refer to NN algorithms as "ordered FFTs."

5.3   Shorthand Notation for the DIF$_{RN}$ Algorithm

For the case $N = 8$, a shorthand notation describing the three-stage process, together with the initial permutation to bit-reversed order, is the sequence

$$\boxed{i_2i_1i_0 \quad \overset{\blacktriangledown}{i_0i_1i_2} \quad \overset{\blacktriangledown}{i_0i_1\tau_2} \quad \overset{\blacktriangledown}{i_0\tau_1\tau_2}}$$

Here the sequence begins with $i_2i_1i_0$, which is intended to imply that $x_{i_2i_1i_0}$ is assumed to be in $a[i_2i_1i_0]$ as before; the notation $\overset{\blacktriangledown}{i_0i_1i_2}$ is intended to imply that $x_{i_2i_1i_0}$ has been permuted to $a[i_0i_1i_2]$ before the first butterfly computation is performed. That is, $i_2i_1i_0$ *always represents the binary representation of the subscripts of* x; the order in which the bits appear, or are permuted during the computation, refer to movements that $x_{i_2i_1i_0}$ or its derivatives undergo in a during the computation.

For $N = 32$, the sequence describing the five-stage process, together with the initial permutation to bit-reversed order, is shown below.

$$\boxed{i_4i_3i_2i_1i_0 \quad \overset{\blacktriangledown}{i_0i_1i_2i_3i_4} \quad \overset{\blacktriangledown}{i_0i_1i_2i_3\tau_4} \quad \overset{\blacktriangledown}{i_0i_1i_2\tau_3\tau_4} \quad \overset{\blacktriangledown}{i_0i_1\tau_2\tau_3\tau_4} \quad \overset{\blacktriangledown}{i_0\tau_1\tau_2\tau_3\tau_4}}$$

5.3.1 Shorthand notation for the twiddle factors

To express the twiddle factors corresponding to the n-stage sequence shown in Section 5.3, one only needs to recall that $a[i_0i_1i_2i_3i_4]$ contains $x_{i_4i_3i_2i_1i_0}$ or its derivative. For example, when the content of $a[i_0i_1i_2i_31]$ is modified during stage $\overset{\blacktriangledown}{i_0i_1i_2i_3i_4}$, the element being updated is $x_{1i_3i_2i_1i_0}$; when the content of $a[i_0i_1i_21\tau_4]$ is modified during stage $\overset{\blacktriangledown}{i_0i_1i_2i_3\tau_4}$, the element being updated is the derivative $x^{(1)}_{\tau_41i_2i_1i_0}$, and so on.

Therefore, the shorthand notation for the twiddle factors is exactly the same as those derived in Chapter 4 for $x_{i_4i_3i_2i_1i_0}$ and its derivatives, namely,

$$\omega_N^{i_3i_2i_1i_0}, \ \omega_N^{i_2i_1i_00}, \ \omega_N^{i_1i_000}, \ \omega_N^{i_0000}, \ \omega_N^0 = 1 \ .$$

5.3.2 Applying algorithm 5.2 to a $N = 32$ example.

The complete process of applying Algorithm 5.2 to a $N = 32$ example is depicted in Figure 5.3. It is important to recognize that the five-stage sequence shown in Section 5.3, together with the twiddle factors shown above, capture all of the details shown in Figure 5.3.

For easy comparison with Figure 4.4, the pairs of subproblems resulting from every stage of butterfly computation are again identified by highlighting a particular pair which involves $a[0] = x_0$ in Figure 5.3—the two subproblems forming the pair are shaded in different grey tones so one can be easily distinguished from the other.

Comparing Figure 5.3 with Figure 4.4, one sees that "the same pair of subproblems" are located in different parts of the data array due to different initial and intermediate orderings. Thus, the computations performed DIF_{NR} and DIF_{RN} algorithms are essentially identical, although they must access the data arrays in different manners.

Figure 5.3 Butterflies for the DIF_{RN} FFT algorithm.

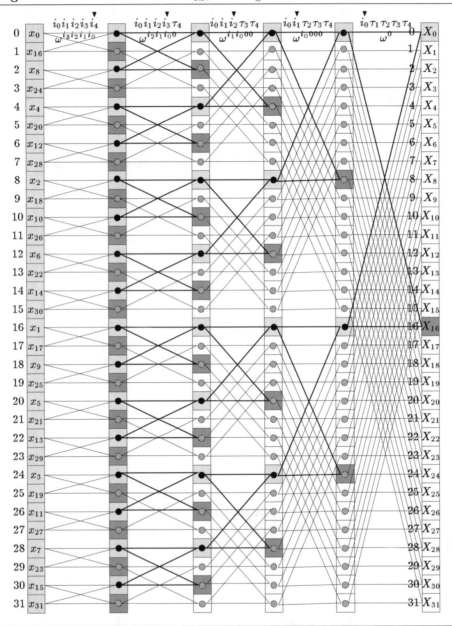

5.4 Using Scrambled Output for Input to the Inverse FFT

As demonstrated in Chapter 1, the DFT and the IDFT (inverse discrete Fourier transform) are essentially the same computation, apart from a scale factor and conjugation of the twiddle factors. Thus, algorithms essentially identical to the DIT and DIF methods already introduced, including their "data scrambling" characteristics, can be used to implement the IDFT as well. One may combine forward DIF_{NR} with inverse FFT using $IDIF_{RN}$, and the latter will automatically transform the bit-reversed frequency output back into naturally ordered time series. This process is depicted for a $N = 8$ example in Figure 5.4.

Figure 5.4 Using DIF_{NR} and DIF_{RN} in computing forward and inverse FFT.

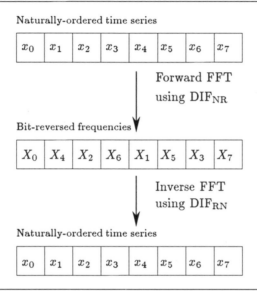

Naturally-ordered time series

| x_0 | x_1 | x_2 | x_3 | x_4 | x_5 | x_6 | x_7 |

Forward FFT
using DIF_{NR}

Bit-reversed frequencies

| X_0 | X_4 | X_2 | X_6 | X_1 | X_5 | X_3 | X_7 |

Inverse FFT
using DIF_{RN}

Naturally-ordered time series

| x_0 | x_1 | x_2 | x_3 | x_4 | x_5 | x_6 | x_7 |

On the other hand, if the input data to the forward FFT are given in bit-reversed order, one has the flexibility of of using an RN algorithm in the forward FFT and a complementary NR algorithm for the inverse FFT. Again, the final time series will be in the same (bit-reversed) order as the given input.

In any case, by appropriate combination of RN and NR algorithms for the forward and inverse FFT, the user may be completely shielded from the complications caused by the different ordering of data in the middle of the computation. Thus, it is useful to have the two different implementations.

5.5 Notes and References

Many "bit-reversal algorithms" that reorder data corresponding to reversing the order of bits in the array index have been proposed in the FFT literature. In [60], Karp reviewed, tested, and compared 30 methods he had recoded in a uniform style on computers with different memory systems. Two new algorithms that perform substantially better than the others were also proposed in [60].

Chapter 6

Performing Bit-Reversal by Repeated Permutation of Intermediate Results

It has been shown in the previous chapter that if the input data are first permuted into bit-reversed order, then the radix-2 DIF_{RN} FFT can be used to obtain naturally ordered output. This process is depicted for a $N = 8$ example in Figure 6.1. When the static permutation step is not performed in place, the bit-reversed input data are available in array b after the reordering.

Figure 6.1 Bit-reversing the input *before* performing in-place DIF_{RN} FFT.

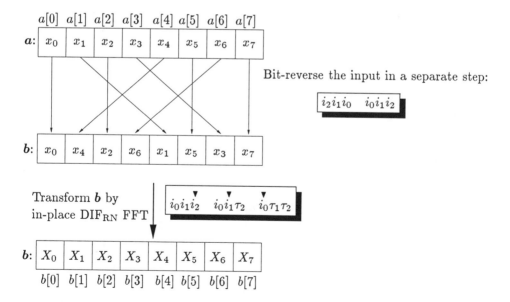

Of course, the same result can be accomplished by bit-reversing the output from an NR FFT algorithm as depicted in Figure 6.2 for the same example.

Figure 6.2 Bit-reversing the output *after* performing in-place DIF$_{NR}$ FFT.

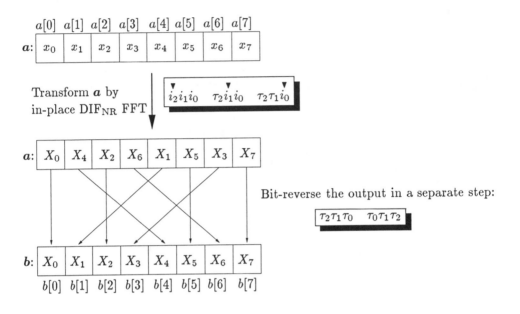

6.1   Combining Permutation with Butterfly Computation

The cost of extra memory accesses in a separate bit-reversing phase can be completely eliminated if data permutation is combined with the butterfly computation at each step. Such an alternative is presented in this section.

6.1.1   The ordered radix-2 DIF$_{NN}$ FFT

When input x and output X are both in natural order, the algorithm is referred to as an "ordered" FFT in the literature. The ordered radix-2 DIF FFT procedure was originally proposed by Stockham [30, 89]. The key to understanding what is required is to view each butterfly computation as consisting of one permutation step followed by one in-place computation step. These permutation steps reorder the initial input as well as the input to each subsequent subproblem, and the notation introduced in the previous chapters can be used to describe this process in a natural way.

Again using the $N = 8$ example above, with the input in the natural order, i.e., $a[i_2 i_1 i_0] = x_{i_2 i_1 i_0}$, the first in-place butterfly is denoted by $i_2 i_1 i_0$. If this butterfly operation is preceded by permuting the data in $a[i_2 i_1 i_0]$ to $b[i_1 i_0 i_2]$, it is natural to use

$$\boxed{\begin{array}{cc} i_2 i_1 i_0 & i_1 i_0 i_2 \\ a & b \end{array}}$$

to denote the permutation, which is followed by in-place butterfly computation denoted by

$$\boxed{\begin{array}{cc} i_1 i_0 i_2 & i_1 i_0 i_2 \ {\scriptstyle\blacktriangledown} \end{array}}$$

To show the combined effect, the two sequences above are condensed into

If the next step involves permuting the derivative $x^{(1)}_{i_2 i_1 i_0}$ in $b[i_1 i_0 i_2]$ to $a[i_0 i_1 i_2]$, then the derivatives $x^{(2)}_{i_2 i_1 i_0}$ and $x^{(3)}_{i_2 i_1 i_0}$ can both be computed *in-place* in $a[i_0 i_1 i_2]$. Since $x^{(3)}_{i_2 i_1 i_0} = X_{i_0 i_1 i_2}$ is contained in $a[i_0 i_1 i_2]$, the output frequencies X_m's are naturally ordered in array a as desired.

However, the easiest way to "understand" an algorithm may not be the most efficient way to "implement" an algorithm. For example, two implementations of a single butterfly computation step involving naturally ordered input elements $a[2] = x_2$ and $a[6] = x_6$ are depicted in Figures 6.3 and 6.4.

In Figure 6.3, the ordered DIF FFT is implemented as one understands it; i.e., a permutation step actually precedes the butterfly computation. As reflected by the fragment of pseudo-code displayed in Figure 6.3, memory locations $b[4]$ and $b[5]$ are each modified twice.

In Figure 6.4, the ordered DIF FFT is implemented without first permuting $a[2]$ to $b[4]$, $a[6]$ to $b[5]$, ..., etc. Instead, the derivative $x^{(1)}_2$ is computed and stored directly into $b[4]$, and so on. As reflected by the fragment of pseudo-code displayed in Figure 6.4, memory locations $b[4]$ and $b[5]$ are each modified only once. Since the same memory accessing pattern applies to all butterflies in every stage, this implementation eliminates all extra memory accesses in reordering intermediate results, and it is a more efficient way to implement the ordered DIF FFT algorithm. The complete pseudo-code program is given as Algorithm 6.1 below.

Figure 6.3 Naive Implementation of the (ordered) $\mathrm{DIF_{NN}}$ FFT.

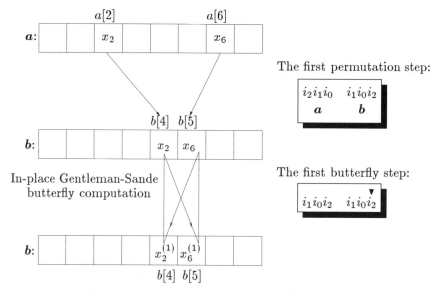

The first permutation step:

$$i_2 i_1 i_0 \qquad i_1 i_0 i_2$$
$$a \qquad\qquad b$$

The first butterfly step:

$$i_1 i_0 i_2 \qquad i_1 i_0 i_2$$

Pseudo-code executed:

$W :=$ Appropriate twiddle factor from $\boldsymbol{w}$
$b[4] := a[2]$ (Writing into memory)
$b[5] := a[6]$ (Writing into memory)

$Temp := b[4]$
$b[4] := Temp + b[5]$ (Writing into memory)
$b[5] := (Temp - b[5]) * W$ (Writing into memory)

Figure 6.4 Implement the (ordered) DIF$_{\text{NN}}$ FFT with no extra memory access.

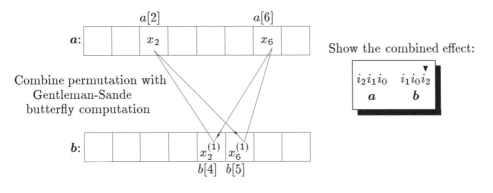

Show the combined effect:

$$i_2 i_1 i_0 \quad i_1 i_0 i_2$$
$$\quad a \qquad\quad b$$

Combine permutation with
Gentleman-Sande
butterfly computation

Pseudo-code executed:

$W :=$ Appropriate twiddle factor from $\boldsymbol{w}$
$b[4] := a[2] + a[6]$ (Writing into memory)
$b[5] := (a[2] - a[6]) * W$ (Writing into memory)

Algorithm 6.1 The (ordered) radix-2 DIF$_{NN}$ FFT algorithm.

<u>begin</u>

    $NumOfProblems := 1$ Initially: One problems of size N

    $ProblemSize := N$ $HalfSize = ProblemSize/2$

    $Distance := 1$

    $NotSwitchInput := true$

    <u>while</u> $ProblemSize > 1$ <u>do</u>   Halve each problem

        <u>if</u> $NotSwitchInput$ Array **a** contains input; array **b** contains output

            <u>for</u> $JFirst := 0$ <u>to</u> $NumOfProblems - 1$ <u>do</u>

                $J := JFirst$; $Jtwiddle := 0$

                $K := JFirst$

                <u>while</u> $J < N - 1$ <u>do</u>

                    $W := w[Jtwiddle]$

                    $b[J] := a[K] + a[K + N/2]$

                    $b[J + Distance] := (a[K] - a[K+N/2]) * W$

                    $Jtwiddle := Jtwiddle + NumOfProblems$ Assume $w[\ell] = \omega_N^\ell$

                    $J := J + 2 * NumOfProblems$

                    $K := K + NumOfProblems$

                <u>end while</u>

            <u>end for</u>

        $NotSwitchInput := false$

      <u>else</u>   Array **b** contains input; array **a** contains output

            <u>for</u> $JFirst := 0$ <u>to</u> $NumOfProblems - 1$ <u>do</u>

                $J := JFirst$; $Jtwiddle := 0$

                $K := JFirst$

                <u>while</u> $J < N - 1$ <u>do</u>

                    $W := w[Jtwiddle]$

                    $a[J] := b[K] + b[K + N/2]$

                    $a[J + Distance] := (b[K] - b[K + N/2]) * W$

                    $Jtwiddle := Jtwiddle + NumOfProblems$ Assume $w[\ell] = \omega_N^\ell$

                    $J := J + 2 * NumOfProblems$

                    $K := K + NumOfProblems$

                <u>end while</u>

            <u>end for</u>

        $NotSwitchInput := true$

      <u>end if</u>

      $NumOfProblems := NumOfProblems * 2$

      $ProblemSize := ProblemSize/2$

      $Distance := Distance * 2$

    <u>end while</u>

<u>end</u>

6.1.2   The shorthand notation

As usual, assuming that x is initially contained in a in the natural order, a second array b would alternately contain the data. The entire computation process, along with the use of the two arrays, is depicted below.

$$
\begin{array}{cccc}
i_2 i_1 i_0 & i_1 i_0 i_2 & i_0 i_1 \tau_2 & i_0 \tau_1 \tau_2 \\
a & b & a & b
\end{array}
$$

Note that the corresponding twiddle factors are

$$
w_N^{i_1 i_0}, \quad w_N^{i_0 0}, \quad w_N^0 = 1,
$$

because DIF_{NR}, DIF_{RN}, and DIF_{NN} FFT algorithms all transform the same element $x_{i_2 i_1 i_0}$, although they refer to the *different addresses* of $x_{i_2 i_1 i_0}$ in expressing the same algorithm.

Once again, all details of the (ordered) DIF_{NN} FFT can be captured by a shorthand notation together with the twiddle factors.

6.2   Applying the Ordered DIF FFT to a $N = 32$ Example

Generalizing the shorthand notation for $N = 32$, the following sequence represents all five stages of permutation and computation depicted in Figure 6.5.

$$
\begin{array}{cccccc}
i_4 i_3 i_2 i_1 i_0 & i_3 i_2 i_1 i_0 i_4 & i_2 i_1 i_0 i_3 \tau_4 & i_1 i_0 i_2 \tau_3 \tau_4 & i_0 i_1 \tau_2 \tau_3 \tau_4 & i_0 \tau_1 \tau_2 \tau_3 \tau_4 \\
a & b & a & b & a & b
\end{array}
$$

The corresponding twiddle factors are

$$
w_N^{i_3 i_2 i_1 i_0}, \quad w_N^{i_2 i_1 i_0 0}, \quad w_N^{i_1 i_0 0 0}, \quad w_N^{i_0 0 0 0}, \quad w_N^0 = 1.
$$

By comparing Figure 6.6, where the butterflies associated with a particular pair of resulting subproblems are shown without the cluttering of others, with the two unordered DIF FFT in Figures 4.4 and 5.3, one immediately observes that

> all three variants of the DIF FFT treat exactly the same pairs of subproblems during each stage of the computation.

Thus they all implement the same radix-2 DIF FFT algorithm.

Figure 6.5 Butterflies of the (ordered) $\mathrm{DIF_{NN}}$ FFT algorithm.

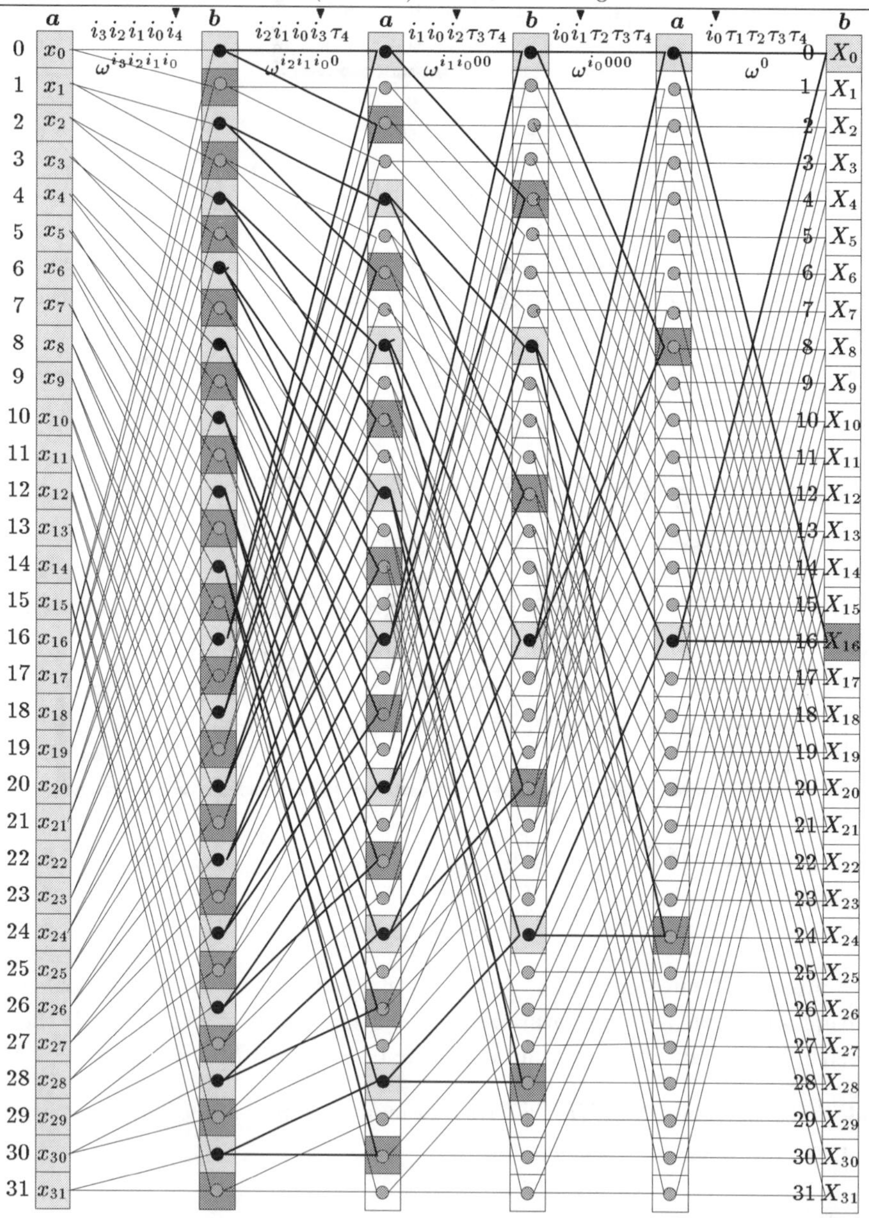

Figure 6.6 Identifying the subproblems paired up by the (ordered) DIF$_{\text{NN}}$ FFT.

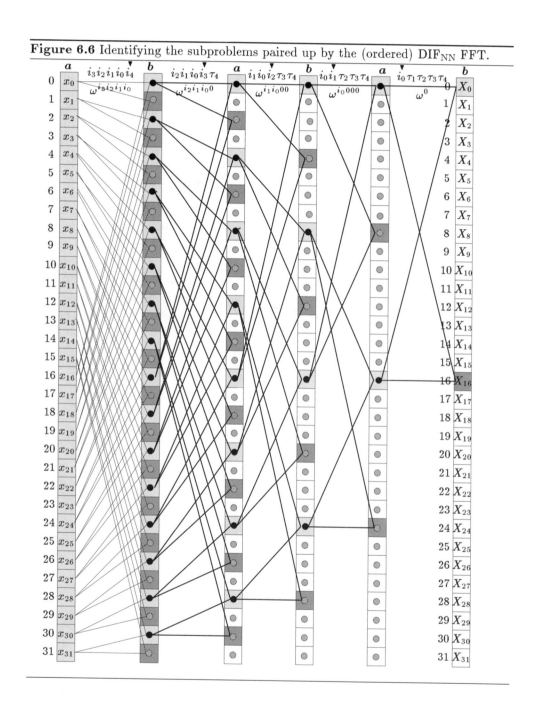

6.3   In-Place Ordered (or Self-Sorting) Radix-2 FFT Algorithms.

Another class of "ordered" FFTs performs in-place permutation and consequently does not need a second array; they are the so-called "self-sorting in-place" algorithms. This class contains variants of the prime-factor algorithms [20, 81, 99] and a radix-2 FFT [58]. This class has been further extended to include self-sorting in-place radix-3, radix-4, radix-5, and finally mixed-radix FFTs [101]. The radix-2 algorithm is relevant to the discussion in this chapter. Using the notation developed earlier, the process of applying the self-sorting in-place radix-2 DIF FFT to array a, which contains naturally ordered x, is depicted below for $N = 32$.

$$
\underset{a}{i_4 i_3 i_2 i_1 i_0} \quad \underset{a}{i_0 i_3 i_2 i_1 i_4} \quad \underset{a}{i_0 i_1 i_2 i_3 \tau_4} \quad \underset{a}{i_0 i_1 i_2 \tau_3 \tau_4} \quad \underset{a}{i_0 i_1 \tau_2 \tau_3 \tau_4} \quad \underset{a}{i_0 \tau_1 \tau_2 \tau_3 \tau_4}
$$

Observe that the permutation always involves bits in symmetric positions: e.g., in step 1, the left-most bit i_4 switches with the right-most bit i_0 and in step 2, bit i_3, the second bit from the left end, switches with bit i_1, the second bit from the right end. Accordingly, the ordering of the bits is "reversed" after only two steps, and the permutation can be implemented using "pairwise" interchanges. The contents in $a[0i_3i_2i_1 1]$ and $a[1i_3i_2i_1 0]$ are switched in step 1 and the contents in $a[i_0 0i_2 1\tau_4]$ and $a[i_0 1i_2 0\tau_4]$ are switched in step 2. Since each "pairwise" interchange can be done using a single temporary location, the array b is not needed.

Chapter 7

An In-Place Radix-2 DIT FFT for Input in Natural Order

The NR, RN, and NN algorithms implementing DIF (*decimation-in-frequency*) FFT were presented in Chapters 4, 5, and 6. Corresponding to them, there are also three variants of the DIT (*decimation-in-time*) FFT, and they are developed in this and the following two chapters.

The three DIT FFT algorithms will be presented using the notation developed in the previous chapters. Accordingly, they are referred to as DIT$_{\text{NR}}$, DIT$_{\text{RN}}$, and DIT$_{\text{NN}}$ FFT algorithms. The DIT$_{\text{NR}}$ and DIT$_{\text{RN}}$ algorithms implement in-place DIT FFT on naturally ordered and bit-reversed input data, whereas the DIT$_{\text{NN}}$ algorithm allows repeated permutation of the intermediate results and can thus produce naturally ordered output from naturally ordered input.

Since both DIF FFT and DIT FFT implement the same Discrete Fourier Transform, one may argue intuitively that the final result which overwrites an input element x_k must remain unchanged in either implementation, and that the many results obtained previously in Chapters 4, 5, and 6 for the three DIF FFT algorithms should apply to the corresponding DIT FFT. However, to make this chapter self-contained, it is useful to develop these iterative DIT FFT algorithms from its recursive definition, and this approach is adopted here.

Since the concepts introduced before for the DIF FFT will not be repeated, it is recommended that Chapters 4, 5, and 6 be studied before Chapter 7.

7.1 Understanding the Recursive DIT FFT and its In-Place Implementation

Recall that the recursive radix-2 DIT FFT algorithm was derived in Chapter 3, in which the Cooley-Tukey butterfly in Figure 7.1 represents the last combination step in computing the transform

$$(7.1) \qquad X_r = \sum_{k=0}^{\frac{N}{2}-1} x_{2k}\omega_N^{r(2k)} + \omega_N^r \sum_{k=0}^{\frac{N}{2}-1} x_{2k+1}\omega_N^{r(2k)}, \qquad r = 0, 1, \ldots, N-1.$$

Figure 7.1 The Cooley-Tukey butterfly.

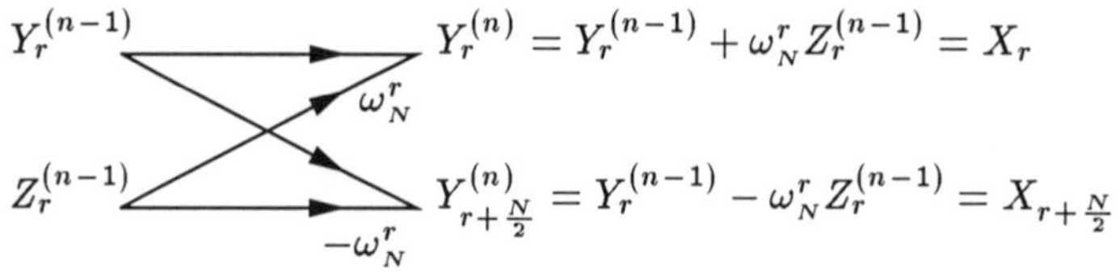

$$Y_r^{(n-1)} \longrightarrow Y_r^{(n)} = Y_r^{(n-1)} + \omega_N^r Z_r^{(n-1)} = X_r$$

$$\omega_N^r$$

$$Z_r^{(n-1)} \longrightarrow Y_{r+\frac{N}{2}}^{(n)} = Y_r^{(n-1)} - \omega_N^r Z_r^{(n-1)} = X_{r+\frac{N}{2}}$$

$$-\omega_N^r$$

Assuming $N = 2^n$, $Y_r^{(n-1)}$ and $Z_r^{(n-1)}$ in Figure 7.1 are solutions to the two half-size subproblems defined by

$$(7.2) \qquad Y_r^{(n-1)} = \sum_{k=0}^{\frac{N}{2}-1} x_{2k} \omega_N^{r(2k)} = \sum_{k=0}^{\frac{N}{2}-1} y_k^{(n-1)} \omega_{\frac{N}{2}}^{rk}, \quad r = 0, 1, \ldots, N/2 - 1$$

and

$$(7.3) \qquad Z_r^{(n-1)} = \sum_{k=0}^{\frac{N}{2}-1} x_{2k+1} \omega_N^{r(2k)} = \sum_{k=0}^{\frac{N}{2}-1} z_k^{(n-1)} \omega_{\frac{N}{2}}^{rk}, \quad r = 0, 1, \ldots, N/2 - 1.$$

Note that $y_k^{(n-1)} \equiv x_{2k}$ and $z_k^{(n-1)} \equiv x_{2k+1}$, $k = 0, 1, \ldots, 2^{n-1} - 1$, identify input elements to the two subproblems of size 2^{n-1} as depicted in Figure 7.2 for $N = 8$.

Since each subproblem is to be solved by the same DIT FFT algorithm recursively, two subproblems of size $N = 2^{n-2}$ are defined by the even and odd elements from $y_k^{(n-1)}$, namely, $y_k^{(n-2)} \equiv y_{2k}^{(n-1)}$, and $z_k^{(n-2)} \equiv y_{2k+1}^{(n-1)}$ as depicted in Figure 7.3; the other two subproblems of size $N = 2^{n-2}$ are defined by the even and odd elements from $z_k^{(n-1)}$, namely, $y_k^{(n-2)} \equiv z_{2k}^{(n-1)}$, and $z_k^{(n-2)} \equiv z_{2k+1}^{(n-1)}$ as depicted in Figure 7.4. The subdivision steps continue until the subproblem size becomes one, and solution to each subproblem is simply itself; i.e., $Y_0^{(0)} = y_0^{(0)}$ and $Z_0^{(0)} = z_0^{(0)}$.

Therefore, the DIT FFT algorithm begins its computation by combining all pairs of (properly identified) $y_0^{(0)}$ and $z_0^{(0)}$ to form solutions to subproblems of size two, and so on. For $N = 8$, the three combination steps are depicted in Figures 7.3, 7.4, and 7.5. As expected, the in-place DIT FFT overwrites the naturally ordered input by bit-reversed output after the last combination step in Figure 7.5.

Figure 7.2 The first division step of in-place $\mathrm{DIT_{NR}}$ FFT.

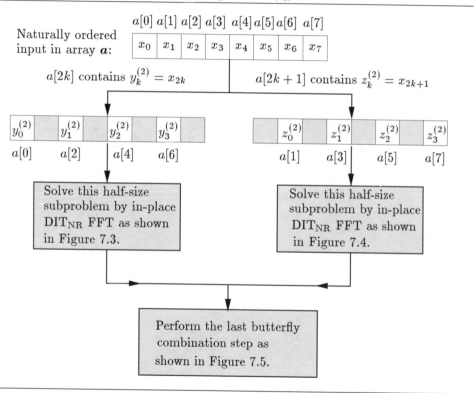

Figure 7.3 Solve one half-size subproblem by in-place DIT_{NR} FFT.

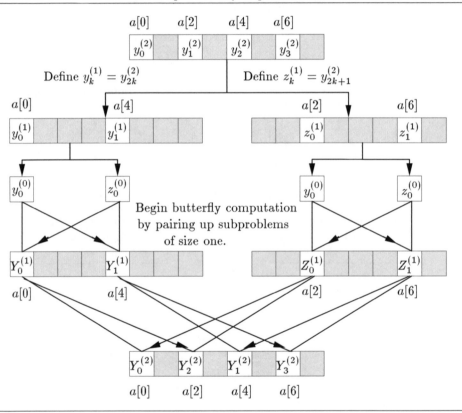

Figure 7.4 Solve the second half-size subproblem by in-place DIT_{NR} FFT.

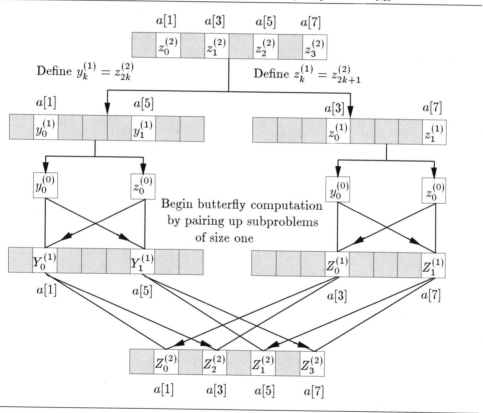

Figure 7.5 Solve the entire problem by in-place DIT$_{\text{NR}}$ FFT.

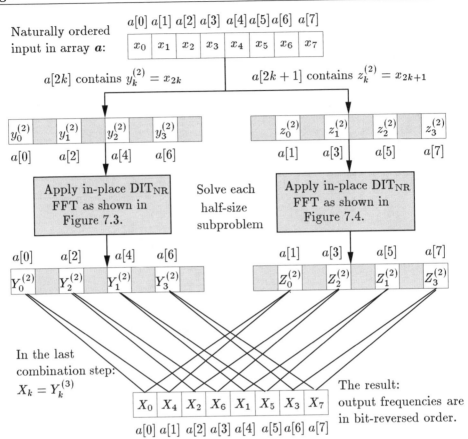

7.2 Developing the Iterative In-Place DIT FFT

Since the input data are in natural order, array element $a[m]$ is assumed to contain x_m in the development of the pseudo-code program in this section. For easy comparison with the DIF$_{\text{NR}}$ FFT specified in Table 4.2, the butterfly computation steps depicted in Figures 7.3 – 7.5 are summarized in Table 7.1, where all butterfly pairs are identified by the addresses of $a[J]$ and $a[J + Distance]$.

Table 7.1 Decimal and binary addresses for $a[J]$ and $a[J+Distance]$ in Figures 7.3–7.5.

Butterfly Computation	Current HalfSize	Decimal a[J]	Binary J	Decimal a[J + Distance]	Binary J + Distance
Stage 1	$HalfSize = N/2^3 = 1$	$a[0]$	000	$a[4]$	100
	($Distance = 2^2 = 4$)	$a[2]$	010	$a[6]$	110
		$a[1]$	001	$a[5]$	101
		$a[3]$	011	$a[7]$	111
Stage 2	$HalfSize = N/2^2 = 2$	$a[0]$	000	$a[2]$	010
	($Distance = 2$)	$a[4]$	100	$a[6]$	110
		$a[1]$	001	$a[3]$	011
		$a[5]$	101	$a[7]$	111
Stage 3	$HalfSize = N/2 = 4$	$a[0]$	000	$a[1]$	001
	($Distance = 1$)	$a[2]$	010	$a[3]$	011
		$a[4]$	100	$a[5]$	101
		$a[6]$	110	$a[7]$	111

Observe that the distance between $a[J]$ and $a[J + Distance]$ is a power of 2, so the binary number property 5 (page 34) applies here as before. That is, the addresses $a[J]$ and $a[J + Distance]$, denoted as *binary numbers* $i_{n-1} \cdots i_1 i_0$, differ only in bit i_k when $Distance = 2^k$ and the algorithm can be expressed in terms of binary addresses as shown below.

Algorithm 7.1 The radix-2 DIT FFT algorithm in terms of binary addresses:

begin
    $k := n - 1$ Assume problem size $N = 2^n$
    while $k \geq 0$ **do**   Combine two subproblems
            Apply Cooley-Tukey butterfly computation
            to "combine" all pairs of array elements whose
            binary addresses differ in bit i_k
            $k := k - 1$
    end while
end

Accordingly, the shorthand notation developed for the DIF_{NR} FFT applies here. That is, for $N = 8$, the three DIT butterfly computation steps are also represented by the sequence

$$\boxed{\quad \overset{\blacktriangledown}{i_2 i_1 i_0} \quad \overset{\blacktriangledown}{\tau_2 i_1 i_0} \quad \overset{\blacktriangledown}{\tau_2 \tau_1 i_0} \quad}$$

However, the corresponding twiddle factors are now different, and they are identified in the next section.

7.2.1 Identifying the twiddle factors in the DIT FFT

Observe that in Table 7.1, each butterfly step *combines* pairs of subproblems of *HalfSize* beginning with *HalfSize* $= 1$ and twiddle factor $\pm w_N^0 = \pm 1$. Following the definition of the Cooley-Tukey butterfly in Figure 7.1, the second butterfly step combines pairs of subproblems of *HalfSize* $= 2$, and the corresponding twiddle factors are $\pm w_4^0 = \pm w_N^0 = \pm 1$, and $\pm w_4^1 = \pm w_N^{N/4}$, and so on. For $N = 8$, the twiddle factors corresponding to all combination steps are identified in Table 7.2.

Table 7.2 Relating twiddle factors to binary addresses of the pair $a[J]$ and $a[J + Distance]$ $(N = 8)$.

Cooley-Tukey Butterfly	Actual Modification of $a[J]$ & $a[J + Distance]$	Butterfly Groups	A Binary Address Based Notation
Stage $i_2 i_1 i_0$ (Distance = 4)	$a[0] = a[0] + w_N^0 a[4]$ $a[4] = a[0] - w_N^0 a[4]$ $a[1] = a[1] + w_N^0 a[5]$ $a[5] = a[1] - w_N^0 a[5]$ $a[2] = a[2] + w_N^0 a[6]$ $a[6] = a[2] - w_N^0 a[6]$ $a[3] = a[3] + w_N^0 a[7]$ $a[7] = a[3] - w_N^0 a[7]$	Group 0 (4 Pairs)	$a[0 i_1 i_0] = a[0 i_1 i_0] + w_N^0 a[1 i_1 i_0]$ $a[1 i_1 i_0] = a[0 i_1 i_0] - w_N^0 a[1 i_1 i_0]$
Stage $\tau_2 i_1 i_0$ (Distance = 2)	$a[0] = a[0] + w_N^0 a[2]$ $a[2] = a[0] - w_N^0 a[2]$ $a[1] = a[1] + w_N^0 a[3]$ $a[3] = a[1] - w_N^0 a[3]$	Group 0 (2 pairs)	$a[i_2 0 i_0] = a[i_2 0 i_0] + w_N^{i_2 0} a[i_2 1 i_0]$ $a[i_2 1 i_0] = a[i_2 0 i_0] - w_N^{i_2 0} a[i_2 1 i_0]$
	$a[4] = a[4] + w_N^2 a[6]$ $a[6] = a[4] - w_N^2 a[6]$ $a[5] = a[5] + w_N^2 a[7]$ $a[7] = a[5] - w_N^2 a[7]$	Group 1 (2 pairs)	
Stage $\tau_2 \tau_1 i_0$ (Distance = 1)	$a[0] = a[0] + w_N^0 a[1]$ $a[1] = a[0] - w_N^0 a[1]$	Group 0 (1 pair)	$a[i_2 i_1 0] = a[i_2 i_1 0] + w_N^{i_1 i_2} a[i_2 i_1 1]$ $a[i_2 i_1 1] = a[i_2 i_1 0] - w_N^{i_1 i_2} a[i_2 i_1 1]$
	$a[2] = a[2] + w_N^2 a[3]$ $a[3] = a[2] - w_N^2 a[3]$	Group 1 (1 pair)	
	$a[4] = a[4] + w_N^1 a[5]$ $a[5] = a[4] - w_N^1 a[5]$	Group 2 (1 pair)	
	$a[6] = a[6] + w_N^3 a[7]$ $a[7] = a[6] - w_N^3 a[7]$	Group 3 (1 pair)	

For easy comparison with Table 4.3 for the DIF_{NR} FFT, the corresponding twiddle factors and their shorthand notation are summarized in Table 7.3. In order not to invoke a bit-reversing subroutine in the program, it is assumed that the pre-computed twiddle factors are stored in bit-reversed order although the input data are in natural order. Therefore, $w[i_2 i_1] = w_N^{i_1 i_2}$ is used in Table 7.3.

Table 7.3 Identifying twiddle factors ($N = 8$) for DIT_{NR} FFT.

Butterfly Computation	Current HalfSize	Decimal $\pm w[Jtwiddle]$	Binary Jtwiddle w.r.t modified $x_{i_2 i_1 i_0}$	Shorthand Notation w.r.t. modified $x_{i_2 i_1 i_0}$
Stage $i_2 i_1 i_0$	$HalfSize = 1$	$\pm w[0] = \pm w_N^0$	00	$\pm w_N^0 = \pm 1$
Stage $\tau_2 i_1 i_0$	$HalfSize = 2$	$\pm w[0] = \pm w_N^0$	$0 i_2 = 00$	$\pm w_N^{i_2 0}$
		$\pm w[1] = \pm w_N^2$	$0 i_2 = 01$	
Stage $\tau_2 \tau_1 i_0$	$HalfSize = 4$	$\pm w[0] = \pm w_N^0$	$i_2 i_1 = 00$	$\pm w_N^{i_1 i_2}$
		$\pm w[1] = \pm w_N^2$	$i_2 i_1 = 01$	
		$\pm w[2] = \pm w_N^1$	$i_2 i_1 = 10$	
		$\pm w[3] = \pm w_N^3$	$i_2 i_1 = 11$	

7.2.2 The pseudo-code program for the DIT_{NR} FFT algorithm

In contrast to the DIF FFT pseudo-code program given by Algorithm 4.2, the complete DIT FFT version is given below as Algorithm 7.2. For simplicity, the twiddle factors are assumed to have been stored in the array w in bit-reversed order. For example, for $N = 8$, the $N/2 = 4$ pre-computed twiddle factors are stored as $w[i_2 i_1] = w_N^{i_1 i_2}$; for $N = 32$, the $N/2 = 16$ pre-computed twiddle factors are stored as $w[i_4 i_3 i_2 i_1] = w_N^{i_1 i_2 i_3 i_4}$.

7.3 Shorthand Notation and a $N = 32$ Example

As derived in the previous section, the butterfly computation steps in Figures 7.3–7.5 (for $N = 8$) are fully specified by the sequence

$$i_2 i_1 i_0 \qquad \tau_2 i_1 i_0 \qquad \tau_2 \tau_1 i_0$$

together with the corresponding twiddle factors

$$\pm w_N^0 = \pm 1, \quad \pm w_N^{i_2 0}, \quad \pm w_N^{i_1 i_2}.$$

For $N = 32$, the entire Figure 7.6 can be replaced by the sequence

$$i_4 i_3 i_2 i_1 i_0 \qquad \tau_4 i_3 i_2 i_1 i_0 \qquad \tau_4 \tau_3 i_2 i_1 i_0 \qquad \tau_4 \tau_3 \tau_2 i_1 i_0 \qquad \tau_4 \tau_3 \tau_2 \tau_1 i_0$$

Algorithm 7.2 The iterative radix-2 DIT FFT algorithm in pseudo-code.

---

begin
  $PairsInGroup := N/2$ Begin with $N/2$ butterflies in one group
  $NumOfGroups := 1$ Same twiddle factor is employed in a group
  $Distance := N/2$
  while $NumOfGroups < N$ **do**
        for $K := 0$ **to** $NumOfGroups - 1$ **do**   Combine pairs in each group
          $JFirst := 2 * K * PairsInGroup$
          $JLast := JFirst + PairsInGroup - 1$
          $Jtwiddle := K$ Access consecutive $w[m]$
          $W := w[Jtwiddle]$ Assume $w[m] = w_N^\ell$, m bit-reverses ℓ
          for $J := JFirst$ **to** $JLast$ **do**
            $Temp := W * a[J + Distance]$
            $a[J + Distance] := a[J] - Temp$
            $a[J] := a[J] + Temp$
          end for
        end for
        $PairsInGroup := PairsInGroup/2$
        $NumOfGroups := NumOfGroups * 2$
        $Distance := Distance/2$
  end while
end

---

with the understanding that on input, $a[i_4 i_3 i_2 i_1 i_0]$ contains $x_{i_4 i_3 i_2 i_1 i_0}$, and on output, $a[i_4 i_3 i_2 i_1 i_0]$ contains the bit-reversed $X_{i_0 i_1 i_2 i_3 i_4}$. Refer to Figure 7.6 for the decimal subscripts of all 32 bit-reversed output elements $X_{i_0 i_1 i_2 i_3 i_4}$.

The corresponding twiddle factors are

$$\pm w_N^0 = \pm 1, \ \pm w_N^{i_4 000}, \ \pm w_N^{i_3 i_4 00}, \ \pm w_N^{i_2 i_3 i_4 0}, \ \pm w_N^{i_1 i_2 i_3 i_4}.$$

As identified in Table 7.3 and assumed in Algorithm 7.2, these twiddle factors are pre-computed and stored in w in bit-reversed order. Hence, for $N = 32$, $w[i_4 i_3 i_2 i_1] = w_N^{i_1 i_2 i_3 i_4}$, which implies $w[0] = w_N^0$, $w[000 i_4] = w_N^{i_4 000}$, $w[00 i_4 i_3] = w_N^{i_3 i_4 00}$, and $w[0 i_4 i_3 i_2] = w_N^{i_2 i_3 i_4 0}$.

Figure 7.6 Butterflies of the in-place DIT$_{NR}$ FFT algorithm.

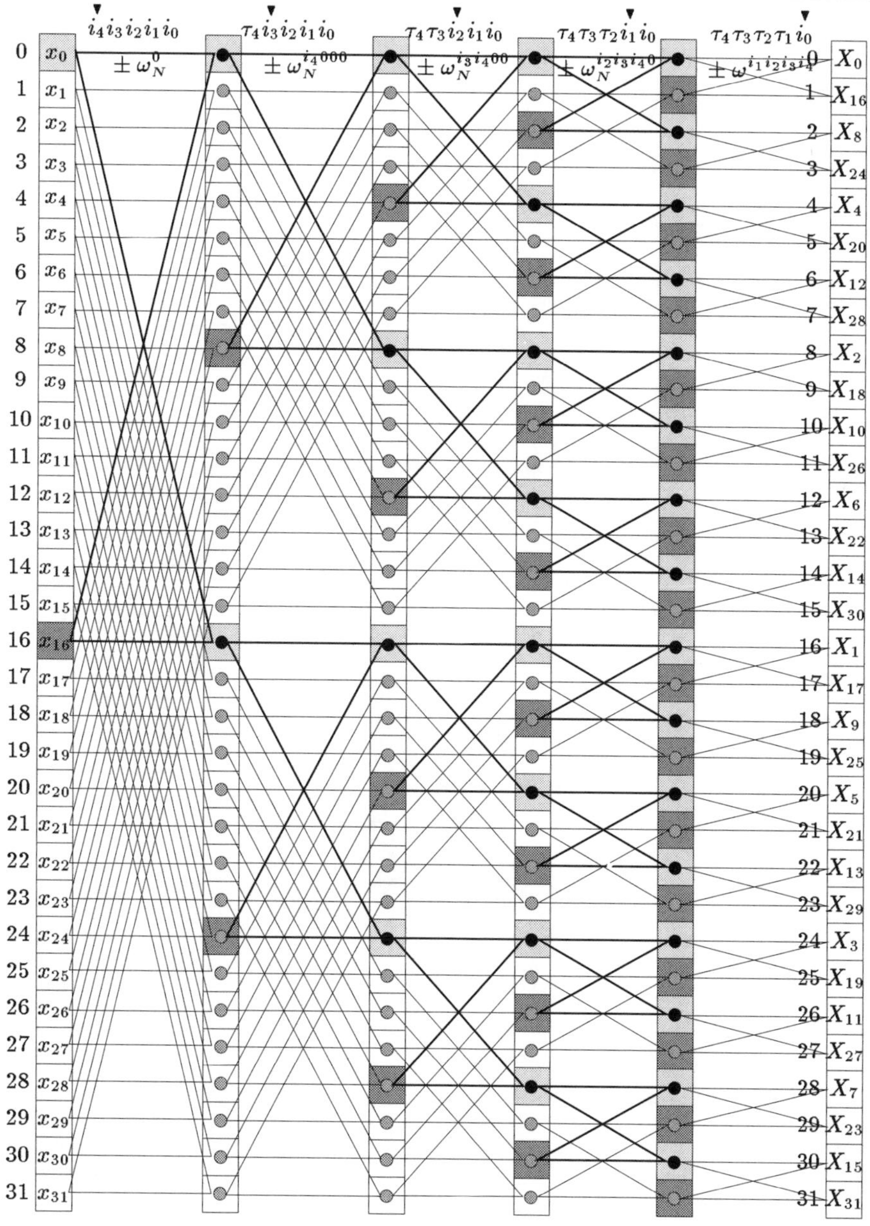

Chapter 8

An In-Place Radix-2 DIT FFT for Input in Bit-Reversed Order

This chapter is a sequel of Chapter 7, so readers are assumed to be familiar with applying the in-place DIT FFT to naturally ordered input as shown in Figures 7.2–7.5.

Recall that $x_{i_{n-1}\cdots i_1 i_0}$ was assumed to be contained in $a[i_{n-1}\cdots i_1 i_0]$ when the in-place DIT FFT algorithm was expressed in terms of the binary addresses of array a in Chapter 7. Accordingly, the following algorithm is correct regardless of where x_k is located.

Algorithm 8.1 The radix-2 DIT FFT algorithm in terms of binary subscripts.

---

begin
    $k := n - 1$ Assume problem size $N = 2^n$
    while $k \geq 0$ **do**   Combine two subproblems
            Apply Cooley-Tukey butterfly computation
            to "combine" all pairs of elements of x *whose*
            binary subscripts differ in bit i_k
            $k := k - 1$
    end while
end

Now, if the input data are in bit-reversed order, the same computation must be performed on the same elements although they are stored in bit-reversed array locations. The application of the in-place DIT_{RN} FFT to bit-reversed input of size $N = 8$ is depicted in Figures 8.1–8.4. As expected, the bit-reversed input is overwritten by naturally ordered output.

Figure 8.1 The first division step of the DIT_{RN} FFT.

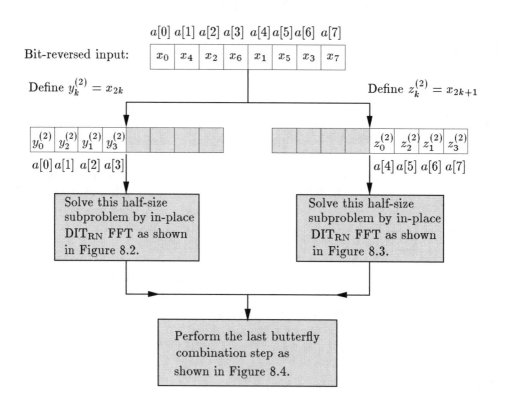

Figure 8.2 Solve one half-size subproblem by in-place DIT_{RN} FFT.

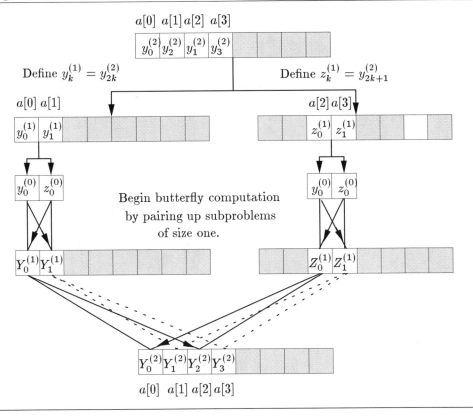

Figure 8.3 Solve one half-size subproblem by in-place DIT_{RN} FFT.

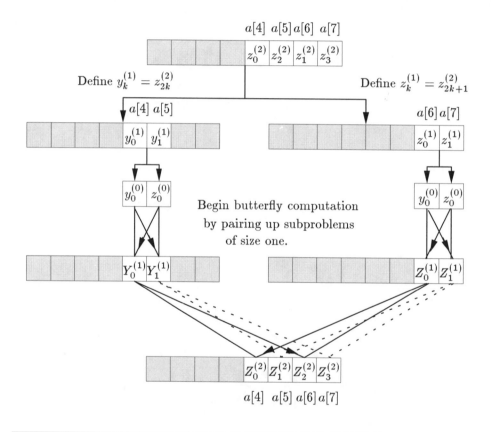

Figure 8.4 Solve the entire problem by in-place DIT_{RN} FFT.

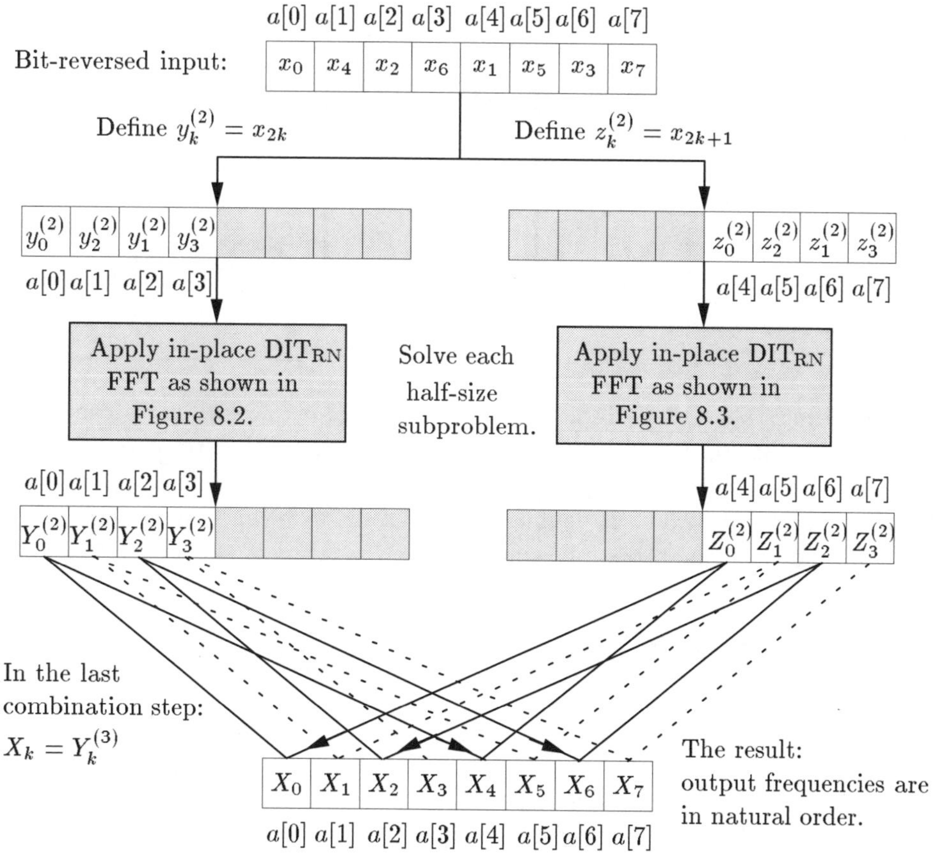

$a[0]\ a[1]\ a[2]\ a[3]\ a[4]\ a[5]\ a[6]\ a[7]$

Bit-reversed input:

x_0	x_4	x_2	x_6	x_1	x_5	x_3	x_7

Define $y_k^{(2)} = x_{2k}$ Define $z_k^{(2)} = x_{2k+1}$

$y_0^{(2)}$	$y_2^{(2)}$	$y_1^{(2)}$	$y_3^{(2)}$				

$a[0]\ a[1]\ a[2]\ a[3]$

				$z_0^{(2)}$	$z_2^{(2)}$	$z_1^{(2)}$	$z_3^{(2)}$

$a[4]\ a[5]\ a[6]\ a[7]$

Apply in-place DIT_{RN} FFT as shown in Figure 8.2.

Solve each half-size subproblem.

Apply in-place DIT_{RN} FFT as shown in Figure 8.3.

$a[0]\ a[1]\ a[2]\ a[3]$

$Y_0^{(2)}$	$Y_1^{(2)}$	$Y_2^{(2)}$	$Y_3^{(2)}$				

$a[4]\ a[5]\ a[6]\ a[7]$

				$Z_0^{(2)}$	$Z_1^{(2)}$	$Z_2^{(2)}$	$Z_3^{(2)}$

In the last combination step:

$X_k = Y_k^{(3)}$

X_0	X_1	X_2	X_3	X_4	X_5	X_6	X_7

The result: output frequencies are in natural order.

$a[0]\ a[1]\ a[2]\ a[3]\ a[4]\ a[5]\ a[6]\ a[7]$

8.1 Developing the Iterative In-Place DIT$_{RN}$ FFT

To facilitate the development of an iterative algorithm, the butterfly computation steps depicted in Figures 8.2–8.4 are summarized in Table 8.1, where all butterfly pairs are identified by the addresses of $a[J]$ and $a[J + Distance]$.

Table 8.1 Decimal and binary addresses for $a[J]$ and $a[J+Distance]$ in Figures 8.2–8.4.

Butterfly Computation	Current HalfSize	Decimal $a[J]$	Binary J	Decimal $a[J + Distance]$	Binary $J + Distance$
Stage 1	$HalfSize = N/2^3 = 1$ ($Distance = 1$)	$a[0]$	000	$a[1]$	001
		$a[2]$	010	$a[3]$	011
		$a[4]$	100	$a[5]$	101
		$a[6]$	110	$a[7]$	111
Stage 2	$HalfSize = N/2^2 = 2$ ($Distance = 2$)	$a[0]$	000	$a[2]$	010
		$a[1]$	001	$a[3]$	011
		$a[4]$	100	$a[6]$	110
		$a[5]$	101	$a[7]$	111
Stage 3	$HalfSize = N/2 = 4$ ($Distance = 2^2 = 4$)	$a[0]$	000	$a[4]$	100
		$a[1]$	001	$a[5]$	101
		$a[2]$	010	$a[6]$	110
		$a[3]$	011	$a[7]$	111

Comparing Table 8.1 to Table 7.1, it is clear that because $x_{i_2i_1i_0}$ is contained in $a[i_0i_1i_2]$, elements $x_{0i_1i_0}$ and $x_{1i_1i_0}$, which form pairs during the first stage of butterfly computation, are found in $a[i_0i_10]$ and $a[i_0i_11]$, and so on. Consequently, the value of $Distance$ starts from 1, and it doubles in each of the following stages. This observation immediately leads to the following algorithm in terms of the binary addresses of elements in array a.

Algorithm 8.2 The in-place DIT$_{RN}$ FFT algorithm in terms of binary addresses:

begin
$\quad \ell := 0$ Assume problem size $N = 2^n$
$\quad$ **while** $\ell \le n - 1$ **do**   Combine two subproblems
$\quad\quad$ *Apply Cooley-Tukey butterfly computation*
$\quad\quad$ *to "combine" all pairs of array elements whose*
$\quad\quad$ *binary addresses differ in the ℓth bit from the right*
$\quad\quad \ell := \ell + 1$
$\quad$ **end while**
end

Accordingly, the shorthand notation developed for the DIF$_{RN}$ FFT, which transforms bit-reversed input, applies here. That is, for $N = 8$, the three DIT butterfly computation steps are also represented by the sequence

$$i_2i_1i_0 \quad \overset{\blacktriangledown}{i_0i_1i_2} \quad \overset{\blacktriangledown}{i_0i_1\tau_2} \quad \overset{\blacktriangledown}{i_0\tau_1\tau_2}$$

Once again, the bit-reversed ordering of the input is reflected by beginning the sequence with permuting $x_{i_2 i_1 i_0}$ from the assumed location $a[i_2 i_1 i_0]$ to $a[i_0 i_1 i_2]$.

However, the corresponding twiddle factors are different from the $\mathrm{DIF_{RN}}$ FFT, as explained in the next section.

8.1.1 Identifying the twiddle factors in the $\mathrm{DIT_{RN}}$ FFT

Observe that in Table 8.1, each butterfly step *combines* pairs of subproblems of *HalfSize* beginning with *HalfSize* $= 1$ and twiddle factor $\pm w_N^0 = \pm 1$. Following the definition of the Cooley-Tukey butterfly in Figure 7.1, the second butterfly step combines pairs of subproblems of *HalfSize* $= 2$, and the corresponding twiddle factors are $\pm w_4^0 = \pm w_N^0 = \pm 1$, and $\pm w_4^1 = \pm w_N^{N/4}$, and so on. For $N = 8$, the twiddle factors corresponding to all combination steps are identified in Table 8.2.

Table 8.2 Relating twiddle factors to binary addresses of the pair $a[J]$ and $a[J + Distance]$.

Cooley-Tukey Butterfly	Actual Modification of $a[J]$ & $a[J + Distance]$	Butterfly Groups	A Binary Address Based Notation
Stage $i_0 i_1 i_2$ $(Distance = 1)$	$a[0] = a[0] + w_N^0 a[1]$ $a[1] = a[0] - w_N^0 a[1]$ $a[2] = a[2] + w_N^0 a[3]$ $a[3] = a[2] - w_N^0 a[3]$ $a[4] = a[4] + w_N^0 a[5]$ $a[5] = a[4] - w_N^0 a[5]$ $a[6] = a[6] + w_N^0 a[7]$ $a[7] = a[6] - w_N^0 a[7]$	Group 0 (4 Pairs)	$a[i_0 i_1 0] = a[i_0 i_1 0] + w_N^0 a[i_0 i_1 1]$ $a[i_0 i_1 1] = a[i_0 i_1 0] - w_N^0 a[i_0 i_1 1]$
Stage $i_0 i_1 \tau_2$ $(Distance = 2)$	$a[0] = a[0] + w_N^0 a[2]$ $a[2] = a[0] - w_N^0 a[2]$ $a[4] = a[4] + w_N^0 a[6]$ $a[6] = a[4] - w_N^0 a[6]$	Group 0 (2 pairs)	$a[i_0 0 i_2] = a[i_0 0 i_2] + w_N^{i_2 0} a[i_0 1 i_2]$ $a[i_0 1 i_2] = a[i_0 0 i_2] - w_N^{i_2 0} a[i_0 1 i_2]$
	$a[1] = a[1] + w_N^2 a[1]$ $a[3] = a[3] - w_N^2 a[3]$ $a[5] = a[5] + w_N^2 a[7]$ $a[7] = a[5] - w_N^2 a[7]$	Group 1 (2 pairs)	
Stage $i_0 \tau_1 \tau_2$ $(Distance = 4)$	$a[0] = a[0] + w_N^0 a[4]$ $a[4] = a[0] - w_N^0 a[4]$	Group 0 (1 pair)	$a[0 i_1 i_2] = a[0 i_1 i_2] + w_N^{i_1 i_2} a[1 i_1 i_2]$ $a[1 i_1 i_2] = a[0 i_1 i_2] - w_N^{i_1 i_2} a[1 i_1 i_2]$
	$a[1] = a[1] + w_N^1 a[5]$ $a[5] = a[1] - w_N^1 a[5]$	Group 1 (1 pair)	
	$a[2] = a[2] + w_N^2 a[6]$ $a[6] = a[4] - w_N^2 a[6]$	Group 2 (1 pair)	
	$a[3] = a[3] + w_N^3 a[7]$ $a[7] = a[3] - w_N^3 a[7]$	Group 3 (1 pair)	

For easy comparison with Table 7.3 for the DIT_{NR} FFT the corresponding twiddle factors and its shorthand notations are summarized in Table 8.3. Note that although the input data are in bit-reversed order, it is assumed that the pre-computed twiddle factors are stored in natural order in the array w. That is, $w[\ell] = w_N^\ell$, $\ell = 0, 1, \ldots, N/2 - 1$. Therefore, $w[i_1 i_2] = w_N^{i_1 i_2}$ is used in Table 8.3.

Table 8.3 Identifying twiddle factors ($N = 8$) for DIT_{RN} FFT.

Butterfly Computation	Current HalfSize	Decimal $\pm w[Jtwiddle]$	Binary $Jtwiddle$ w.r.t modified $x_{i_2 i_1 i_0}$	Shorthand Notation w.r.t. modified $x_{i_2 i_1 i_0}$
Stage $i_0 i_1 i_2$	$HalfSize = 1$	$\pm w[0] = \pm w_N^0$	00	$\pm w_N^0 = \pm 1$
Stage $i_0 i_1 \tau_2$	$HalfSize = 2$	$\pm w[0] = \pm w_N^0$	$i_2 0 = 00$	$\pm w_N^{i_2 0}$
		$\pm w[2] = \pm w_N^2$	$i_2 0 = 10$	
Stage $i_0 \tau_1 \tau_2$	$HalfSize = 4$	$\pm w[0] = \pm w_N^0$	$i_1 i_2 = 00$	$\pm w_N^{i_1 i_2}$
		$\pm w[1] = \pm w_N^1$	$i_1 i_2 = 01$	
		$\pm w[2] = \pm w_N^2$	$i_1 i_2 = 10$	
		$\pm w[3] = \pm w_N^3$	$i_1 i_2 = 11$	

8.1.2   The pseudo-code program for the DIT$_{RN}$ FFT

In contrast to the DIF$_{RN}$ FFT pseudo-code program given by Algorithm 4.2, the complete DIT$_{RN}$ version is given below as Algorithm 8.3. As mentioned in the last section, the twiddle factors are stored in natural order, i.e., $w[\ell] = \omega_N^\ell$, $\ell = 0, 1, \ldots, N/2 - 1$.

Algorithm 8.3 The iterative radix-2 DIT$_{RN}$ FFT algorithm in pseudo-code.

---

begin
    PairsInGroup := *N*/2   Begin with *N*/2 butterflies in one group
    NumOfGroups := 1   Same twiddle factor is employed in a group
    Distance := 1
    while *NumOfGroups* < *N* **do**
        GapToNextPair := 2 * *NumOfGroups*
        GapToLastPair := *GapToNextPair* * (*PairsInGroup* − 1)
        for *K* := 0 **to** *NumOfGroups* − 1 **do**   Modify one group at a time
            J := *K*   Address of the first pair
            JLast := *K* + *GapToLastPair*   Address of the last pair
            Jtwiddle := *K* * *PairsInGroup*
            W := *w*[*Jtwiddle*]   Assume $w[\ell] = \omega_N^\ell$
            while *J* ≤ *JLast* **do**   Modify all pairs in the same group
                Temp := *W* * *a*[*J* + *Distance*]
                a[*J* + *Distance*] := *a*[*J*] − *Temp*
                a[*J*] := *a*[*J*] + *Temp*
                J := *J* + *GapToNextPair*   Process next pair in the group
            end while
        end for
        PairsInGroup := *PairsInGroup*/2
        NumOfGroups := *NumOfGroups* * 2
        Distance := *Distance* * 2
    end while
end

8.2   Shorthand Notation and a $N = 32$ Example

As derived in the previous section, the butterfly computation steps in Figures 8.2–8.4 (for $N = 8$) are fully specified by the sequence

$$\boxed{i_2 i_1 i_0 \quad i_0 i_1 i_2 \quad i_0 i_1 \tau_2 \quad i_0 \tau_1 \tau_2}$$

together with the corresponding twiddle factors

$$\pm \omega_N^0 = \pm 1, \ \pm \omega_N^{i_2 0}, \ \pm \omega_N^{i_1 i_2}.$$

For $N = 32$, the entire Figure 8.5 can be replaced by the sequence

$$\boxed{i_4 i_3 i_2 i_1 i_0 \quad i_0 i_1 i_2 i_3 i_4 \quad i_0 i_1 i_2 i_3 \tau_4 \quad i_0 i_1 i_2 \tau_3 \tau_4 \quad i_0 i_1 \tau_2 \tau_3 \tau_4}$$

with the understanding that on input, $a[i_0 i_1 i_2 i_3 i_4]$ contains the bit-reversed $x_{i_4 i_3 i_2 i_1 i_0}$, and on output, $a[i_0 i_1 i_2 i_3 i_4]$ contains $X_{i_0 i_1 i_2 i_3 i_4}$. Refer to Figure 8.5 for the decimal subscripts of all 32 bit-reversed input elements in $a[i_0 i_1 i_2 i_3 i_4]$.

The corresponding twiddle factors are

$$\pm \omega_N^0 = \pm 1, \ \pm \omega_N^{i_4 000}, \ \pm \omega_N^{i_3 i_4 00}, \ \pm \omega_N^{i_2 i_3 i_4 0}, \ \pm \omega_N^{i_1 i_2 i_3 i_4},$$

and $w[i_1 i_2 i_3 i_4] = \omega_N^{i_1 i_2 i_3 i_4}$ is assumed in Algorithm 8.3.

Figure 8.5 Butterflies of the in-place DIT_{RN} FFT algorithm.

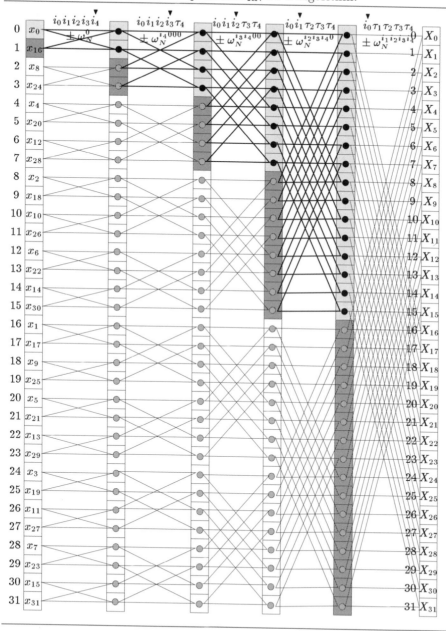

Chapter 9

An Ordered Radix-2 DIT FFT

This chapter presents the sixth variant of the FFT algorithm, which applies *decimation-in-time* FFT to transform naturally ordered input time series to naturally ordered output frequencies. As expected, this ordered DIT$_{NN}$ FFT corresponds to the ordered DIF$_{NN}$ FFT presented in Chapter 6. Readers of this chapter are thus assumed to be familiar with the concept of combining permutation with butterfly computation in the ordered DIF FFT, which allows the implementation of repeated permutation of intermediate results without extra accesses to memory. Since it would be helpful to look at the same from a different angle, this chapter begins with the following generic description of the DIT FFT algorithm, which is correct regardless of where the input element x_m, $m = 0, 1, \ldots, N - 1$, and its derivatives are stored in array $\boldsymbol{a}$.

Algorithm 9.1 The radix-2 DIT FFT algorithm in terms of binary subscripts.

---

begin
    $k := n - 1$ Assume problem size $N = 2^n$
    while $k \geq 0$ **do**   Combine two subproblems
        Apply Cooley-Tukey butterfly computation
        to "combine" all pairs of elements of $\boldsymbol{x}$ whose
        binary subscripts differ in bit i_k
        $k := k - 1$
    end while
end

Accordingly, the two in-place DIT FFT algorithms presented in Chapters 7 and 8 and the ordered DIT$_{NN}$ FFT algorithm presented in this chapter must all implement exactly the same computation, and they each can be understood by relating the locations where x_m and its derivatives are stored in the input and the working arrays as shown in Table 9.1 for the familiar $N = 8$ example.

Table 9.1 Relating the NR, RN, and NN variants of the DIT FFT.

Actual Modification of x_m & its derivatives	DIT$_{NR}$ FFT	DIT$_{RN}$ FFT	DIT$_{NN}$ FFT
$x^{(1)}_{0i_1i_0} = x_{0i_1i_0} + \omega_N^0 x_{1i_1i_0}$	$a[i_2i_1i_0] = x^{(1)}_{i_2i_1i_0}$	$a[i_0i_1i_2] = x^{(1)}_{i_2i_1i_0}$	$b[i_2i_1i_0] = x^{(1)}_{i_2i_1i_0}$
$x^{(1)}_{1i_1i_0} = x_{0i_1i_0} - \omega_N^0 x_{1i_1i_0}$	(in-place)	(in-place)	$(a[i_2i_1i_0] = x_{i_2i_1i_0})$
$x^{(2)}_{i_20i_0} = x^{(1)}_{i_20i_0} + \omega_N^{i_20} x^{(1)}_{i_21i_0}$	$a[i_2i_1i_0] = x^{(2)}_{i_2i_1i_0}$	$a[i_0i_1i_2] = x^{(2)}_{i_2i_1i_0}$	$a[i_1i_2i_0] = x^{(2)}_{i_2i_1i_0}$
$x^{(2)}_{i_21i_0} = x^{(1)}_{i_20i_0} - \omega_N^{i_20} x^{(1)}_{i_21i_0}$	(in-place)	(in-place)	(permuted)
$x^{(3)}_{i_2i_10} = x^{(2)}_{i_2i_10} + \omega_N^{i_1i_2} x^{(2)}_{i_2i_11}$	$a[i_2i_1i_0] = x^{(3)}_{i_2i_1i_0}$	$a[i_0i_1i_2] = x^{(3)}_{i_2i_1i_0}$	$b[i_0i_1i_2] = x^{(3)}_{i_2i_1i_0}$
$x^{(3)}_{i_2i_11} = x^{(2)}_{i_2i_10} - \omega_N^{i_1i_2} x^{(2)}_{i_2i_11}$	(in-place)	(in-place)	(permuted)

Recall that $x^{(3)}_{i_2i_1i_0} = X_{i_0i_1i_2}$. Therefore, the result $b[i_0i_1i_2] = x^{(3)}_{i_2i_1i_0}$ from the DIT$_{NN}$ algorithm ensures that the output is in natural order. Based on Table 9.1, the ordered DIT$_{NN}$ FFT can immediately be described in full for the $N = 8$ example in Table 9.2, which can be compared with Table 7.2 for the DIT$_{NR}$ FFT and Table 8.2 for the DIT$_{RN}$ FFT.

The effect of combining butterfly computation with permutation from array a to b, or vice versa, is reflected by the shorthand notation

$$
\begin{array}{cccc}
\blacktriangledown & \blacktriangledown & \blacktriangledown & \\
i_2i_1i_0 & i_2i_1i_0 & i_1\tau_2i_0 & i_0\tau_1\tau_2 \\
a & b & a & b
\end{array}
$$

.

Observe that the twiddle factors remain the same for all three DIT FFT algorithms, i.e., they are

$$\pm\omega_N^0 = \pm 1, \quad \pm\omega_N^{i_20}, \quad \pm\omega_N^{i_1i_2}.$$

Table 9.2 The NN DIT FFT applied to a $N = 8$ example.

Cooley-Tukey Butterfly Combined with Permutation	Actual Modification & Permutation from a to b or Vice Versa	Butterfly Groups	A Binary Address Based Notation to Show the Combined Effect
Stage 1: $i_2 i_1 i_0$	$b[0] = a[0] + \omega_N^0 a[4]$	Group 0	$b[0 i_1 i_0] = a[0 i_1 i_0] + \omega_N^0 a[1 i_1 i_0]$
	$b[4] = a[0] - \omega_N^0 a[4]$	(4 Pairs)	$b[1 i_1 i_0] = a[0 i_1 i_0] - \omega_N^0 a[1 i_1 i_0]$
	$b[1] = a[1] + \omega_N^0 a[5]$		
	$b[5] = a[1] - \omega_N^0 a[5]$		Note: $b[i_2 i_1 i_0] = x_{i_2 i_1 i_0}^{(1)}$
	$b[2] = a[2] + \omega_N^0 a[6]$		$(a[i_2 i_1 i_0] = x_{i_2 i_1 i_0})$
	$b[6] = a[2] - \omega_N^0 a[6]$		
	$b[3] = a[3] + \omega_N^0 a[7]$		
	$b[7] = a[3] - \omega_N^0 a[7]$		
Stage 2: $i_1 \tau_2 i_0$	$a[0] = b[0] + \omega_N^0 b[2]$	Group 0	$a[0 i_2 i_0] = b[i_2 0 i_0] + \omega_N^{i_2 0} b[i_2 1 i_0]$
	$a[4] = b[0] - \omega_N^0 b[2]$	(2 pairs)	$a[1 i_2 i_0] = b[i_2 0 i_0] - \omega_N^{i_2 0} b[i_2 1 i_0]$
	$a[1] = b[1] + \omega_N^0 b[3]$		
	$a[5] = b[1] - \omega_N^0 b[3]$		Note: $a[i_1 i_2 i_0] = x_{i_2 i_1 i_0}^{(2)}$
	$a[2] = b[4] + \omega_N^2 b[6]$	Group 1	
	$a[6] = b[4] - \omega_N^2 b[6]$	(2 pairs)	
	$a[3] = b[5] + \omega_N^2 b[7]$		
	$a[7] = b[5] - \omega_N^2 b[7]$		
Stage 3: $i_0 \tau_1 \tau_2$	$b[0] = a[0] + \omega_N^0 a[1]$	Group 0	$b[0 i_1 i_2] = a[i_1 i_2 0] + \omega_N^{i_1 i_2} a[i_1 i_2 1]$
	$b[4] = a[0] - \omega_N^0 a[1]$	(1 pair)	$b[1 i_1 i_2] = a[i_1 i_2 0] - \omega_N^{i_1 i_2} a[i_1 i_2 1]$
	$b[1] = a[2] + \omega_N^1 a[3]$	Group 1	
	$b[5] = a[2] - \omega_N^1 a[3]$	(1 pair)	Note: $b[i_0 i_1 i_2] = x_{i_2 i_1 i_0}^{(3)} = X_{i_0 i_1 i_2}$
	$b[2] = a[4] + \omega_N^2 a[5]$	Group 2	
	$b[6] = a[4] - \omega_N^2 a[5]$	(1 pair)	
	$b[3] = a[6] + \omega_N^3 a[7]$	Group 3	
	$b[7] = a[6] - \omega_N^3 a[7]$	(1 pair)	

9.1   Deriving the (Ordered) DIT_{NN} FFT From Its Recursive Definition

For completeness, the derivation of the ordered DIT FFT from its recursive definition is also depicted for the $N = 8$ example in Figures 9.1–9.4.

Figure 9.1 The first division step of the (ordered) DIT_{NN} FFT.

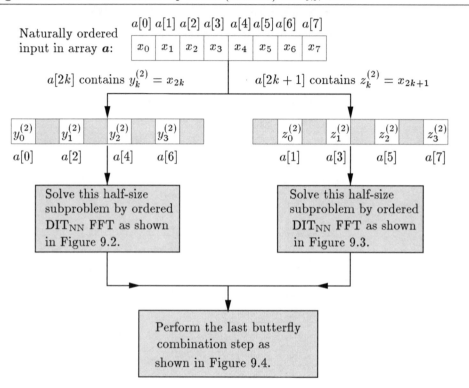

Figure 9.2 Solve one half-size subproblem by (ordered) DIT$_{\text{NN}}$ FFT.

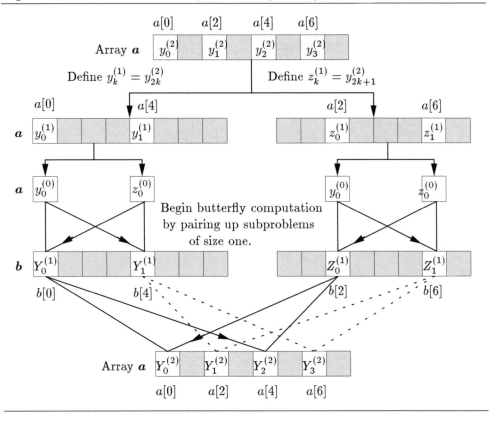

Figure 9.3 Solve one half-size subproblem by (ordered) DIT$_{\text{NN}}$ FFT.

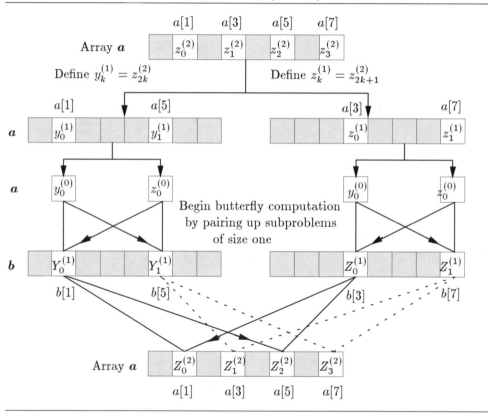

Figure 9.4 Solve the entire problem by (ordered) DIT_{NN} FFT.

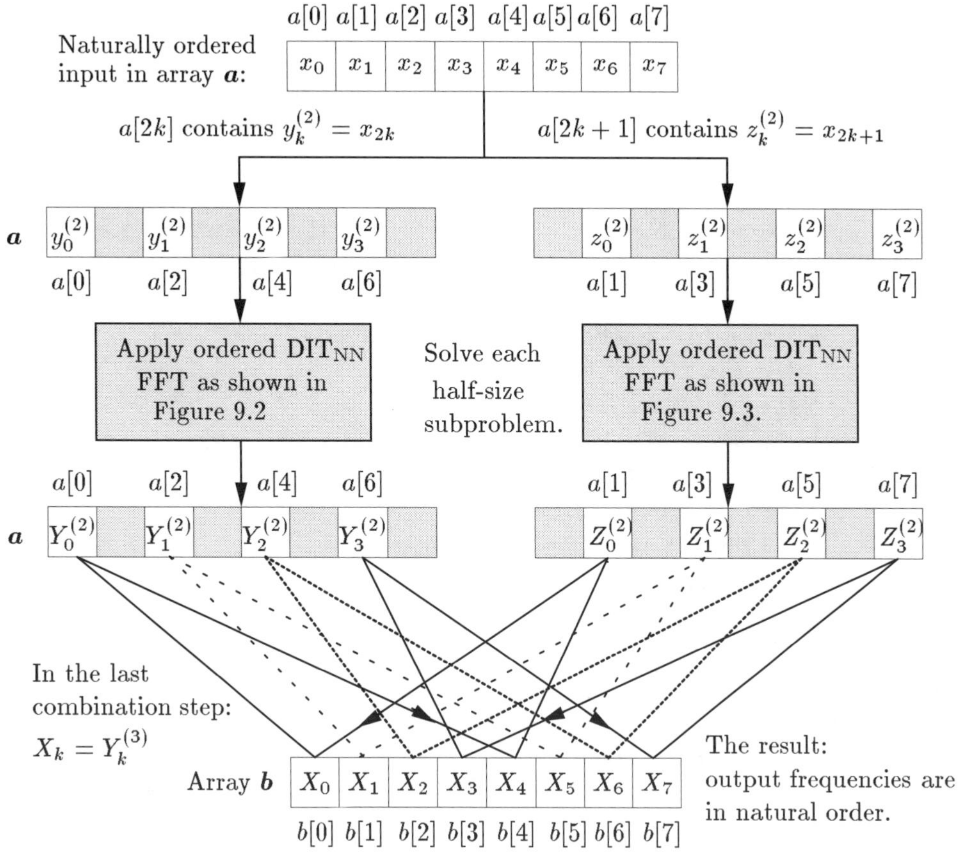

9.2 The Pseudo-code Program for the DIT_{NN} FFT

In contrast to the DIF_{NN} FFT pseudo-code program given by Algorithm 6.1, the DIT_{NN} version is given below as Algorithm 9.2. Note that the twiddle factors are assumed to be stored in natural order, i.e., $w[\ell] = \omega_N^\ell$, $\ell = 0, 1, \ldots, N/2 - 1$.

Algorithm 9.2 The (ordered) radix-2 DIT$_{\text{NN}}$ FFT algorithm.

<u>begin</u>
   $PairsInGroup := N/2$ Initially: N/2 pairs in one group
   $NumOfGroups := 1$
   $Distance := N/2$
   $NotSwitchInput := true$
   <u>while</u> $NumOfGroups < N$ <u>do</u>   Combine pairs in each group
      <u>if</u> $NotSwitchInput$ Array **a** contains input; array **b** contains output
        $L := 0$
        <u>for</u> $K := 0$ <u>to</u> $NumOfGroups - 1$ <u>do</u>
           $JFirst := 2 * K * PairsInGroup$
           $JLast := JFirst + PairsInGroup - 1$
           $Jtwiddle := K * PairsInGroup$ Assume $w[\ell] = \omega_N^\ell$
           $W := w[Jtwiddle]$ Same twiddle factor in each group
           <u>for</u> $J := JFirst$ <u>to</u> $JLast$ <u>do</u>
              $Temp := W * a[J + Distance]$
              $b[L] := a[J] + Temp$
              $b[L + N/2] := a[J] - Temp$
              $L := L + 1$
           <u>end for</u>
        <u>end for</u>
        $NotSwitchInput := false$
      <u>else</u>   Array **b** contains input; array **a** contains output
        $L := 0$
        <u>for</u> $K := 0$ <u>to</u> $NumOfGroups - 1$ <u>do</u>
           $JFirst := 2 * K * PairsInGroup$
           $JLast := JFirst + PairsInGroup - 1$
           $Jtwiddle := K * PairsInGroup$ Assume $w[\ell] = \omega_N^\ell$
           $W := w[Jtwiddle]$ Same twiddle factor in each group
           <u>for</u> $J := JFirst$ <u>to</u> $JLast$ <u>do</u>
              $Temp := W * a[J + Distance]$
               $a[L] := b[J] + Temp$
               $a[L + N/2] := b[J] - Temp$
               $L := L + 1$
           <u>end for</u>
        <u>end for</u>
        $NotSwitchInput := true$
      <u>end if</u>
      $PairsInGroup := PairsInGroup/2$
      $NumOfGroups := NumOfGroups * 2$
      $Distance := Distance/2$
   <u>end while</u>
<u>end</u>

9.3 Applying the (Ordered) DIT_{NN} FFT to a $N = 32$ Example

Generalizing the shorthand notation for $N = 32$, the following sequence represents all five stages of permutation and computation depicted in Figure 9.5.

$\blacktriangledown$	$\blacktriangledown$	$\blacktriangledown$	$\blacktriangledown$	$\blacktriangledown$	
$i_4 i_3 i_2 i_1 i_0$	$i_4 i_3 i_2 i_1 i_0$	$i_3 \tau_4 i_2 i_1 i_0$	$i_2 \tau_3 \tau_4 i_1 i_0$	$i_1 \tau_2 \tau_3 \tau_4 i_0$	$i_0 \tau_1 \tau_2 \tau_3 \tau_4$
a	b	a	b	a	b

The corresponding twiddle factors are

$$\pm w_N^0 = \pm 1, \ \pm w_N^{i_4 000}, \ \pm w_N^{i_3 i_4 00}, \ \pm w_N^{i_2 i_3 i_4 0}, \ \pm w_N^{i_1 i_2 i_3 i_4};$$

and $w[i_1 i_2 i_3 i_4] = w_N^{i_1 i_2 i_3 i_4}$ is assumed in Algorithm 9.2.

By comparing Figure 9.6, where the butterflies associated with a particular pair of resulting subproblems are shown without the cluttering of others, with the two unordered DIT FFT in Figures 7.6 and 8.5, one once again observes that

all three DIT algorithms treat exactly the same pairs of subproblems during each stage of the computation.

Thus, they all implement the same radix-2 DIT FFT algorithm.

Figure 9.5 Butterflies of the (ordered) DIT_{NN} FFT algorithm.

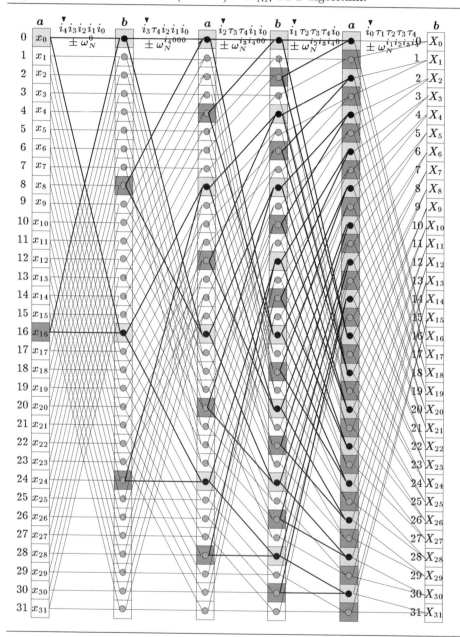

Figure 9.6 Identifying the subproblem pairs in the (ordered) $\mathrm{DIT_{NN}}$ FFT.

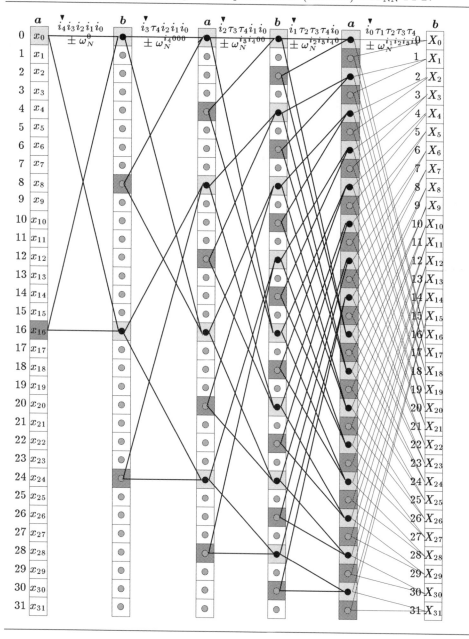

Chapter 10

Ordering Algorithms and Computer Implementation of Radix-2 FFTs

This chapter examines the different roles that ordering algorithms may play in the computer implementation of the fast Fourier transforms. For example, a bit-reversal algorithm may be used to reorder input or output data, and a perfect shuffle algorithm and its variants may be used in a parallel processor to maintain data adjacency throughout the computation.

10.1   Bit-Reversal and Ordered FFTs

The six radix-2 FFT variants have each been discussed in detail in the preceding six chapters. Although it appears that only two specific algorithms are suitable for performing "ordered" transforms at the cost of doubling the primary storage, the other four variants can be used to perform ordered FFT if one is willing to apply a bit-reversal algorithm to either the naturally ordered input sequence (for use with the DIF_{RN} or DIT_{RN} algorithm) or the bit-reversed output sequence (from the DIF_{NR} or DIT_{NR} algorithm). The four possible combinations are shown in Figure 10.1. In any case, the following remarks should be noted at the outset.

Since the bit-reversal algorithm is often not performed in-place [60], the primary storage of the FFT is doubled in the reordering phase. In addition, the pre-processing of input data or the post-processing of output data incurs arithmetic and memory-accessing cost. Such extra cost could account for a significant part of the total execution time if the ordered FFT must be obtained via one of the four unordered variants. (Recall that the additional memory accessing cost can be avoided using the DIF_{NN} or the DIT_{NN} FFT.)

As reported by Karp in his review article [60], there had been a surprisingly large number of papers on the subject of bit-reversal in the literature. He collected and tested 30 different methods for bit reversing an array on several uniprocessor machines, each with a different memory system. He also proposed two new bit-reversal algorithms to make better use of the interleaved memory machines and the hierarchical memories on modern high-performance uniprocessors. Clearly, the design and implementation of bit-reversal algorithms is interesting and important in its own right.

Figure 10.1 Combining bit reversal and unordered DIF and DIT FFTs.

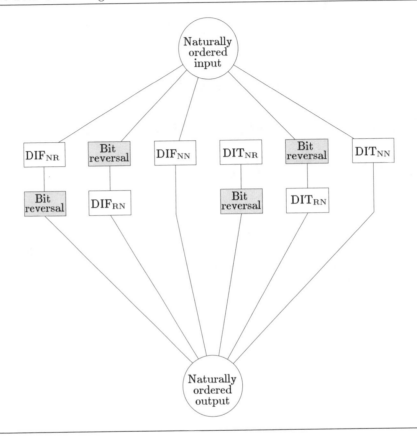

10.2   Perfect Shuffle and In-Place FFTs

The *perfect shuffle* was originally proposed by Stone [90] in 1971 as an interconnection scheme for a parallel processor. The thesis of Stone's paper was that the perfect shuffle was not only a "good" pattern for particular parallel algorithms but had a wide variety of applications. Parallelizing the FFT with the perfect shuffle was among the series of examples described in [90]. Other interesting applications include polynomial evaluation, sorting, and matrix transposition.

The "effect" of a perfect shuffle can be most intuitively understood by applying it to a deck of cards—the cards in the first half of the deck will be interlaced with cards in the second half of the deck as a result.

Viewing the scheme as a computer algorithm, a perfect shuffle of an N element vector simply permutes the elements according to an interlacing pattern identical to the shuffled deck of cards. Following the binary address based notations introduced in

Chapter 4, if element $x_{i_2 i_1 i_0}$ is in $a[i_2 i_1 i_0]$ before the perfect shuffle, it will be relocated to $b[i_1 i_0 i_2]$, and such a permutation is represented by the sequence

$i_2 i_1 i_0$	$i_1 i_0 i_2$
a	b

Keep in mind that the shorthand notations are always formed by the addresses of element $x_{i_2 i_1 i_0}$ and its derivatives.

Just like any other computer algorithm, the perfect shuffle can be implemented *either* by software *or* by hardware as originally proposed.

10.2.1 Combining a software implementation with the FFT

Recall that the unordered DIF_{NR} algorithm and DIT_{NR} algorithm perform in-place butterfly computations on an array of $N = 2^n$ complex numbers, and the series of butterflies involve pairing up elements located 2^k places apart for k values decreasing from $n-1$ to 0. Using the shorthand notations developed in Chapters 4 and 7, the DIF_{NR} and DIT_{NR} FFTs are depicted side by side for $N = 8$ in Table 10.1. Clearly, the two variants differ only in the application of twiddle factors.

Table 10.1 In-place DIF_{NR} and DIT_{NR} FFTs with twiddle factors for $N = 8$.

Input: $a[i_2 i_1 i_0] = x_{i_2 i_1 i_0}$; Output: $a[i_2 i_1 i_0] = x^{(3)}_{i_2 i_1 i_0} = X_{i_0 i_1 i_2}$

DIF_{NR} FFT Algorithm	DIT_{NR} FFT Algorithm
$i_2 i_1 i_0 \quad \tau_2 i_1 i_0 \quad \tau_2 \tau_1 i_0$	$i_2 i_1 i_0 \quad \tau_2 i_1 i_0 \quad \tau_2 \tau_1 i_0$
Twiddle Factors:	Twiddle Factors:
$\omega_N^{i_1 i_0}, \quad \omega_N^{i_0 0}, \quad \omega_N^0 = 1$	$\pm\omega_N^0 = \pm 1, \quad \pm\omega_N^{i_2 0}, \quad \omega_N^{i_1 i_2}$

Now if a perfect shuffle precedes each stage of in-place butterfly computation, either DIF_{NR} or DIT_{NR} FFT in Table 10.1 can be represented by the addresses of relocated $x_{i_2 i_1 i_0}$ and its derivatives as shown in the following sequence:

$i_2 i_1 i_0$	$i_1 i_0 i_2$	$i_0 \tau_2 i_1$	$\tau_2 \tau_1 i_0$
a	b	a	b

Therefore, such a software implementation could provide another sequential iterative program for the recursively-defined FFT algorithm, although the primary storage must be doubled to accommodate the perfect shuffle. Observe that because $b[i_2 i_1 i_0] = x^{(3)}_{i_2 i_1 i_0} = X_{i_0 i_1 i_2}$ on output, there is no change to the bit-reversed output ordering. In fact, there does not appear to be any advantage in combining the perfect shuffle with the FFT in a software implementation.

10.2.2   Data adjacency afforded by a hardware implementation

The combination of the perfect shuffle with the in-place NR algorithms as discussed
above leads to the following observation:

> The addresses of the two elements involved in each butterfly computation
> always differ in their right-most bit; they are thus located in consecutive
> positions in the array.

This suggests that the same set of butterfly computing modules, with each module
pairing up and modifying data in two neighboring locations, can be connected to the
output of the perfect shuffle for multiple stages of butterfly computation. After that,
the updated data output from the butterfly modules can be transferred back into the
same perfect shuffle to produce data for the next stage of butterfly computation.

A hardware implementation allowing all $N/2$ butterflies to be performed simultaneously by $N/2$ butterfly modules will thus result in a dedicated parallel processor for
the FFT. Such a parallel processor with four butterfly modules is shown in Figure 10.2.

Figure 10.2 A parallel processor with perfect shuffle and four butterfly modules.

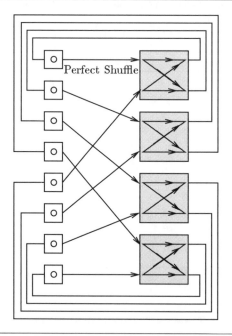

Early FFT hardware implementations were discussed in the overview and survey articles by Bergland [6, 7]. More recent development in digital signal processors and
dedicated VLSI implementation was addressed in [41, 103].

10.3 Reverse Perfect Shuffle and In-Place FFTs

As its name implies, the *reverse* perfect shuffle is expected to "undo" the perfect shuffle when it is applied to a shuffled deck of cards. That is, the data in the even-numbered locations of an N element vector are permuted to the top half of the vector, and the data in the odd-numbered locations are permuted to the bottom half of the vector.

Since it has already been established in the previous section that the perfect shuffle moves data to new locations which are obtained by cyclic-rotating the current binary address of the data one bit to the left (with the left-most bit moved to the vacated right-most position), the reverse perfect shuffle can be readily obtained by cyclically rotating the binary address one bit to the right (with the right-most bit moved to the now vacated left-most position).

Recall that algorithms DIF_{RN} and DIT_{RN} both implement unordered FFT on a bit-reversed $N = 2^n$ element vector as shown in Table 10.2 below.

Table 10.2 In-place DIF_{RN} and DIT_{RN} FFTs with twiddle factors for $N = 8$.

Input: $a[i_0 i_1 i_2] = x_{i_2 i_1 i_0}$; Output: $a[i_0 i_1 i_2] = x^{(3)}_{i_2 i_1 i_0} = X_{i_0 i_1 i_2}$

DIF_{RN} FFT Algorithm	DIT_{RN} FFT Algorithm
$i_2 i_1 i_0 \quad i_0 i_1 i_2 \quad i_0 i_1 \tau_2 \quad i_0 \tau_1 \tau_2$	$i_2 i_1 i_0 \quad i_0 i_1 i_2 \quad i_0 i_1 \tau_2 \quad i_0 \tau_1 \tau_2$
Twiddle Factors:	Twiddle Factors:
$\omega_N^{i_1 i_0}, \quad \omega_N^{i_0 0}, \quad \omega_N^0 = 1$	$\pm \omega_N^0 = \pm 1, \quad \pm \omega_N^{i_2 0}, \quad \omega_N^{i_1 i_2}$

One can immediately see that by reverse perfect shuffling the butterfly output at the end of every stage, one can again be sure that all butterflies will involve only neighboring elements. The combined effect from a software implementation can be depicted by the following sequence:

$i_2 i_1 i_0$	$i_0 i_1 i_2$	$\tau_2 i_0 i_1$	$\tau_1 \tau_2 i_0$	$\tau_0 \tau_1 \tau_2$
b	a	b	a	b

A parallel processor combining $N/2$ butterfly modules with the reverse perfect shuffle network will allow simultaneous computation of all butterflies during each stage. A parallel processor suitable for computing the DIF_{RN} or DIT_{RN} algorithm with four butterfly modules is shown in Figure 10.3.

10.4 Fictitious Block Perfect Shuffle and Ordered FFTs

It is interesting to "interpret" the permutation and butterflies in the two NN algorithms under the perfect shuffle framework, although it was concluded in Chapters 6 and 9 that an efficient implementation of the ordered DIF_{NN} or DIT_{NN} FFTs should not have a separate permutation step. For purposes of exposition, the discussion below discusses permutations and computation as separate operations. However, as noted

Figure 10.3 A parallel processor with reverse perfect shuffle and four butterfly modules.

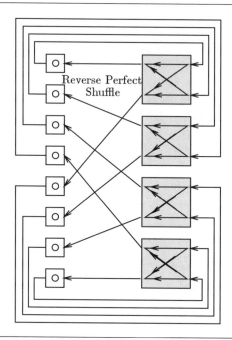

in Chapters 6 and 9, the two operations are usually combined to achieve efficiency by avoiding unnecessary memory accesses.

10.4.1   Interpreting the ordered DIF$_{\text{NN}}$ FFT algorithm

Recall from Chapter 6 that the following sequence represents the five stages of permutation and computation depicted in Figure 6.5 for a $N = 32$ example.

$$
\begin{array}{cccccc}
i_4 i_3 i_2 i_1 i_0 & i_3 i_2 i_1 i_0 i_4 & i_2 i_1 i_0 i_3 \tau_4 & i_1 i_0 i_2 \tau_3 \tau_4 & i_0 i_1 \tau_2 \tau_3 \tau_4 & i_0 \tau_1 \tau_2 \tau_3 \tau_4 \\
a & b & a & b & a & b
\end{array}
$$

Each step can be viewed as consisting of a separate permutation step followed by an in-place butterfly computation. Clearly, the permutation preceding the first butterfly involves a perfect shuffle.

$$
\begin{array}{cc}
i_4 i_3 i_2 i_1 i_0 & i_3 i_2 i_1 i_0 i_4 \\
a & b
\end{array}
$$

After the in-place butterfly computation involving adjacent elements in array b, the derivative $x^{(1)}_{i_4 i_3 i_2 i_1 i_0}$ is located in $b[i_3 i_2 i_1 i_0 \tau_4]$. Now, if array b is partitioned into blocks of size 2, then the leftmost four bits, namely $i_3 i_2 i_1 i_0$, in the address field may be used to address the blocks. That is, an array of N elements is now viewed as an array containing $N/2$ blocks. If each block is treated as a unit, then the following permutation is a perfect shuffle of the $N/2$ blocks.

$$\boxed{\begin{array}{cc} i_3 i_2 i_1 i_0 & i_2 i_1 i_0 i_3 \\ b & a \end{array}}$$

Since the block is permuted as a whole, an individual element is permuted from $b[i_3 i_2 i_1 i_0 \tau_4]$ to $a[i_2 i_1 i_0 i_3 \tau_4]$. As expected, only the address of the block is changed after the block perfect shuffle. Observe that the in-place butterfly computation now involves elements in "adjacent" blocks, and the next derivative $x^{(2)}_{i_4 i_3 i_2 i_1 i_0}$ is therefore contained in $a[i_2 i_1 i_0 \tau_3 \tau_4]$.

To continue the analysis, imagine that array a is partitioned into blocks of size 4. Then the leftmost three bits, namely $i_2 i_1 i_0$, in the address may be used to address the blocks. Note that an array of N elements is now viewed as an array of $N/4$ blocks, with each block containing four consecutive elements. Each block is again treated as a unit, and the following permutation is a perfect shuffle of the $N/4$ blocks.

$$\boxed{\begin{array}{cc} i_2 i_1 i_0 & i_1 i_0 i_2 \\ a & b \end{array}}$$

Since the block is permuted as whole, only the address of the block is changed after the perfect shuffle, and the individual element in $a[i_2 i_1 i_0 \tau_3 \tau_4]$ is relocated to $b[i_1 i_0 i_2 \tau_3 \tau_4]$. Observe again that the in-place butterfly computation may be viewed as pairing up elements in "adjacent" blocks.

It is now obvious that the perfect shuffle in each stage that follow will permute blocks which double the size until the blocksize reaches $N/2$.

10.4.2 Interpreting the ordered DIT$_{\text{NN}}$ FFT algorithm

Recall from Chapter 9 that the following sequence represents the five stages of computation and permutation depicted in Figure 9.5 for transforming a naturally ordered input series of length $N = 32$.

$$\boxed{\begin{array}{cccccc} \blacktriangledown & \blacktriangledown & \blacktriangledown & \blacktriangledown & \blacktriangledown & \\ i_4 i_3 i_2 i_1 i_0 & \tau_4 i_3 i_2 i_1 i_0 & \tau_3 \tau_4 i_2 i_1 i_0 & \tau_2 \tau_3 \tau_4 i_1 i_0 & \tau_1 \tau_2 \tau_3 \tau_4 i_0 & \tau_0 \tau_1 \tau_2 \tau_3 \tau_4 \\ a & b & a & b & a & b \end{array}}$$

As expected, the analysis of the ordered DIT$_{\text{NN}}$ FFT mirrors that of the DIF$_{\text{NN}}$ in the previous section. That is, each stage is viewed as consisting of an in-place butterfly computation involving elements in "adjacent" blocks, which is followed by a *reverse* block perfect shuffle. In contrast to the DIF$_{\text{NN}}$ FFT, the first DIT$_{\text{NN}}$ butterfly pairs up elements in two adjacent blocks of size $N/2$, which is followed by a reverse block perfect shuffle which happens to effect no change when there are only two blocks.

$$\boxed{\begin{array}{cc} \tau_4 & \tau_4 \\ a & b \end{array}}$$

The next reverse perfect shuffle permutes four blocks of size $N/4$.

$$\begin{array}{cc} T_4 T_3 & T_3 T_4 \\ b & a \end{array}$$

Eight blocks of size $N/8$ are shuffled in the step below, and so on.

$$\begin{array}{cc} T_3 T_4 T_2 & T_2 T_3 T_4 \\ a & b \end{array}$$

The reverse block perfect shuffle is repeated at the end of every stage of the in-place butterfly computation. Note that the blocksize is halved each time, and the last shuffle involves blocks of size one.

Chapter 11

The Radix-4 and the Class of Radix-2^S FFTs

The divide-and-conquer paradigm introduced in Chapter 3 is not restricted to dividing a problem into two subproblems. In fact, as explained in Section 2.4 and Appendix B, the recurrence equation

$$T(N) = \begin{cases} \alpha T\left(\frac{N}{\alpha}\right) + bN & \text{if } N = \alpha^k > 1, \\ \gamma & \text{if } N = 1, \end{cases}$$

represents the arithmetic cost of an algorithm which solves the original problem of size N by combining the results from (recursively) solving α problems of size N/α. In this chapter, the cases for $\alpha = 4$ as well as $\alpha = 2^s$ are considered.

11.1   The Radix-4 DIT FFTs

The DFT of a time series consisting of $N = 4^n$ discrete samples is considered in this section. Since $N = 4^n = 2^{2n}$, any version of the radix-2 FFTs introduced in Sections 3.1 and 3.2 can certainly be used to compute the transform. The reason it is worthwhile to develop a radix-4 implementation instead of simply using the radix-2 FFTs is that the arithmetic cost can be further reduced, and this advantage is carried over to the design of parallel FFTs. In fact, both radix-2 and radix-4 FFTs are special cases in the class of radix-2^s FFTs.

The radix-4 DIT FFT [5, 70, 84] is derived from equation (3.1), which defines the discrete Fourier transform of a complex time series. With the help of the identities $\omega_N^{\frac{N}{4}} = -j$ and $\omega_N^4 = \omega_{\frac{N}{4}}$, equation (3.1) can be rewritten in terms of four partial sums;

101

that is,

(11.1)

$$X_r = \sum_{\ell=0}^{N-1} x_\ell \omega_N^{r\ell}, \quad r = 0, 1, \ldots, N-1$$

$$= \sum_{k=0}^{\frac{N}{4}-1} x_{4k} \omega_N^{r(4k)} + \sum_{k=0}^{\frac{N}{4}-1} x_{4k+1} \omega_N^{r(4k+1)} + \sum_{k=0}^{\frac{N}{4}-1} x_{4k+2} \omega_N^{r(4k+2)} + \sum_{k=0}^{\frac{N}{4}-1} x_{4k+3} \omega_N^{r(4k+3)}$$

$$= \sum_{k=0}^{\frac{N}{4}-1} x_{4k} \omega_N^{r(4k)} + \omega_N^r \sum_{k=0}^{\frac{N}{4}-1} x_{4k+1} \omega_N^{r(4k)} + \omega_N^{2r} \sum_{k=0}^{\frac{N}{4}-1} x_{4k+2} \omega_N^{r(4k)} + \omega_N^{3r} \sum_{k=0}^{\frac{N}{4}-1} x_{4k+3} \omega_N^{r(4k)}.$$

By *decimating* the time series into four sets, namely the set $\{y_k | y_k = x_{4k}, \ 0 \le k \le N/4 - 1\}$, the set $\{z_k | z_k = x_{4k+1}, \ 0 \le k \le N/4 - 1\}$, the set $\{g_k | g_k = x_{4k+2}, \ 0 \le k \le N/4 - 1\}$, and the set $\{h_k | h_k = x_{4k+3}, \ 0 \le k \le N/4 - 1\}$, the four subproblems with period of $N/4$ can be defined after the appropriate twiddle factor $\omega_{\frac{N}{4}} = \omega_N^4$ is identified. The four subproblems are

(11.2)

$$Y_r = \sum_{k=0}^{\frac{N}{4}-1} x_{4k} \omega_N^{r(4k)} = \sum_{k=0}^{\frac{N}{4}-1} x_{4k} \left(\omega_N^4\right)^{rk} = \sum_{k=0}^{\frac{N}{4}-1} y_k \omega_{\frac{N}{4}}^{rk}, \quad r = 0, 1, \ldots, N/4 - 1.$$

(11.3)

$$Z_r = \sum_{k=0}^{\frac{N}{2}-1} x_{4k+1} \omega_N^{r(4k)} = \sum_{k=0}^{\frac{N}{4}-1} x_{4k+1} \left(\omega_N^4\right)^{rk} = \sum_{k=0}^{\frac{N}{4}-1} z_k \omega_{\frac{N}{4}}^{rk}, \quad r = 0, 1, \ldots, N/4 - 1.$$

(11.4)

$$G_r = \sum_{k=0}^{\frac{N}{2}-1} x_{4k+2} \omega_N^{r(4k)} = \sum_{k=0}^{\frac{N}{4}-1} x_{4k+2} \left(\omega_N^4\right)^{rk} = \sum_{k=0}^{\frac{N}{4}-1} g_k \omega_{\frac{N}{4}}^{rk}, \quad r = 0, 1, \ldots, N/4 - 1.$$

(11.5)

$$H_r = \sum_{k=0}^{\frac{N}{2}-1} h_{4k+3} \omega_N^{r(4k)} = \sum_{k=0}^{\frac{N}{4}-1} x_{4k+3} \left(\omega_N^4\right)^{rk} = \sum_{k=0}^{\frac{N}{4}-1} h_k \omega_{\frac{N}{4}}^{rk}, \quad r = 0, 1, \ldots, N/4 - 1.$$

The size of each subproblem is thus $N/4$, which is equal to the number of input data points or the number of computed output data points in one period. After these four subproblems are each (recursively) solved, the solution to the original problem of size N can be obtained according to (11.1) for $r = 0, 1, \ldots, N-1$. Since the series Y_r has a period of $N/4$, $Y_r = Y_{r+\frac{N}{4}} = Y_{r+\frac{N}{2}} = Y_{r+\frac{3N}{4}}$, and the same applies to the series Z_r, G_r, and H_r. The output X_r may be expressed in terms of Y_r, Z_r, G_r, and H_r for

$r = 0, 1, \ldots, N/4 - 1$ as shown below.

(11.6) $$X_r = Y_r + w_N^r Z_r + w_N^{2r} G_r + w_N^{3r} H_r .$$

(11.7) $$X_{r+\frac{N}{4}} = Y_r + w_N^{r+\frac{N}{4}} Z_r + w_N^{2(r+\frac{N}{4})} G_r + w_N^{3(r+\frac{N}{4})} H_r .$$

(11.8) $$X_{r+\frac{N}{2}} = Y_r + w_N^{r+\frac{N}{2}} Z_r + w_N^{2(r+\frac{N}{2})} G_r + w_N^{3(r+\frac{N}{2})} H_r .$$

(11.9) $$X_{r+\frac{3N}{4}} = Y_r + w_N^{r+\frac{3N}{4}} Z_r + w_N^{2(r+\frac{3N}{4})} G_r + w_N^{3(r+\frac{3N}{4})} H_r .$$

By noting that the twiddle factors $w_N^{\frac{N}{4}} = w_4 = e^{-j\frac{2\pi}{4}} = -j$, $w_N^{\frac{N}{2}} = (-j)^2 = -1$, and $w_N^{\frac{3N}{4}} = (-j)^3 = j$, the four equations above can be simplified to

(11.10) $$X_r = Y_r + w_N^r Z_r + w_N^{2r} G_r + w_N^{3r} H_r ,$$

(11.11) $$X_{r+\frac{N}{4}} = Y_r - jw_N^r Z_r - w_N^{2r} G_r + jw_N^{3r} H_r ,$$

(11.12) $$X_{r+\frac{N}{2}} = Y_r - w_N^r Z_r + w_N^{2r} G_r - w_N^{3r} H_r ,$$

(11.13) $$X_{r+\frac{3N}{4}} = Y_r + jw_N^r Z_r - w_N^{2r} G_r - jw_N^{3r} H_r ,$$

where $r = 0, 1, \ldots, N/4 - 1$.

If the radix-4 algorithm is implemented based on equations (11.10), (11.11), (11.12), and (11.13), one step of the radix-4 algorithm will require more arithmetic operations than two steps of the radix-2 algorithm, because some partial results were computed more than once. However, if such partial results can be identified and computed only once, one step of the radix-4 algorithm can require fewer arithmetic operations than two steps of the radix-2 algorithm, and the total cost of the radix-4 algorithm can be lower than the radix-2 algorithm. The four recurrent partial results are shown below inside each pair of parentheses.

(11.14) $$X_r = \left(Y_r + w_N^{2r} G_r\right) + \left(w_N^r Z_r + w_N^{3r} H_r\right),$$

(11.15) $$X_{r+\frac{N}{4}} = \left(Y_r - w_N^{2r} G_r\right) - j\left(w_N^r Z_r - w_N^{3r} H_r\right),$$

(11.16) $$X_{r+\frac{N}{2}} = -\left(w_N^r Z_r + w_N^{3r} H_r\right) + \left(Y_r + w_N^{2r} G_r\right),$$

(11.17) $$X_{r+\frac{3N}{4}} = j\left(w_N^r Z_r - w_N^{3r} H_r\right) + \left(Y_r - w_N^{2r} G_r\right),$$

where $r = 0, 1, \ldots, N/4 - 1$. The computation represented by (11.14), (11.15), (11.16), and (11.17) can now be represented by the two stages of butterfly computation in Figure 11.1.

11.1.1 Analyzing the arithmetic cost

To determine the arithmetic cost of the radix-4 FFT algorithm, observe that $w_N^{2r} G_r$, $w_N^r Z_r$ and $w_N^{3r} H_r$ need to be computed before the four partial sums can be obtained. Since the size of each subproblem is $N/4$, $3N/4$ complex multiplications and N complex additions are performed during the first stage of butterfly computation. The second stage of butterfly computation involves no multiplication by the twiddle factors, so only N complex additions are needed. Thus, $3N/4$ complex multiplications and $2N$ complex additions are required to implement the butterfly computation in Figure 11.1.

Figure 11.1 The radix-4 DIT FFT butterflies.

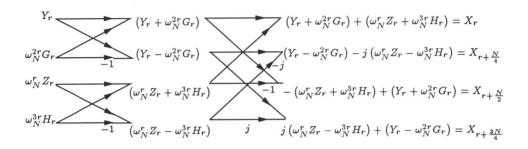

Recall again that the arithmetic cost of computer algorithms is measured by the number of real arithmetic operations, and that one complex addition incurs two real additions according to (2.1), and one complex multiplication (with pre-computed intermediate results involving the real and imaginary parts of a twiddle factor) incurs three real multiplications and three real additions according to (2.4). Accordingly, $9N/4$ real multiplications and $25N/4$ real additions are required per step of radix-4 DIT FFT. Thus, one step of the radix-4 DIT FFT algorithm requires $17N/2$ flops in total.

Since the objective of developing the radix-4 algorithm is to minimize the essential real operations, a careful analysis of the cost should exclude the trivial multiplication by $w_N^0 = 1$ and $w_4^\ell = (-j)^\ell = \pm 1$ or $\pm j$, since they will certainly not be done in an efficient implementation. Furthermore, note that the cost of multiplication by a twiddle factor which is an odd power of $w_8 = (1-j)/\sqrt{2}$ is less than the cost of a complex multiplication because

$$(11.18) \qquad w_8^{2\ell+1} = w_4^\ell \times w_8 = (-j)^\ell \left(\frac{1-j}{\sqrt{2}}\right) = \pm \left(\frac{1-j}{\sqrt{2}}\right) \quad \text{or} \quad \pm \left(\frac{1+j}{\sqrt{2}}\right).$$

These special factors are identified from the computation of $w_N^{2r} G_r$, $w_N^r Z_r$, and $w_N^{3r} H_r$ for $r = 0, 1, \ldots, \frac{N}{4} - 1$ below.

Table 11.1 Special cases of twiddle-factor multiplication in the radix-4 algorithm.

$0 \leq r \leq \frac{N}{4} - 1$	$r = 0$	$r = \left(\frac{1}{2}\right)\frac{N}{4}$	$r = \left(\frac{1}{4}\right)\frac{N}{4}$	$r = \left(\frac{3}{4}\right)\frac{N}{4}$
$w_N^{2r} G_r$	$1 \times G_r = G_r$	$w_4 G_r = -j G_r$	$w_8 G_r$	$w_8^3 G_r$
$w_N^r Z_r$	$1 \times Z_r = Z_r$	$w_8 Z_r$	–	–
$w_N^{3r} H_r$	$1 \times H_r = H_r$	$w_8^3 H_r$	–	–

Thus, there are eight special cases: the four cases involving multiplication by 1 and $-j$ are trivial, and the other four cases involving the multiplication by an odd power of w_8 are to be treated specially. The total nontrivial complex multiplications thus is reduced to $\frac{3}{4}N - 8$. Since only 4 flops are needed to compute each of $w_8 G_r$, $w_8^3 G_r$,

$\omega_8 Z_r$, $\omega_8^3 H_r$, the total flop count becomes

$$(11.19) \qquad (3+3) \times \left(\frac{3}{4}N - 8\right) + 4 \times 4 + 2 \times (2N) = \frac{17}{2}N - 32 .$$

Because these special factors also occur in every subsequent step, the savings can be incorporated in setting up the recurrence equation. For completeness, assuming that the problem size is $N = 4^n$, these special factors are identified for $r = 0, 1, \ldots, N/\left(4^{i+1}\right)$ in the ith step for $i = 0, 1, \ldots, n-2$ in Table 11.2.

Table 11.2 The recurrent special twiddle-factor multiplications in the radix-4 algorithm.

$0 \le r \le \frac{N}{4^{i+1}} - 1$	$r = 0$	$r = \left(\frac{1}{2}\right)\frac{N}{4^{i+1}}$	$r = \left(\frac{1}{4}\right)\frac{N}{4^{i+1}}$	$r = \left(\frac{3}{4}\right)\frac{N}{4^{i+1}}$
$\omega_{N/4^i}^{2r} G_r$	$1 \times G_r = G_r$	$\omega_4 G_r = -jG_r$	$\omega_8 G_r$	$\omega_8^3 G_r$
$\omega_{N/4^i}^{r} Z_r$	$1 \times Z_r = Z_r$	$\omega_8 Z_r$	–	–
$\omega_{N/4^i}^{3r} H_r$	$1 \times H_r = H_r$	$\omega_8^3 H_r$		–	–

To set up the recurrence equation, the boundary condition for $N = 4$ is needed. Recall that when $N = 4$, the twiddle factors are the four primitive roots of unity, namely 1, -1, j and $-j$, so the first stage of butterfly computation involves no nontrivial complex multiplications. Therefore, when $N = 4$, only $2 \times N = 8$ complex additions or $4 \times N = 16$ real arithmetic operations are required.

The cost of the radix-4 FFT algorithm can now be represented by the following recurrence:

$$(11.20) \qquad T(N) = \begin{cases} 4T\left(\frac{N}{4}\right) + \frac{17}{2}N - 32 & \text{if } N = 4^n > 4 , \\ 16 & \text{if } N = 4 . \end{cases}$$

Solving (11.20) (see Appendix B), one obtains

$$(11.21) \qquad T(N) = 4\frac{1}{4}N \log_2 N - \frac{43}{6}N + \frac{32}{3} .$$

The derivation above confirms similar results given in [70, 1981] and [46, 1996].

Therefore, compared to the arithmetic cost of $T(N) = 5N \log_2 N$ of the radix-2 algorithm in (3.10), the saving by the radix-4 algorithm is 15 percent. It will be shown in Chapter 12 that the split-radix algorithm can further reduce the arithmetic cost to $T(N) = 4N \log_2 N + \Theta(N)$, which represents a saving of 25 percent compared to the radix-2 algorithm.

11.2 The Radix-4 DIF FFTs

A radix-4 DIF FFT algorithm can be derived from recursively *decimating* the frequency series into four subsets, i.e., the set denoted by $Y_k = X_{4k}$ for $0 \le k \le N/4 - 1$, the set denoted by $Z_k = X_{4k+1}$ for $0 \le k \le N/4 - 1$, the set denoted by $G_k = X_{4k+2}$ for

$0 \le k \le N/4 - 1$, and the set denoted by $H_k = X_{4k+3}$ for $0 \le k \le N/4 - 1$ as shown below. The derivation again begins with the DFT definition from (3.1).

(11.22)

$$X_r = \sum_{\ell=0}^{N-1} x_\ell \omega_N^{r\ell}, \quad r = 0, 1, \ldots, N-1,$$

$$= \sum_{\ell=0}^{\frac{N}{4}-1} x_\ell \omega_N^{r\ell} + \sum_{\ell=\frac{N}{4}}^{\frac{N}{2}-1} x_\ell \omega_N^{r\ell} + \sum_{\ell=\frac{N}{2}}^{\frac{3N}{4}-1} x_\ell \omega_N^{r\ell} + \sum_{\ell=\frac{3N}{4}}^{N-1} x_\ell \omega_N^{r\ell}$$

$$= \sum_{\ell=0}^{\frac{N}{4}-1} x_\ell \omega_N^{r\ell} + \sum_{\ell=0}^{\frac{N}{4}-1} x_{\ell+\frac{N}{4}} \omega_N^{r\left(\ell+\frac{N}{4}\right)} + \sum_{\ell=0}^{\frac{N}{4}-1} x_{\ell+\frac{N}{2}} \omega_N^{r\left(\ell+\frac{N}{2}\right)} + \sum_{\ell=0}^{\frac{N}{4}-1} x_{\ell+\frac{3N}{4}} \omega_N^{r\left(\ell+\frac{3N}{4}\right)}$$

$$= \sum_{\ell=0}^{\frac{N}{4}-1} x_\ell \omega_N^{r\ell} + \omega_4^r \sum_{\ell=0}^{\frac{N}{4}-1} x_{\ell+\frac{N}{4}} \omega_N^{r\ell} + \omega_4^{2r} \sum_{\ell=0}^{\frac{N}{4}-1} x_{\ell+\frac{N}{2}} \omega_N^{r\ell} + \omega_4^{3r} \sum_{\ell=0}^{\frac{N}{4}-1} x_{\ell+\frac{3N}{4}} \omega_N^{r\ell}$$

$$= \sum_{\ell=0}^{\frac{N}{4}-1} \left(x_\ell + x_{\ell+\frac{N}{4}} \omega_4^r + x_{\ell+\frac{N}{2}} \omega_4^{2r} + x_{\ell+\frac{3N}{4}} \omega_4^{3r} \right) \omega_N^{r\ell}.$$

The four subproblems can thus be constructed by substituting $r = 4k$, $r = 4k+1$, $r = 4k+2$, and $r = 4k+3$ into the equation above.

$$Y_k = X_{4k} = \sum_{\ell=0}^{\frac{N}{4}-1} \left(x_\ell + x_{\ell+\frac{N}{4}} \omega_4^{4k} + x_{\ell+\frac{N}{2}} \omega_4^{2\times4k} + x_{\ell+\frac{3N}{4}} \omega_4^{3\times4k} \right) \omega_N^{4k\ell}$$

$$= \sum_{\ell=0}^{\frac{N}{4}-1} \left(x_\ell + x_{\ell+\frac{N}{4}} + x_{\ell+\frac{N}{2}} + x_{\ell+\frac{3N}{4}} \right) \omega_{\frac{N}{4}}^{k\ell}$$

(11.23)

$$= \sum_{\ell=0}^{\frac{N}{4}-1} \left(\left(x_\ell + x_{\ell+\frac{N}{2}} \right) + \left(x_{\ell+\frac{N}{4}} + x_{\ell+\frac{3N}{4}} \right) \right) \omega_{\frac{N}{4}}^{k\ell}$$

$$= \sum_{\ell=0}^{\frac{N}{4}-1} y_\ell \omega_{\frac{N}{4}}^{k\ell}, \quad k = 0, 1, \ldots, N/4 - 1.$$

(11.24)

$$Z_k = X_{4k+1} = \sum_{\ell=0}^{\frac{N}{4}-1} \left(x_\ell + x_{\ell+\frac{N}{4}} \omega_4^{4k+1} + x_{\ell+\frac{N}{2}} \omega_4^{2(4k+1)} + x_{\ell+\frac{3N}{4}} \omega_4^{3(4k+1)} \right) \omega_N^{(4k+1)\ell}$$

$$= \sum_{\ell=0}^{\frac{N}{4}-1} \left(\left(x_\ell - x_{\ell+\frac{N}{2}} \right) - j \left(x_{\ell+\frac{N}{4}} - x_{\ell+\frac{3N}{4}} \right) \right) \omega_N^\ell \omega_{\frac{N}{4}}^{k\ell}$$

$$= \sum_{\ell=0}^{\frac{N}{4}-1} z_\ell \omega_{\frac{N}{4}}^{k\ell}, \quad k = 0, 1, \ldots, N/4 - 1.$$

(11.25)

$$G_k = X_{4k+2} = \sum_{\ell=0}^{\frac{N}{4}-1} \left(x_\ell + x_{\ell+\frac{N}{4}} \omega_4^{4k+2} + x_{\ell+\frac{N}{2}} \omega_4^{2(4k+2)} + x_{\ell+\frac{3N}{4}} \omega_4^{3(4k+2)} \right) \omega_N^{(4k+2)\ell}$$

$$= \sum_{\ell=0}^{\frac{N}{4}-1} \left(\left(x_\ell + x_{\ell+\frac{N}{2}} \right) - \left(x_{\ell+\frac{N}{4}} + x_{\ell+\frac{3N}{4}} \right) \right) \omega_N^{2\ell} \omega_{\frac{N}{4}}^{k\ell}$$

$$= \sum_{\ell=0}^{\frac{N}{4}-1} g_\ell\, \omega_{\frac{N}{4}}^{k\ell}, \quad k = 0, 1, \ldots, N/4 - 1 .$$

(11.26)

$$H_k = X_{4k+3} = \sum_{\ell=0}^{\frac{N}{4}-1} \left(x_\ell + x_{\ell+\frac{N}{4}} \omega_4^{4k+3} + x_{\ell+\frac{N}{2}} \omega_4^{2(4k+3)} + x_{\ell+\frac{3N}{4}} \omega_4^{3(4k+3)} \right) \omega_N^{(4k+3)\ell}$$

$$= \sum_{\ell=0}^{\frac{N}{4}-1} \left(\left(x_\ell - x_{\ell+\frac{N}{2}} \right) + j \left(x_{\ell+\frac{N}{4}} - x_{\ell+\frac{3N}{4}} \right) \right) \omega_N^{3\ell} \omega_{\frac{N}{4}}^{k\ell}$$

$$= \sum_{\ell=0}^{\frac{N}{4}-1} h_\ell\, \omega_{\frac{N}{4}}^{k\ell}, \quad k = 0, 1, \ldots, N/4 - 1.$$

To form these four subproblems using two stages of butterfly computation, the partial sums identified above are first rearranged to facilitate the butterfly computation as shown below.

(11.27) $$y_\ell = \left(x_\ell + x_{\ell+\frac{N}{2}} \right) + \left(x_{\ell+\frac{N}{4}} + x_{\ell+\frac{3N}{4}} \right), \quad 0 \le \ell \le \frac{N}{4} - 1.$$

(11.28) $$z_\ell = \left(\left(x_\ell - x_{\ell+\frac{N}{2}} \right) - j \left(x_{\ell+\frac{N}{4}} - x_{\ell+\frac{3N}{4}} \right) \right) \omega_N^\ell, \quad 0 \le \ell \le \frac{N}{4} - 1.$$

(11.29) $$g_\ell = \left(- \left(x_{\ell+\frac{N}{4}} + x_{\ell+\frac{3N}{4}} \right) + \left(x_\ell + x_{\ell+\frac{N}{2}} \right) \right) \omega_N^{2\ell}, \quad 0 \le \ell \le \frac{N}{4} - 1.$$

(11.30) $$h_\ell = \left(j \left(x_{\ell+\frac{N}{4}} - x_{\ell+\frac{3N}{4}} \right) + \left(x_\ell - x_{\ell+\frac{N}{2}} \right) \right) \omega_N^{3\ell}, \quad 0 \le \ell \le \frac{N}{4} - 1.$$

The computation represented by (11.27), (11.28), (11.29), and (11.30) can now be represented by the two stages of butterfly computation in Figure 11.2.

Figure 11.2 The radix-4 DIF FFT butterflies.

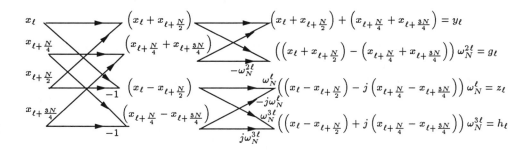

11.3   The Class of Radix-2^s DIT and DIF FFTs

The techniques used to develop the radix-2 and the radix-4 FFT algorithms can be
generalized to develop the entire class of radix-2^s FFTs. Setting $q = 2^s$, a radix-q DIT
FFT algorithm may be developed from decomposing (3.1) into q partial sums:

$$X_r = \sum_{\ell=0}^{N-1} x_\ell \omega_N^{r\ell}, \quad r = 0, 1, \ldots, N-1,$$

$$= \sum_{u=0}^{q-1} \sum_{k=0}^{\frac{N}{q}-1} x_{qk+u} \omega_N^{r(qk+u)}$$

(11.31)
$$= \sum_{u=0}^{q-1} \omega_N^{ur} \sum_{k=0}^{\frac{N}{q}-1} x_{qk+u} \omega_N^{r(qk)}$$

$$= \sum_{u=0}^{q-1} \omega_N^{ur} \sum_{k=0}^{\frac{N}{q}-1} x_{qk+u} \left(\omega_N^q\right)^{rk}$$

$$= \sum_{u=0}^{q-1} \omega_N^{ur} \left(\sum_{k=0}^{\frac{N}{q}-1} x_{qk+u} \omega_{\frac{N}{q}}^{rk}\right).$$

Observe that (3.3) and (11.1) are special cases of the equation above when $q = 2$ and
$q = 4$. According to (11.31), the time series can be *decimated* into $q = 2^s$ sets so that
each of the q partial sums represented by $\sum_{k=0}^{\frac{N}{q}-1} x_{qk+u} \omega_{\frac{N}{q}}^{rk}$ for $u = 0, 1, \ldots, q-1$, can
be recursively computed independent of each other. Each partial sum represents the
DFT of a subproblem of size N/q.

The output frequencies are computed as q separate segments, and each segment de-
noted by $X_{\ell+\lambda\frac{N}{q}}$ has N/q consecutive elements indexed by ℓ, where $0 \leq \ell \leq N/q-1$ and
$0 \leq \lambda \leq q-1$. By substituting $r = \ell + \lambda\frac{N}{q}$ in (11.31), one obtains the equation for
computing the output frequencies in each of the q segments. The equation for the λth
segment is shown below, where $0 \leq \lambda \leq q-1$.

$$X_{\ell+\lambda\frac{N}{q}} = \sum_{u=0}^{q-1} \omega_N^{u\ell} \left(\omega_N^{\frac{N}{q}}\right)^{u\lambda} \left(\sum_{k=0}^{\frac{N}{q}-1} x_{qk+u} \omega_{\frac{N}{q}}^{k\left(\ell+\lambda\frac{N}{q}\right)}\right)$$

(11.32)
$$= \sum_{u=0}^{q-1} \omega_N^{u\ell} \omega_q^{u\lambda} \left(\sum_{k=0}^{\frac{N}{q}-1} x_{qk+u} \omega_{\frac{N}{q}}^{k\ell}\right), \quad \ell = 0, 1, \ldots, N/q-1.$$

Of course, to minimize the arithmetic cost, the computation of the q frequency segments
should be reorganized to avoid redundant computation as demonstrated earlier in the
derivation of the radix-4 DIT FFT algorithm. Observe also that by substituting $q =$
2, $\lambda = 0, 1$ in (11.32), one obtains (3.7) and (3.8) for computing the two frequency
segments in the radix-2 DIT FFT algorithm; on the other hand, by substituting $q = 4$,
$\lambda = 0, 1, 2, 3$ in (11.32), one obtains the four equations (11.6), (11.7), (11.8), and (11.9)
for computing the four frequency segments in the radix-4 DIT FFT algorithm.

To develop a radix-q DIF FFT algorithm, one would simply *decimate* the frequency
series X_r into q sets with each set containing $\{Y_k^{(u)} | Y_k^{(u)} = X_{qk+u}, 0 \leq k \leq N/q-1\}$

for $u = 0, 1, \ldots, q-1$. Each of the q subproblems is thus of size N/q, and is defined by substituting $r = qk + u$ in (11.33) as shown for $q = 2$ and $q = 4$ in developing the radix-2 and radix-4 DIF FFT algorithms in Sections 3.2 and 11.2. For completeness, a brief derivation, which illuminates the generalization from the radix-2 and radix-4 algorithms, is provided below.

$$X_r = \sum_{\ell=0}^{N-1} x_\ell \omega_N^{r\ell}, \quad r = 0, 1, \ldots, N-1,$$

$$= \sum_{\lambda=0}^{q-1} \sum_{\ell=0}^{\frac{N}{q}-1} x_{\ell+\lambda\frac{N}{q}} \omega_N^{r\left(\ell+\lambda\frac{N}{q}\right)}$$

(11.33)
$$= \sum_{\ell=0}^{\frac{N}{q}-1} \sum_{\lambda=0}^{q-1} x_{\ell+\lambda\frac{N}{q}} \omega_N^{r\left(\ell+\lambda\frac{N}{q}\right)}$$

$$= \sum_{\ell=0}^{\frac{N}{q}-1} \sum_{\lambda=0}^{q-1} x_{\ell+\lambda\frac{N}{q}} \omega_N^{r\lambda\frac{N}{q}} \omega_N^{r\ell}$$

$$= \sum_{\ell=0}^{\frac{N}{q}-1} \left(\sum_{\lambda=0}^{q-1} x_{\ell+\lambda\frac{N}{q}} \omega_q^{\lambda r} \right) \omega_N^{r\ell}.$$

The q subproblems can thus be constructed by substituting $r = qk + u$ in (11.33) for $u = 0, 1, \ldots, q-1$.

$$Y_k^{(u)} = X_{qk+u} = \sum_{\ell=0}^{\frac{N}{q}-1} \left(\sum_{\lambda=0}^{q-1} x_{\ell+\lambda\frac{N}{q}} \omega_q^{\lambda(qk+u)} \right) \omega_N^{(qk+u)\ell}$$

(11.34)
$$= \sum_{\ell=0}^{\frac{N}{q}-1} \left\{ \left(\sum_{\lambda=0}^{q-1} x_{\ell+\lambda\frac{N}{q}} \omega_q^{\lambda(qk+u)} \right) \omega_N^{u\ell} \right\} \omega_{\frac{N}{q}}^{k\ell},$$

$$= \sum_{\ell=0}^{\frac{N}{q}-1} y_\ell^{(u)} \omega_{\frac{N}{q}}^{k\ell}, \quad k = 0, 1, \ldots, N/q - 1.$$

Note that the N/q input data points to each subproblems are labelled by $y_\ell^{(u)}$ for $u = 0, 1, \ldots, q-1$. To show that the radix-2 and radix-4 DIF FFT are special cases when $q = 2$ and $q = 4$, the generalized formulae for forming each of the $q = 2^s \geq 2$ subproblems are explicitly identified from (11.34) and it is displayed once again below. Observe that when $q = 2$, y_ℓ in (3.13) and z_ℓ in (3.15) correspond to $y_\ell^{(0)}$ and $y_\ell^{(1)}$ in the generalized formula; when $q = 4$, y_ℓ, z_ℓ, g_ℓ and h_ℓ in equations (11.23) to (11.26) correspond to $y_\ell^{(0)}$, $y_\ell^{(1)}$, $y_\ell^{(2)}$ and $y_\ell^{(3)}$ in this generalized formula.

(11.35)
$$y_\ell^{(u)} = \left(\sum_{\lambda=0}^{q-1} x_{\ell+\lambda\frac{N}{q}} \omega_q^{\lambda(qk+u)} \right) \omega_N^{u\ell}, \quad \ell = 0, 1, \ldots, N/q - 1.$$

Since it is known that a radix-q FFT for $q = 2^s > 4$ is less efficient than the probably optimal split-radix algorithm which recursively applies both radix-2 and radix-4 algorithms to solve each subproblem [86], further details on higher radix algorithms are omitted here, and readers are referred to [5, 84] for details about the popular radix-8 and radix-16 FFT algorithms.

Chapter 12

The Mixed-Radix and Split-Radix FFTs

12.1 The Mixed-Radix FFTs

There are two kinds of mixed-radix FFT algorithms. The first kind refers to a situation arising naturally when a radix-q algorithm, where $q = 2^m > 2$, is applied to an input series consisting of $N = 2^k \times q^s$ equally spaced points, where $1 \leq k < m$. In this case, out of necessity, k steps of radix-2 algorithm are applied either at the beginning or at the end of the transform, while the rest of the transform is carried out by s steps of the radix-q algorithm. For example, if $N = 2^{2m+1} = 2 \times 4^m$, the mixed-radix algorithm combines one step of the radix-2 algorithm and m steps of the radix-4 algorithm.[1]

The second kind of mixed-radix algorithms in the literature refers to those specialized for a composite $N = N_0 \times N_1 \times \cdots \times N_k$. Different algorithms may be used depending on whether the factors satisfy certain restrictions. The FFT algorithms for composite N will be treated in Chapter 15.

12.2 The Split-Radix DIT FFTs

After one has studied the radix-2 and radix-4 FFT algorithms in Chapters 3 and 11, it is interesting to see that the computing cost of the FFT algorithm can be further reduced by combining the two in a split-radix algorithm. The split-radix approach was first proposed by Duhamel and Hollmann in 1984 [39]. There are again DIT versions and DIF versions of the algorithm, depending on whether the input time series or the output frequency series is decimated.

The split-radix DIT algorithm is derived from (3.1), which defines the discrete

---

[1]It is of historical interest to note that a program for $N = 2^{2m+1}$ was written by Gentleman and Sande [47] in 1966, where they claimed a doubling of efficiency by this approach. However, Singleton observed in [83] that when computing with all the data stored in memory, a good radix-2 program was nearly as efficient as a radix-4 plus one step of radix-2 program and was simpler.

Fourier transform of a complex time series:

$$X_r = \sum_{\ell=0}^{N-1} x_\ell w_N^{r\ell}, \quad r = 0, 1, \ldots, N - 1,$$

$$(12.1) \qquad = \sum_{k=0}^{\frac{N}{2}-1} x_{2k} w_N^{r(2k)} + \sum_{k=0}^{\frac{N}{4}-1} x_{4k+1} w_N^{r(4k+1)} + \sum_{k=0}^{\frac{N}{4}-1} x_{4k+3} w_N^{r(4k+3)}$$

$$= \sum_{k=0}^{\frac{N}{2}-1} x_{2k} w_N^{r(2k)} + w_N^r \sum_{k=0}^{\frac{N}{4}-1} x_{4k+1} w_N^{r(4k)} + w_N^{3r} \sum_{k=0}^{\frac{N}{4}-1} x_{4k+3} w_N^{r(4k)}.$$

By *decimating* the time series into three sets, namely the set $\{y_k | y_k = x_{2k}, \ 0 \le k \le N/2 - 1\}$, the set $\{z_k | z_k = x_{4k+1}, \ 0 \le k \le N/4 - 1\}$, and the set $\{h_k | h_k = x_{4k+3}, \ 0 \le k \le N/4 - 1\}$, the three subproblems are defined after the appropriate twiddle factors $w_{\frac{N}{2}} = w_N^2$ and $w_{\frac{N}{4}} = w_N^4$ are identified.

(12.2)

$$Y_r = \sum_{k=0}^{\frac{N}{2}-1} x_{2k} w_N^{r(2k)} = \sum_{k=0}^{\frac{N}{2}-1} x_{2k} \left(w_N^2\right)^{rk} = \sum_{k=0}^{\frac{N}{2}-1} y_k w_{\frac{N}{2}}^{rk}, \quad r = 0, 1, \ldots, N/2 - 1.$$

(12.3)

$$Z_r = \sum_{k=0}^{\frac{N}{2}-1} x_{4k+1} w_N^{r(4k)} = \sum_{k=0}^{\frac{N}{4}-1} x_{4k+1} \left(w_N^4\right)^{rk} = \sum_{k=0}^{\frac{N}{4}-1} z_k w_{\frac{N}{4}}^{rk}, \quad r = 0, 1, \ldots, N/4 - 1.$$

(12.4)

$$H_r = \sum_{k=0}^{\frac{N}{2}-1} x_{4k+3} w_N^{r(4k)} = \sum_{k=0}^{\frac{N}{4}-1} x_{4k+3} \left(w_N^4\right)^{rk} = \sum_{k=0}^{\frac{N}{4}-1} h_k w_{\frac{N}{4}}^{rk}, \quad r = 0, 1, \ldots, N/4 - 1.$$

After these three subproblems are each (recursively) solved by the split-radix algorithm, the solution to the original problem of size N can be obtained according to (12.1) for $r = 0, 1, \ldots, N - 1$. Because $Y_{r+k\frac{N}{2}} = Y_r$ for $0 \le r \le N/2 - 1$, $Z_{r+k\frac{N}{4}} = Z_r$ for $0 \le r \le N/4 - 1$, and $H_{r+k\frac{N}{4}} = H_r$ for $0 \le r \le N/4 - 1$, equation (12.1) can be rewritten in terms of Y_r, $Y_{r+\frac{N}{4}}$, Z_r, and H_r for $0 \leq r \leq N/4 - 1$ as shown below.

$$X_r = Y_r + \omega_N^r Z_r + \omega_N^{3r} H_r$$

(12.5)
$$= Y_r + \left(\omega_N^r Z_r + \omega_N^{3r} H_r\right), \quad 0 \leq r \leq \frac{N}{4} - 1,$$

$$X_{r+\frac{N}{4}} = Y_{r+\frac{N}{4}} + \omega_N^{r+\frac{N}{4}} Z_r + \omega_N^{3\left(r+\frac{N}{4}\right)} H_r$$

(12.6)
$$= Y_{r+\frac{N}{4}} - j\left(\omega_N^r Z_r - \omega_N^{3r} H_r\right), \quad 0 \leq r \leq \frac{N}{4} - 1,$$

$$X_{r+\frac{N}{2}} = Y_r + \omega_N^{r+\frac{N}{2}} Z_r + \omega_N^{3\left(r+\frac{N}{2}\right)} H_r$$

(12.7)
$$= Y_r - \left(\omega_N^r Z_r + \omega_N^{3r} H_r\right), \quad 0 \leq r \leq \frac{N}{4} - 1,$$

$$X_{r+\frac{3N}{4}} = Y_{r+\frac{N}{4}} + \omega_N^{r+\frac{3N}{4}} Z_r + \omega_N^{3\left(r+\frac{3N}{4}\right)} H_r$$

(12.8)
$$= Y_{r+\frac{N}{4}} + j\left(\omega_N^r Z_r - \omega_N^{3r} H_r\right), \quad 0 \leq r \leq \frac{N}{4} - 1.$$

The computation represented by (12.5), (12.6), (12.7), and (12.8) is referred to as an *unsymmetric* DIT butterfly computation in the literature as shown in Figure 12.1.

Figure 12.1 The split-radix DIT FFT butterflies.

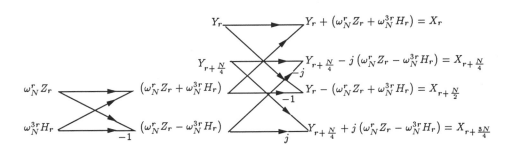

12.2.1 Analyzing the arithmetic cost

To determine the arithmetic cost of the split-radix DIT FFT algorithm, observe that $\omega_N^r Z_r$ and $\omega_N^{3r} H_r$ must be computed before the two partial sums can be formed. Since these two subproblems are each of size $N/4$, $N/2$ complex multiplications and $N/2$ complex additions are required in order to obtain the partial sums. Among the $N/2$ complex multiplications, there are four special cases which were already identified in the earlier discussion of the radix-4 algorithm: they are two cases of multiplication by 1 and two cases of multiplication by an odd power of ω_8. Recall that the former two cases are trivial, and only four real operations rather than six would be used in the latter two cases. Thus, $(3 + 3) \times (N/2 - 4) + 4 \times 2 + 2 \times (N/2) = 4N - 16$ nontrivial real operations are performed in the first stage of butterfly computation. In the second stage of the butterfly computation, only N complex additions or $2N$ real operations are required. The total cost for a single split-radix step thus involves $6N - 16$ nontrivial real operations (flops).

To set up the recurrence equation, the boundary conditions for both $N = 4$ and $N = 2$ are needed; when the size of a subproblem is reduced to 8, the three subsequent subproblems are of sizes 4, 2 and 2. As noted earlier, $T(4) = 16$ flops, and $T(2) = 4$ flops. The cost of the split-radix FFT algorithm (in terms of nontrivial flops) can be represented by the following recurrence:

$$(12.9) \qquad T(N) = \begin{cases} T\left(\frac{N}{2}\right) + 2T\left(\frac{N}{4}\right) + 6N - 16 & \text{if } N = 4^n > 4, \\ 16 & \text{if } N = 4, \\ 4 & \text{if } N = 2. \end{cases}$$

Solving (12.9) (see Appendix B), one obtains the solution

$$(12.10) \qquad T(N) = 4N \log_2 N - 6N + 8 .$$

12.3   The Split-Radix DIF FFTs

A split-radix DIF FFT algorithm can be derived by recursively applying both radix-2 and radix-4 DIF FFT algorithm to solve each subproblem resulting from *decimating* the output frequency series in a similar fashion. That is, the frequency series is recursively decimated into three subsets, i.e., the set denoted by $Y_k = X_{2k}$ for $0 \le k \le N/2-1$, the set denoted by $Z_k = X_{4k+1}$ for $0 \le k \le N/4 - 1$, and the set denoted by $H_k = X_{4k+3}$ for $0 \le k \le N/4 - 1$ as shown below. The derivation begins with the discrete Fourier transform defined by (3.1). Using the results developed earlier for the radix-2 DIF algorithm in (3.11), one obtains

$$(12.11) \qquad \begin{aligned} X_r &= \sum_{\ell=0}^{\frac{N}{2}-1} x_\ell \omega_N^{r\ell} + \sum_{\ell=\frac{N}{2}}^{N-1} x_\ell \omega_N^{r\ell} \\ &= \sum_{\ell=0}^{\frac{N}{2}-1} \left(x_\ell + x_{\ell+\frac{N}{2}} \omega_N^{r\frac{N}{2}} \right) \omega_N^{r\ell}, \quad r = 0, 1, \ldots, N - 1. \end{aligned}$$

By letting $Y_k = X_{2k}$, $y_\ell = x_\ell + x_{\ell+\frac{N}{2}}$, one subproblem of half the size is defined by

$$(12.12) \qquad \begin{aligned} Y_k = X_{2k} &= \sum_{\ell=0}^{\frac{N}{2}-1} \left(x_\ell + x_{\ell+\frac{N}{2}} \right) \omega_{\frac{N}{2}}^{k\ell} \\ &= \sum_{\ell=0}^{\frac{N}{2}-1} y_\ell \, \omega_{\frac{N}{2}}^{k\ell}, \quad k = 0, 1, \ldots, N/2 - 1. \end{aligned}$$

To construct the other two subproblems of size $N/4$, begin with the DFT definition in (3.1) and use the results developed earlier for the DIF radix-4 algorithm in (11.22).

$$(12.13) \qquad \begin{aligned} X_r &= \sum_{\ell=0}^{N-1} x_\ell \omega_N^{r\ell}, \quad r = 0, 1, \ldots, N - 1, \\ &= \sum_{\ell=0}^{\frac{N}{4}-1} \left(x_\ell + x_{\ell+\frac{N}{4}} \omega_4^{r} + x_{\ell+\frac{N}{2}} \omega_4^{2r} + x_{\ell+\frac{3N}{4}} \omega_4^{3r} \right) \omega_N^{r\ell} . \end{aligned}$$

By substituting $r = 4k + 1$ and $r = 4k + 3$, one again obtains

(12.14)

$$
Z_k = X_{4k+1} = \sum_{\ell=0}^{\frac{N}{4}-1} \left(x_\ell + x_{\ell+\frac{N}{4}} w_4^{4k+1} + x_{\ell+\frac{N}{2}} w_4^{2(4k+1)} + x_{\ell+\frac{3N}{4}} w_4^{3(4k+1)} \right) w_N^{(4k+1)\ell}
$$

$$
= \sum_{\ell=0}^{\frac{N}{4}-1} \left(\left(x_\ell - x_{\ell+\frac{N}{2}} \right) - j \left(x_{\ell+\frac{N}{4}} - x_{\ell+\frac{3N}{4}} \right) \right) w_N^\ell w_{\frac{N}{4}}^{k\ell}
$$

$$
= \sum_{\ell=0}^{\frac{N}{4}-1} z_\ell \, w_{\frac{N}{4}}^{k\ell}, \quad k = 0, 1, \ldots, N/4 - 1 \,.
$$

(12.15)

$$
H_k = X_{4k+3} = \sum_{\ell=0}^{\frac{N}{4}-1} \left(x_\ell + x_{\ell+\frac{N}{4}} w_4^{4k+3} + x_{\ell+\frac{N}{2}} w_4^{2(4k+3)} + x_{\ell+\frac{3N}{4}} w_4^{3(4k+3)} \right) w_N^{(4k+3)\ell}
$$

$$
= \sum_{\ell=0}^{\frac{N}{4}-1} \left(\left(x_\ell - x_{\ell+\frac{N}{2}} \right) + j \left(x_{\ell+\frac{N}{4}} - x_{\ell+\frac{3N}{4}} \right) \right) w_N^{3\ell} w_{\frac{N}{4}}^{k\ell}
$$

$$
= \sum_{\ell=0}^{\frac{N}{4}-1} h_\ell \, w_{\frac{N}{4}}^{k\ell}, \quad k = 0, 1, \ldots, N/4 - 1 \,.
$$

To form these three subproblems using two stages of *unsymmetric* butterfly computation, the computation of the partial sums is again rearranged to facilitate the butterfly computation.

(12.16) $\quad y_\ell = \left(x_\ell + x_{\ell+\frac{N}{2}} \right), \quad 0 \le \ell \le \dfrac{N}{4} - 1 \,.$

(12.17) $\quad y_{\ell+\frac{N}{4}} = \left(x_{\ell+\frac{N}{4}} + x_{\ell+\frac{3N}{4}} \right), \quad 0 \le \ell \le \dfrac{N}{4} - 1 \,.$

(12.18) $\quad z_\ell = \left(\left(x_\ell - x_{\ell+\frac{N}{2}} \right) - j \left(x_{\ell+\frac{N}{4}} - x_{\ell+\frac{3N}{4}} \right) \right) w_N^\ell, \quad \le \ell \le \dfrac{N}{4} - 1 \,.$

(12.19) $\quad h_\ell = \left(j \left(x_{\ell+\frac{N}{4}} - x_{\ell+\frac{3N}{4}} \right) + \left(x_\ell - x_{\ell+\frac{N}{2}} \right) \right) w_N^{3\ell}, \quad 0 \le \ell \le \dfrac{N}{4} - 1 \,.$

The computation represented by (12.16), (12.17), (12.18), and (12.19) again yields an *unsymmetric* DIF butterfly computation as depicted in Figure 12.2.

Figure 12.2 The split-radix DIF FFT butterflies.

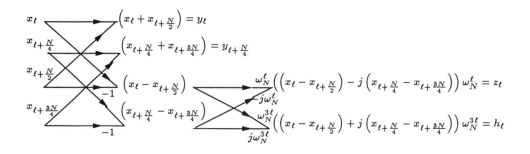

12.4   Notes and References

The split-radix FFT was originally developed by Duhamel and Hollmann [39] in 1984, and it was subsequently extended and implemented for complex, real and real-symmetric data by Duhamel in [38]. In 1986, Sorensen, Heideman, and Burrus presented an indexing scheme which efficiently implemented the Duhamel-Hollmann split-radix FFT [86]. Both DIF and DIT Fortran programs were presented in [86]. The history of the ideas on the fast Fourier transforms from Gauss to the split-radix algorithm is presented in [41].

Chapter 13

FFTs for Arbitrary N

For some applications, one can always choose $N = 2^n$ and employ the radix-2 algorithm. For example, if one is sampling N discrete date points from an (essentially) infinite set, one can certainly choose $N = 2^n$ points. This is the normal situation in the practice of signal processing. As one more example, it is well-known that the FFT and its inverse transformation can be used to effect polynomial evaluation and interpolation very efficiently. Indeed, one can multiply two (degree $N-1$) polynomials in $\Theta(N \log_2 N)$ time. Since the degree of the polynomial can be increased by adding higher-order terms with zero coefficients, one can always choose $N = 2^n$ and use the radix-2 algorithm [63].

Consequently, the concern here is what to do when it is necessary to compute the DFT of a given time series consisting of equally spaced $N \neq 2^n$ points: $\{x_0, x_1, ..., x_{N-1}\}$, assuming that the less efficient $\Theta(N^2)$ DFT algorithm is not acceptable. Note that one *cannot* simply add zeros to the data set to make N reach the next power of two. That is, the strategy of padding the input array by extra zero entries is not valid. If N can be chosen to be composite, one may either employ Bergland's algorithm [3] for composite $N = q_0 \times q_1 \times \cdots \times q_k$, where each q_i is a small prime, or employ Bluestein's algorithm [10, 11, 96] which works for arbitrary N (including large primes.) While the former is faster than the latter if N is composite, Bluestein's algorithm is more general because there is no restriction on N.

While many variants of Bergland's mixed-radix algorithm is in wide use on shared-memory multiprocessor computers [94], Bluestein's algorithm was recently shown to incur the least amount of communication on distributed-memory hypercube multiprocessors [96]. Since the FFTs for composite N will be presented in Chapter 15, only Bluestein's algorithm is studied in this chapter, and the material is largely adapted from [96] with added details. To highlight that this algorithm is recommended only when $N \neq 2^n$, the example for $N = 4$ in [96] is replaced by a new example for $N = 6$, and some notations used in [96] are modified to be consistent with the rest of the text.

13.1   The Main Ideas Behind Bluestein's FFT

Bluestein's FFT is a good example to demonstrate that a sophisticated algorithm can be built from "a few elegant ideas." The main ideas employed in the Bluestein's FFT are the following.

- Express the DFT defined by (3.1) as a Toeplitz matrix-vector product.

- Embed the Toeplitz matrix-vector product in a circulant matrix-vector product of order $M = 2^s$.

- Diagonalize a circulant matrix by a DFT matrix.

- Evaluate the diagonally scaled DFT using a radix-2 FFT.

- Perform a radix-2 inverse FFT to undo the diagonalization.

Each idea is explored further below.

13.1.1   DFT and the symmetric Toeplitz matrix-vector product

Definition 13.1 A $N \times N$ matrix T is Toeplitz if, for all i and j, $t_{i,j} = h_{i-j}$ for the given $2N - 1$ scalars $h_{-N+1}, \ldots, h_0, \ldots, h_{N-1}$.

Observe that a $N \times N$ matrix has exactly $2N - 1$ diagonals. Since all $t_{i,j}$ entries along a diagonal have the same $i - j$ value, they are assigned the same constant h_{i-j} in a Toeplitz matrix.

Example 13.1 Given below is a 6×6 Toeplitz Matrix.

$$T = \begin{bmatrix} t_{0,0} & t_{0,1} & t_{0,2} & t_{0,3} & t_{0,4} & t_{0,5} \\ t_{1,0} & t_{1,1} & t_{1,2} & t_{1,3} & t_{1,4} & t_{1,5} \\ t_{2,0} & t_{2,1} & t_{2,2} & t_{2,3} & t_{2,4} & t_{2,5} \\ t_{3,0} & t_{3,1} & t_{3,2} & t_{3,3} & t_{3,4} & t_{3,5} \\ t_{4,0} & t_{4,1} & t_{4,2} & t_{4,3} & t_{4,4} & t_{4,5} \\ t_{5,0} & t_{5,1} & t_{5,2} & t_{5,3} & t_{5,4} & t_{5,5} \end{bmatrix} = \begin{bmatrix} h_0 & h_{-1} & h_{-2} & h_{-3} & h_{-4} & h_{-5} \\ h_1 & h_0 & h_{-1} & h_{-2} & h_{-3} & h_{-4} \\ h_2 & h_1 & h_0 & h_{-1} & h_{-2} & h_{-3} \\ h_3 & h_2 & h_1 & h_0 & h_{-1} & h_{-2} \\ h_4 & h_3 & h_2 & h_1 & h_0 & h_{-1} \\ h_5 & h_4 & h_3 & h_2 & h_1 & h_0 \end{bmatrix}$$

Example 13.2 A Toeplitz matrix is symmetric if $h_{i-j} = h_{j-i}$ as shown below.

$$T = \begin{bmatrix} t_{0,0} & t_{0,1} & t_{0,2} & t_{0,3} & t_{0,4} & t_{0,5} \\ t_{1,0} & t_{1,1} & t_{1,2} & t_{1,3} & t_{1,4} & t_{1,5} \\ t_{2,0} & t_{2,1} & t_{2,2} & t_{2,3} & t_{2,4} & t_{2,5} \\ t_{3,0} & t_{3,1} & t_{3,2} & t_{3,3} & t_{3,4} & t_{3,5} \\ t_{4,0} & t_{4,1} & t_{4,2} & t_{4,3} & t_{4,4} & t_{4,5} \\ t_{5,0} & t_{5,1} & t_{5,2} & t_{5,3} & t_{5,4} & t_{5,5} \end{bmatrix} = \begin{bmatrix} h_0 & h_1 & h_2 & h_3 & h_4 & h_5 \\ h_1 & h_0 & h_1 & h_2 & h_3 & h_4 \\ h_2 & h_1 & h_0 & h_1 & h_2 & h_3 \\ h_3 & h_2 & h_1 & h_0 & h_1 & h_2 \\ h_4 & h_3 & h_2 & h_1 & h_0 & h_1 \\ h_5 & h_4 & h_3 & h_2 & h_1 & h_0 \end{bmatrix}$$

Bluestein's idea was to express the DFT matrix-vector product as a *symmetric* Toeplitz matrix-vector product using the identity

(13.1) $$r\ell = \frac{1}{2}\left(r^2 + \ell^2 - (r - \ell)^2\right).$$

Replacing $r\ell$ in (3.1) by the right side of (13.1) yields

$$X_r = \sum_{\ell=0}^{N-1} x_\ell \omega_N^{r\ell}, \quad r = 0, 1, \ldots, N-1,$$

(13.2)
$$= \sum_{\ell=0}^{N-1} x_\ell \omega_N^{\frac{1}{2}(r^2 + \ell^2 - (r-\ell)^2)}$$

$$= \omega_N^{\frac{1}{2}r^2} \sum_{\ell=0}^{N-1} x_\ell \omega_N^{\frac{1}{2}\ell^2} \omega_N^{-\frac{1}{2}(r-\ell)^2}.$$

To keep only the matrix-vector product on the right-hand side, multiply (13.2) by $\omega_N^{-\frac{1}{2}r^2}$ on both sides, yielding

(13.3)
$$\omega_N^{-\frac{1}{2}r^2} X_r = \sum_{\ell=0}^{N-1} \left(x_\ell \omega_N^{\frac{1}{2}\ell^2} \right) \omega_N^{-\frac{1}{2}(r-\ell)^2}, \quad r = 0, \ldots, N-1.$$

To further simplify (13.3), define

(13.4)
$$Z_r = \omega_N^{-\frac{1}{2}r^2} X_r, \qquad y_\ell = x_\ell \omega_N^{\frac{1}{2}\ell^2}, \qquad h_{r-\ell} = \omega_N^{-\frac{1}{2}(r-\ell)^2},$$

and rewrite (13.3) as

(13.5)
$$Z_r = \sum_{\ell=0}^{N-1} y_\ell h_{r-\ell}, \quad r = 0, 1, \ldots, N-1.$$

For $N = 6$, the corresponding equations expressed as a matrix-vector product are

(13.6)
$$\begin{bmatrix} Z_0 \\ Z_1 \\ Z_2 \\ Z_3 \\ Z_4 \\ Z_5 \end{bmatrix} = \begin{bmatrix} h_0 & h_{-1} & h_{-2} & h_{-3} & h_{-4} & h_{-5} \\ h_1 & h_0 & h_{-1} & h_{-2} & h_{-3} & h_{-4} \\ h_2 & h_1 & h_0 & h_{-1} & h_{-2} & h_{-3} \\ h_3 & h_2 & h_1 & h_0 & h_{-1} & h_{-2} \\ h_4 & h_3 & h_2 & h_1 & h_0 & h_{-1} \\ h_5 & h_4 & h_3 & h_2 & h_1 & h_0 \end{bmatrix} \begin{bmatrix} y_0 \\ y_1 \\ y_2 \\ y_3 \\ y_4 \\ y_5 \end{bmatrix}.$$

According to Definition 13.1, equation (13.6) represents a Toeplitz matrix-vector product. Furthermore, because $(r - \ell)^2 = (\ell - r)^2$, which implies

$$h_{r-\ell} = \omega_N^{-\frac{1}{2}(r-\ell)^2} = \omega_N^{-\frac{1}{2}(\ell-r)^2} = h_{\ell-r}$$

according to (13.4). Thus, the six equations can be expressed as a *symmetric* Toeplitz matrix-vector product as shown in (13.7).

(13.7)
$$\begin{bmatrix} Z_0 \\ Z_1 \\ Z_2 \\ Z_3 \\ Z_4 \\ Z_5 \end{bmatrix} = \begin{bmatrix} h_0 & h_1 & h_2 & h_3 & h_4 & h_5 \\ h_1 & h_0 & h_1 & h_2 & h_3 & h_4 \\ h_2 & h_1 & h_0 & h_1 & h_2 & h_3 \\ h_3 & h_2 & h_1 & h_0 & h_1 & h_2 \\ h_4 & h_3 & h_2 & h_1 & h_0 & h_1 \\ h_5 & h_4 & h_3 & h_2 & h_1 & h_0 \end{bmatrix} \begin{bmatrix} y_0 \\ y_1 \\ y_2 \\ y_3 \\ y_4 \\ y_5 \end{bmatrix}.$$

In summary, instead of computing X_r directly from (3.1) using the $\Theta\left(N^2\right)$ Algorithm, the steps in the alternative Algorithm 13.1 may be performed.

Algorithm 13.1 Computing DFT as a symmetric Toeplitz matrix-vector product.

---

begin
    for $\ell := 0$ **to** $N - 1$ **do**   Preprocessing phase
        $y_\ell := x_\ell * \omega_N^{\frac{1}{2}\ell^2}$; $h_\ell := \omega_N^{-\frac{1}{2}\ell^2}$
    end for
    for $r := 0$ **to** $N - 1$ **do**
        $Z_r := 0.0$ Initialize Z_r
        for $\ell := 0$ **to** $N - 1$ **do**
            $m := \text{abs}(r - \ell)$ $m = |r - \ell|$
            $Z_r := Z_r + y_\ell * h_m$
        end for
    end for
    for $r := 0$ **to** $N - 1$ **do**   Recover phase
        $X_r := Z_r * \omega_N^{\frac{1}{2}r^2}$
    end for

However, nothing is gained. Indeed, the $\Theta\left(N^2\right)$ Toeplitz DFT algorithm 13.1 takes even longer to compute due to the extra work in the *preprocessing* phase and the *recovery* phase. The challenge is to compute the Toeplitz matrix-vector product in $\Theta\left(N\log_2 N\right)$ time when N is *not* a power of two.

Bluestein's second idea is to enlarge the $N \times N$ Toeplitz matrix to a circulant matrix of dimension $2N - 2$. If $2N - 2 = 2^s$, then the matrix-vector product involving this enlarged circulant matrix can be computed next. If $2N - 2 \neq 2^s$, the circulant matrix will be enlarged further to dimension $M = 2^s$. For example, if $N = 5$, then $2N - 2 = 8 = 2^3$, and the matrix-vector product involving the 8×8 circulant matrix is computed next. However, if $N = 6$, then $2N - 2 = 10 \neq 2^s$, the circulant matrix will be further enlarged to dimension $M = 2^4 = 16$, and the matrix-vector product computed next involves the 16×16 circulant matrix. These two separate embedding processes are examined in the next two sections.

13.1.2 Enlarging the Toeplitz matrix to a circulant matrix

Bluestein's "embedding" idea can be best demonstrated by enlarging a Toeplitz matrix resulting from a DFT of length $N = 6$. The 6×6 symmetric Toeplitz matrix $\boldsymbol{H}$ is enlarged to a circulant matrix $\boldsymbol{H}^{(1)}$ of dimension $2N - 2 = 10$ below. The process involves extending the existent diagonals as well as adding new ones to fill the two vacant corners. The added diagonals are formed by the same h_i's excluding h_{N-1} and h_0, and they are properly ordered so that the result is a circulant matrix, which is a symmetric Toeplitz matrix with the additional property: *the elements of each row are cyclic shifted from the elements in the preceding row.*

Example 13.3

$$(13.8) \quad H^{(1)} = \begin{bmatrix} h_0 & h_1 & h_2 & h_3 & h_4 & h_5 \\ h_1 & h_0 & h_1 & h_2 & h_3 & h_4 \\ h_2 & h_1 & h_0 & h_1 & h_2 & h_3 \\ h_3 & h_2 & h_1 & h_0 & h_1 & h_2 \\ h_4 & h_3 & h_2 & h_1 & h_0 & h_1 \\ h_5 & h_4 & h_3 & h_2 & h_1 & h_0 \end{bmatrix}$$

$$+ \begin{bmatrix} & & & & & h_5 & & & & \\ & & & & & h_4 & h_5 & & & \\ & & & & & h_3 & h_4 & h_5 & & \\ & & & & & h_2 & h_3 & h_4 & h_5 & \\ & & & & & h_1 & h_2 & h_3 & h_4 \\ h_5 & h_4 & h_3 & h_2 & h_1 & h_0 & h_1 & h_2 & h_3 \\ & h_5 & h_4 & h_3 & h_2 & h_1 & h_0 & h_1 & h_2 \\ & & h_5 & h_4 & h_3 & h_2 & h_1 & h_0 & h_1 \\ & & & h_5 & h_4 & h_3 & h_2 & h_1 & h_0 \end{bmatrix}$$

$$+ \begin{bmatrix} & & & & & & h_4 & h_3 & h_2 & h_1 \\ & & & & & & & h_4 & h_3 & h_2 \\ & & & & & & & & h_4 & h_3 \\ & & & & & & & & & h_4 \\ & & & & & & & & & \\ h_4 & & & & & & & & & \\ h_3 & h_4 & & & & & & & & \\ h_2 & h_3 & h_4 & & & & & & & \\ h_1 & h_2 & h_3 & h_4 & & & & & & \end{bmatrix}.$$

The result is a circulant matrix:

$$(13.9) \quad H^{(1)} = \begin{bmatrix} h_0 & h_1 & h_2 & h_3 & h_4 & h_5 & h_4 & h_3 & h_2 & h_1 \\ h_1 & h_0 & h_1 & h_2 & h_3 & h_4 & h_5 & h_4 & h_3 & h_2 \\ h_2 & h_1 & h_0 & h_1 & h_2 & h_3 & h_4 & h_5 & h_4 & h_3 \\ h_3 & h_2 & h_1 & h_0 & h_1 & h_2 & h_3 & h_4 & h_5 & h_4 \\ h_4 & h_3 & h_2 & h_1 & h_0 & h_1 & h_2 & h_3 & h_4 & h_5 \\ h_5 & h_4 & h_3 & h_2 & h_1 & h_0 & h_1 & h_2 & h_3 & h_4 \\ h_4 & h_5 & h_4 & h_3 & h_2 & h_1 & h_0 & h_1 & h_2 & h_3 \\ h_3 & h_4 & h_5 & h_4 & h_3 & h_2 & h_1 & h_0 & h_1 & h_2 \\ h_2 & h_3 & h_4 & h_5 & h_4 & h_3 & h_2 & h_1 & h_0 & h_1 \\ h_1 & h_2 & h_3 & h_4 & h_5 & h_4 & h_3 & h_2 & h_1 & h_0 \end{bmatrix}.$$

Formally, the circulant $H^{(1)}$ is a symmetric Toeplitz matrix consisting of $M = 2N - 2$ constant diagonals defined by the set of scalars $\{h_0^{(1)}, h_1^{(1)}, h_2^{(1)}, \ldots, h_{M-1}^{(1)}\}$, which is extended from the h_i's defining H (in a circular fashion) according to the rules given below.

(13.10)

$$H^{(1)}[r, \ell] \equiv h_{|r-\ell|}^{(1)}, \quad 0 \leq r, \ell \leq M - 1, \text{where}$$

$$h_\lambda^{(1)} \equiv \begin{cases} h_\lambda & \text{if } 0 \leq \lambda \leq N - 1, \\ h_{M-\lambda} & \text{if } N \leq \lambda \leq M - 1, \text{ and } M = 2N - 2. \end{cases}$$

Accordingly, $H^{(1)}$ is circulant if and only if $h_\lambda^{(1)} = h_{M-\lambda}^{(1)}$ for $0 \leq \lambda \leq M - 1$.

13.1.3   Enlarging the dimension of a circulant matrix to $M = 2^s$

As explained earlier, when $N = 6$, because $2N - 2 = 10$ is not a power of two, the 10×10 circulant matrix $H^{(1)}$ in (13.9) needs to be further enlarged to dimension $M = 2^s = 16$. Continuing with the matrix $H^{(1)}$ from the previous section, the second embedding process is depicted below. The result is a 16×16 circulant matrix $H^{(2)}$. It can be clearly seen that the "embedding" is done by moving the two previously added corners further away to the corners of this larger matrix, followed by extending the existent diagonals. The unfilled diagonals remain zeros.

Example 13.4

The embedded matrix $H^{(1)}$ is

$$\begin{bmatrix}
h_0 & h_1 & h_2 & h_3 & h_4 & h_5 & & & & & & h_4 & h_3 & h_2 & h_1 \\
h_1 & h_0 & h_1 & h_2 & h_3 & h_4 & h_5 & & & & & & h_4 & h_3 & h_2 \\
h_2 & h_1 & h_0 & h_1 & h_2 & h_3 & h_4 & h_5 & & & & & & h_4 & h_3 \\
h_3 & h_2 & h_1 & h_0 & h_1 & h_2 & h_3 & h_4 & h_5 & & & & & & h_4 \\
h_4 & h_3 & h_2 & h_1 & h_0 & h_1 & h_2 & h_3 & h_4 & h_5 & & & & & \\
h_5 & h_4 & h_3 & h_2 & h_1 & h_0 & h_1 & h_2 & h_3 & h_4 & & & & & \\
 & h_5 & h_4 & h_3 & h_2 & h_1 & h_0 & h_1 & h_2 & h_3 & & & & & \\
 & & h_5 & h_4 & h_3 & h_2 & h_1 & h_0 & h_1 & h_2 & & & & & \\
 & & & h_5 & h_4 & h_3 & h_2 & h_1 & h_0 & h_1 & & & & & \\
 & & & & h_5 & h_4 & h_3 & h_2 & h_1 & h_0 & & & & & \\
 & & & & & & & & & & & & & & \\
h_4 & & & & & & & & & & & & & & \\
h_3 & h_4 & & & & & & & & & & & & & \\
h_2 & h_3 & h_4 & & & & & & & & & & & & \\
h_1 & h_2 & h_3 & h_4 & & & & & & & & & & &
\end{bmatrix}.$$

The extended circulant matrix $H^{(2)}$ is

$$
\begin{bmatrix}
h_0 & h_1 & h_2 & h_3 & h_4 & h_5 &   &   &   &   &   &   & h_4 & h_3 & h_2 & h_1 \\
h_1 & h_0 & h_1 & h_2 & h_3 & h_4 & h_5 &   &   &   &   &   &   & h_4 & h_3 & h_2 \\
h_2 & h_1 & h_0 & h_1 & h_2 & h_3 & h_4 & h_5 &   &   &   &   &   &   & h_4 & h_3 \\
h_3 & h_2 & h_1 & h_0 & h_1 & h_2 & h_3 & h_4 & h_5 &   &   &   &   &   &   & h_4 \\
h_4 & h_3 & h_2 & h_1 & h_0 & h_1 & h_2 & h_3 & h_4 & h_5 &   &   &   &   &   &   \\
h_5 & h_4 & h_3 & h_2 & h_1 & h_0 & h_1 & h_2 & h_3 & h_4 & h_5 &   &   &   &   &   \\
    & h_5 & h_4 & h_3 & h_2 & h_1 & h_0 & h_1 & h_2 & h_3 & h_4 & h_5 &   &   &   &   \\
    &   & h_5 & h_4 & h_3 & h_2 & h_1 & h_0 & h_1 & h_2 & h_3 & h_4 & h_5 &   &   &   \\
    &   &   & h_5 & h_4 & h_3 & h_2 & h_1 & h_0 & h_1 & h_2 & h_3 & h_4 & h_5 &   &   \\
    &   &   &   & h_5 & h_4 & h_3 & h_2 & h_1 & h_0 & h_1 & h_2 & h_3 & h_4 & h_5 &   \\
    &   &   &   &   & h_5 & h_4 & h_3 & h_2 & h_1 & h_0 & h_1 & h_2 & h_3 & h_4 & h_5 \\
    &   &   &   &   &   & h_5 & h_4 & h_3 & h_2 & h_1 & h_0 & h_1 & h_2 & h_3 & h_4 \\
h_4 &   &   &   &   &   &   & h_5 & h_4 & h_3 & h_2 & h_1 & h_0 & h_1 & h_2 & h_3 \\
h_3 & h_4 &   &   &   &   &   &   & h_5 & h_4 & h_3 & h_2 & h_1 & h_0 & h_1 & h_2 \\
h_2 & h_3 & h_4 &   &   &   &   &   &   & h_5 & h_4 & h_3 & h_2 & h_1 & h_0 & h_1 \\
h_1 & h_2 & h_3 & h_4 &   &   &   &   &   &   & h_5 & h_4 & h_3 & h_2 & h_1 & h_0
\end{bmatrix}.
$$

Formally, the circulant $H^{(2)}$ is a symmetric Toeplitz matrix consisting of M constant diagonals defined by the set of scalars $\{h_0^{(2)}, h_1^{(2)}, h_2^{(2)}, \ldots, h_{M-1}^{(2)}\}$, which is extended from the h_i's in H (in a circular fashion) according to the rules given below.

(13.11)
$$
H^{(2)}[r,\ell] \equiv h^{(2)}_{|r-\ell|}, \quad 0 \le r, \ell \le M-1 \text{, where}
$$
$$
h^{(2)}_\lambda \equiv \begin{cases} h_\lambda, & \text{if } 0 \le \lambda \le N-1, \\ 0, & \text{if } N \le \lambda \le M-N+1, \\ h_{M-\lambda}, & \text{if } M-N+2 \le \lambda \le M-1. \end{cases}
$$

Observe again that $H^{(2)}$ is circulant if and only if $h^{(2)}_\lambda = h^{(2)}_{M-\lambda}$ for $0 \le \lambda \le M-1$.

13.1.4   Forming the $M \times M$ circulant matrix-vector product.

Recall that depending on whether $M = 2N - 2$ is a power of two, the Toeplitz matrix-vector product $z = Hy$ is to be embedded in either $z^{(1)} = H^{(1)}y^{(1)}$ or $z^{(2)} = H^{(2)}y^{(2)}$. To continue, $y^{(1)}$ and $y^{(2)}$ must now be determined.

Since H is kept *intact* as an $N \times N$ submatrix in the upper left corner of either $H^{(1)}$ or $H^{(2)}$, an appropriate $y^{(1)}$ or $y^{(2)}$ can be obtained by simply padding y with $M - N$ more zeros, where $2N - 2 \le M = 2^s < 4N$. As depicted below for $N = 6$, the elements of the vector z, namely, $z_0, z_1, \ldots, z_{N-1}$, can then be retrieved from the top N positions in either $z^{(1)}$ or $z^{(2)}$. Note that the numerical values of the remaining entries are irrelevant, and they are marked by "$\times$" symbols for simplicity.

Example 13.5 Depicting the embedded $z = Hy$ in $z^{(1)} = H^{(1)}y^{(1)}$ or $z^{(2)} = H^{(2)}y^{(2)}$ for $N = 6$.

$$
\begin{bmatrix}
z_0 \\
z_1 \\
z_2 \\
z_3 \\
z_4 \\
z_5 \\
\times \\
\vdots \\
\times
\end{bmatrix}
=
\begin{bmatrix}
h_0 & h_1 & h_2 & h_3 & h_4 & h_5 & \times & \cdots & \times \\
h_1 & h_0 & h_1 & h_2 & h_3 & h_4 & \times & \cdots & \times \\
h_2 & h_1 & h_0 & h_1 & h_2 & h_3 & \times & \cdots & \times \\
h_3 & h_2 & h_1 & h_0 & h_1 & h_2 & \times & \cdots & \times \\
h_4 & h_3 & h_2 & h_1 & h_0 & h_1 & \times & \cdots & \times \\
h_5 & h_4 & h_3 & h_2 & h_1 & h_0 & \times & \cdots & \times \\
\times & \times & \times & \times & \times & \times & \times & \cdots & \times \\
\cdots\cdots\cdots\cdots\cdots\cdots \\
\times & \times & \times & \times & \times & \times & \times & \cdots & \times
\end{bmatrix}
\begin{bmatrix}
y_0 \\
y_1 \\
y_2 \\
y_3 \\
y_4 \\
y_5 \\
0 \\
\vdots \\
0
\end{bmatrix}
$$

13.1.5 Diagonalizing a circulant matrix by a DFT matrix

The next step involves the diagonalization of the circulant $H^{(1)}$ or $H^{(2)}$ by a DFT matrix. *The objective is to obtain a matrix-vector product which represents a DFT of length M so that it can be computed by a radix-2 FFT in $\Theta\left(M \log_2 M\right)$ time.* In this section, the diagonalization of $H^{(2)}$ is considered without loss of generality, because $H^{(2)}$ is reduced to $H^{(1)}$ when $M = 2N - 2 = 2^s$.

Recall that a DFT of length M defined by (3.1) can also be written as a matrix-vector product involving an $M \times M$ DFT matrix Ω formed by the *twiddle factors*, with $\omega_M^{r\ell}$ contained in $\Omega[\,r, \ell\,]$ as shown below.

$$
(13.12) \quad
\begin{bmatrix}
Y_0 \\
Y_1 \\
Y_2 \\
\vdots \\
Y_{M-1}
\end{bmatrix}
=
\begin{bmatrix}
1 & 1 & 1 & \cdots & 1 \\
1 & \omega_M^1 & \omega_M^2 & \cdots & \omega_M^{M-1} \\
1 & \omega_M^2 & \omega_N^4 & \cdots & \omega_M^{2(M-1)} \\
\cdots\cdots\cdots\cdots\cdots \\
1 & \omega_M^{M-1} & \omega_M^{2(M-1)} & \cdots & \omega_M^{(M-1)^2}
\end{bmatrix}
\begin{bmatrix}
y_0 \\
y_1 \\
y_2 \\
\vdots \\
y_{M-1}
\end{bmatrix}.
$$

Bluestein's next strategy employs the identity $\Omega H^{(2)} = D\Omega$, where D is a diagonal matrix. The diagonal of D defines a DFT (of length M) of the $h_\ell^{(2)}$'s. This identity is proved in Lemma 13.2.

Lemma 13.2 If $H^{(2)}$ is the circulant matrix of dimension M defined in (13.11), and Ω is the DFT matrix of dimension M defined in (13.12), then $\Omega H^{(2)} = D\Omega$, where

$$
(13.13) \quad
D =
\begin{bmatrix}
\hat{h}_0^{(2)} \\
& \hat{h}_1^{(2)} \\
& & \ddots \\
& & & \hat{h}_M^{(2)}
\end{bmatrix}
$$

and

$$
D[r,r] = \hat{h}_r^{(2)} = \sum_{\lambda=0}^{M-1} h_\lambda^{(2)} \omega_M^{r\lambda}, \quad r = 0, 1, 2, \ldots, M - 1.
$$

Proof: Define $A \equiv \Omega H^{(2)}$ and $B \equiv D\Omega$. In what follows $A[r, \ell] = B[r, \ell]$ is proved for $0 \le r, \ell \le M - 1$. For any such r and ℓ,

$$(13.14) \qquad B[r, \ell] = \sum_{k=0}^{M-1} D[r, k] \times \Omega[k, \ell] = D[r, r] \times \Omega[r, \ell],$$

and

$$(13.15) \qquad \begin{aligned} A[r, \ell] &= \sum_{k=0}^{M-1} \Omega[r, k] \times H^{(2)}[k, \ell] \\ &= \sum_{k=0}^{M-1} \omega_M^{rk} h^{(2)}_{|k-\ell|} \\ &= \sum_{k=0}^{\ell-1} \omega_M^{rk} h^{(2)}_{\ell-k} + \sum_{k=\ell}^{M-1} \omega_M^{rk} h^{(2)}_{k-\ell} \\ &= S_2 + S_1 . \end{aligned}$$

To simplify the summation S_1, define $\lambda \equiv k - \ell$, which implies $k = \lambda + \ell$. Furthermore, when $k = \ell$, $\lambda = 0$; when $k = M - 1$, $\lambda = M - 1 - \ell$. By using these identities, λ may be substituted into S_1 to obtain what follows.

$$(13.16) \qquad \begin{aligned} S_1 &= \sum_{k=\ell}^{M-1} \omega_M^{rk} h^{(2)}_{k-\ell} \\ &= \sum_{\lambda=0}^{M-1-\ell} \omega_M^{r(\lambda+\ell)} h^{(2)}_{\lambda} \\ &= \omega_M^{r\ell} \sum_{\lambda=0}^{M-1-\ell} \omega_M^{r\lambda} h^{(2)}_{\lambda} . \end{aligned}$$

To simplify the summation S_2, define $\lambda \equiv k + M - \ell$, which implies $\ell - k = M - \lambda$, and $k = \lambda - M + \ell$. Furthermore, when $k = 0$, $\lambda = M - \ell$; when $k = \ell - 1$, $\lambda = M - 1$. By using these identities, λ may be substituted into S_2 to obtain

$$(13.17) \qquad \begin{aligned} S_2 &= \sum_{k=0}^{\ell-1} \omega_M^{rk} h^{(2)}_{\ell-k} \\ &= \sum_{\lambda=M-\ell}^{M-1} \omega_M^{r(\lambda-M+\ell)} h^{(2)}_{M-\lambda} \\ &= \omega_M^{-rM} \omega_M^{r\ell} \sum_{\lambda=M-\ell}^{M-1} \omega_M^{r\lambda} h^{(2)}_{M-\lambda} . \end{aligned}$$

To simplify S_2 further, observe that $\omega_M^M \equiv 1$ and $h^{(2)}_{M-\lambda} = h^{(2)}_{\lambda}$ because $H^{(2)}$ is circulant. Accordingly,

$$(13.18) \qquad S_2 = \omega_M^{r\ell} \sum_{\lambda=M-\ell}^{M-1} \omega_M^{r\lambda} h^{(2)}_{\lambda}$$

and

$$A[r, \ell] = S_1 + S_2$$

$$= \omega_M^{r\ell} \left(\sum_{\lambda=0}^{M-1-\ell} \omega_M^{r\lambda} h_\lambda^{(2)} + \sum_{\lambda=M-\ell}^{M-1} \omega_M^{r\lambda} h_\lambda^{(2)} \right)$$

(13.19)
$$= \left(\sum_{\lambda=0}^{M-1} \omega_M^{r\lambda} h_\lambda^{(2)} \right) \omega_M^{r\ell}$$

$$= \left(\boldsymbol{D}[r, r] \right) \omega_M^{r\ell}$$

$$= \boldsymbol{D}[r, r] \times \boldsymbol{\Omega}[r, \ell]$$

$$= \boldsymbol{B}[r, \ell], \quad 0 \le r, \ell \le M - 1.$$

Hence $\boldsymbol{\Omega} H^{(2)} = \boldsymbol{D}\boldsymbol{\Omega}$ follows. ∎

Note that the diagonal of $\boldsymbol{D}$ can be computed by a radix-2 FFT in $\Theta\left(M \log_2 M\right)$ time. Using Lemma 13.2, one may now multiply the product $\boldsymbol{z}^{(2)} = \boldsymbol{H}^{(2)} \boldsymbol{y}^{(2)}$ by the DFT matrix $\boldsymbol{\Omega}$ on both sides to obtain

(13.20)
$$\hat{\boldsymbol{z}} = \boldsymbol{\Omega}\boldsymbol{z}^{(2)} = \boldsymbol{\Omega}\boldsymbol{H}^{(2)}\boldsymbol{y}^{(2)} = \boldsymbol{D}\left(\boldsymbol{\Omega}\boldsymbol{y}^{(2)}\right).$$

Observe that the right-hand side is a diagonally-scaled DFT matrix-product of dimension $M = 2^s$, and can be evaluated by a radix-2 FFT in $\Theta\left(M \log_2 M\right)$ time to obtain $\hat{\boldsymbol{z}}$.

In the next step, observe that

(13.21)
$$\boldsymbol{z}^{(2)} = \boldsymbol{\Omega}^{-1}\hat{\boldsymbol{z}}, \quad \text{where } \boldsymbol{\Omega}^{-1} = \frac{1}{M}\overline{\boldsymbol{\Omega}}.$$

Since $\boldsymbol{\Omega}$ is a DFT matrix of dimension M, $\boldsymbol{z}^{(2)}$ is the inverse DFT of $\hat{\boldsymbol{z}}$, and it can be obtained by applying a radix-2 inverse FFT to $\hat{\boldsymbol{z}}$ in $\Theta\left(M \log_2 M\right)$ time.

In the final step, the N desired X_r's are recovered from the Z_r's found in the top N positions in $\boldsymbol{z}^{(2)}$ using equation (13.4). This completes the DFT for arbitrary $N \ne 2^n$.

13.2 Bluestein's Algorithm for Arbitrary N

As seen above, assuming that the diagonal $\boldsymbol{D}$ is computed in a preprocessing phase, Bluestein's algorithm computes a DFT of arbitrary N by performing one forward and one inverse FFTs for the smallest $M = 2^s \ge 2N - 2$. That is, instead of using the $\Theta\left(N^2\right)$ DFT algorithm, a transform for $N \ne 2^n$ is accomplished by two radix-2 FFTs of length $M = 2^s$ using $\Theta(M \log_2 M)$ arithmetic operations. Since $2N - 2 \le M \le 4N$, the complexity of Bluestein's algorithm remains $\Theta\left(N \log_2 N\right)$ for arbitrary N.

The complete Bluestein's algorithm is summarized below, which is adapted from the description in [96]. Observe that the preprocessing in steps A, B, and C do not have to be repeated if N remains unchanged.

Bluestein's Algorithm for computing the discrete Fourier transform

$$X_r = \sum_{\ell=0}^{N-1} x_\ell \omega_N^{r\ell}, \quad r = 0, 1, \ldots, N - 1, \; N \ne 2^n.$$

A. (Preprocessing) Compute the symmetric Toeplitz H: for given N, compute the scalar constants

(13.22) $$h_\ell = w_N^{-\frac{1}{2}\ell^2}, \quad \ell = 0, 1, \ldots, N-1.$$

B. (Preprocessing) Embed H in the circulant $H^{(2)}$: define M as the smallest power of two that is greater than or equal to $2N-2$, and compute the vector $h^{(2)}$ of length M defined by

(13.23) $$h_\ell^{(2)} = h_\ell, \quad \ell = 0, 1, \ldots, N-1,$$

(13.24) $$h_\ell^{(2)} = h_{M-\ell}, \quad \ell = M - N + 2, \ldots, M-1,$$

and if $M > 2N - 2$,

(13.25) $$h_\ell^{(2)} = 0, \quad \ell = N, \ldots, M - N + 1.$$

Observe that if $M = 2N - 2 = 2^s$, the computation in (13.25) will not be performed, and $H^{(2)}$ is reduced to $H^{(1)}$ in (13.10).

C. (Preprocessing) Compute the diagonal D in Lemma 13.2: use the radix-2 FFT to compute the DFT matrix-vector product defined by

(13.26) $$\hat{h}_r = \sum_{\ell=0}^{M-1} h_\ell^{(2)} w_M^{r\ell}, \quad r = 0, 1, \ldots, M-1.$$

D. Given x_ℓ, define the extended vector $y^{(2)}$ of length M as

(13.27) $$y_\ell^{(2)} = x_\ell w_N^{\frac{1}{2}\ell^2}, \quad \ell = 0, 1, \ldots, N-1,$$

(13.28) $$y_\ell^{(2)} = 0, \quad \ell = N, \ldots, M-1.$$

E. Compute $\Omega y^{(2)}$: use the radix-2 FFT to compute the DFT matrix-vector product defined by

(13.29) $$Y_r = \sum_{\ell=0}^{M-1} y_\ell^{(2)} w_M^{r\ell}, \quad r = 0, 1, \ldots, M-1.$$

F. Compute $\hat{z} = D\left(\Omega y^{(2)}\right)$: scale Y_r by $\hat{h}_r = D[r, r]$; i.e.,

(13.30) $$\hat{Z}_r = \hat{h}_r Y_r, \quad r = 0, 1, \ldots, M-1.$$

G. Compute $z^{(2)} = \Omega^{-1}\hat{z}$: use the radix-2 inverse FFT to compute the inverse DFT matrix-vector product defined by

(13.31) $$Z_r = \frac{1}{M} \sum_{\ell=0}^{M-1} \hat{Z}_\ell w_M^{-r\ell}, \quad r = 0, 1, \ldots, M-1.$$

H. Extract the X_r's from the top N elements in $z^{(2)}$ by

(13.32) $$X_r = Z_r w_N^{\frac{1}{2}r^2}, \quad r = 0, \ldots, N-1.$$

Chapter 14

FFTs for Real Input

In Chapter 3, the discrete Fourier transform on N discrete samples from a *complex* time series is defined by formula (3.1):

(14.1)
$$X_r = \sum_{\ell=0}^{N-1} x_\ell \omega_N^{r\ell}, \quad r = 0, 1, \ldots, N-1,$$
$$= \sum_{\ell=0}^{N-1} \left(\mathrm{Re}\,(x_\ell) + j\, \mathrm{Im}\,(x_\ell) \right) \omega_N^{r\ell}.$$

When the samples come from a *real* time series, they can be treated as complex numbers with zero-valued imaginary part, i.e., $\mathrm{Im}(x_\ell) = 0$ for $0 \leq \ell \leq N-1$. In other words, real data represent a special case when approximately one half of the arithmetic operations are redundantly performed on zeros. Since many FFTs are performed on real-valued time series, it is worthwhile to handle real input more efficiently. Two such algorithms are described below. The first algorithm allows one to compute *two* real FFTs of size N by computing one complex FFT of size N; and the second algorithm allows one to compute a real FFT of size N by computing a complex FFT of size $N/2$.

14.1 Computing Two Real FFTs Simultaneously

In this section, a method which computes two real FFTs of size N by computing one complex FFT of size N is introduced. The two sets of real numbers are denoted by f_ℓ and g_ℓ for $0 \leq \ell \leq N-1$. By setting $\mathrm{Re}(x_\ell) = f_\ell$, and $\mathrm{Im}(x_\ell) = g_\ell$, one obtains a set of N complex numbers $x_\ell = f_\ell + jg_\ell$ for $0 \leq \ell \leq N-1$.

The definition of the DFT implies that

(14.2)
$$F_r = \sum_{\ell=0}^{N-1} f_\ell \omega_N^{r\ell} \quad \text{and} \quad G_r = \sum_{\ell=0}^{N-1} g_\ell \omega_N^{r\ell}, \quad 0 \leq r \leq N-1,$$

129

and

(14.3)
$$X_r = \sum_{\ell=0}^{N-1} x_\ell \omega_N^{r\ell}$$
$$= \sum_{\ell=0}^{N-1} (f_\ell + jg_\ell) \omega_N^{r\ell}$$
$$= \sum_{\ell=0}^{N-1} f_\ell \omega_N^{r\ell} + j \sum_{\ell=0}^{N-1} g_\ell \omega_N^{r\ell}$$
$$= F_r + jG_r .$$

Thus, one complex FFT on the x_ℓ's can be computed to obtain X_r's, and almost half of the arithmetic operations can be saved if the F_r's and the G_r's can be recovered efficiently from the computed X_r's. This can be done by using the *symmetry property* for the DFT of a real-valued series, which was established in Chapter 1. For convenience, the result is rederived here.

The *symmetry property* ensures that the complex conjugate of F_{N-r} is equal to F_r. This property is derived using the DFT definition and the fact that the complex conjugate of a real-valued f_ℓ is equal to itself, $\omega_N^N = 1$, and the complex conjugate of $\omega_N^{-r\ell}$ is equal to $\omega_N^{r\ell}$.

(14.4)
$$\overline{F}_{N-r} = \sum_{\ell=0}^{N-1} \bar{f}_\ell \left(\bar{\omega}_N^{N\ell}\right) \bar{\omega}_N^{-r\ell} = \sum_{\ell=0}^{N-1} f_\ell \omega_N^{r\ell} = F_r .$$

Since the g_ℓ's are also real, $\overline{G}_{N-r} = G_r$. Now the complex conjugate of X_{N-r} can be expressed in terms of F_r and G_r as shown below.

(14.5)
$$\overline{X}_{N-r} = \overline{F}_{N-r} - j\overline{G}_{N-r} = F_r - jG_r .$$

Combining (14.3) and (14.5), one immediately obtains

(14.6)
$$F_r = \frac{1}{2} \left(X_r + \overline{X}_{N-r}\right), \qquad G_r = \frac{j}{2} \left(\overline{X}_{N-r} - X_r\right).$$

Therefore, only $2N$ extra complex additions/subtractions are required to recover the two real FFTs after one complex FFT is performed, which requires $\Theta\left(N \log_2 N\right)$ arithmetic operations as usual.

14.2 Computing a Real FFT

To apply the results in the previous section to transform a single series, the latter is first split into two real series of half the size. The derivation is similar to the work in deriving the DIT FFT algorithm in Section 3.1, namely,

(14.7)
$$X_r = \sum_{\ell=0}^{N-1} x_\ell \omega_N^{r\ell}, \quad r = 0, 1, \ldots, N-1,$$

$$= \sum_{\ell=0}^{\frac{N}{2}-1} x_{2\ell} \omega_N^{r(2\ell)} + \omega_N^r \sum_{\ell=0}^{\frac{N}{2}-1} x_{2\ell+1} \omega_N^{r(2\ell)}$$

$$= \sum_{\ell=0}^{\frac{N}{2}-1} x_{2\ell} \omega_{\frac{N}{2}}^{r\ell} + \omega_N^r \sum_{\ell=0}^{\frac{N}{2}-1} x_{2\ell+1} \omega_{\frac{N}{2}}^{r\ell}.$$

By setting $f_\ell = x_{2\ell}$, $g_\ell = x_{2\ell+1}$ for $0 \leq \ell \leq N/2 - 1$, the DFT of two real series and a DFT of $N/2$ complex numbers $y_\ell = f_\ell + jg_\ell$ are defined below.

(14.8)
$$F_r = \sum_{\ell=0}^{\frac{N}{2}-1} f_\ell \omega_{\frac{N}{2}}^{r\ell}, \qquad G_r = \sum_{\ell=0}^{\frac{N}{2}-1} g_\ell \omega_{\frac{N}{2}}^{r\ell},$$

and

(14.9)
$$Y_r = \sum_{\ell=0}^{\frac{N}{2}-1} y_\ell \omega_{\frac{N}{2}}^{r\ell} = \sum_{\ell=0}^{\frac{N}{2}-1} (f_\ell + jg_\ell) \omega_{\frac{N}{2}}^{r\ell} = \sum_{\ell=0}^{\frac{N}{2}-1} f_\ell \omega_{\frac{N}{2}}^{r\ell} + j \sum_{\ell=0}^{\frac{N}{2}-1} g_\ell \omega_{\frac{N}{2}}^{r\ell} = F_r + jG_r.$$

Using the results from the previous section, one complex FFT on the y_ℓ's can be computed to obtain the Y_r's, and the F_r's and G_r's (for $0 \leq r \leq N/2 - 1$) can be recovered using the following equations.

(14.10)
$$F_r = \frac{1}{2} \left(Y_r + \overline{Y}_{\frac{N}{2}-r} \right), \qquad G_r = \frac{j}{2} \left(\overline{Y}_{\frac{N}{2}-r} - Y_r \right).$$

However, it is no longer sufficient to have successfully recovered the F_r's and G_r's, because the goal is to compute the X_r's defined by equation (14.7), which can now be obtained from the available F_r's and G_r's as shown below.

(14.11)
$$X_r = \sum_{\ell=0}^{\frac{N}{2}-1} x_{2\ell} \omega_{\frac{N}{2}}^{r\ell} + \omega_N^r \sum_{\ell=0}^{\frac{N}{2}-1} x_{2\ell+1} \omega_{\frac{N}{2}}^{r\ell}$$

$$= \sum_{\ell=0}^{\frac{N}{2}-1} f_\ell \omega_{\frac{N}{2}}^{r\ell} + \omega_N^r \sum_{\ell=0}^{\frac{N}{2}-1} g_\ell \omega_{\frac{N}{2}}^{r\ell}$$

$$= F_r + \omega_N^r G_r, \quad r = 0, 1, 2, \ldots, N/2 - 1.$$

Since the x_ℓ's are real, the X_r's have the *symmetry property* derived in (14.4); thus, $X_{\frac{N}{2}+1}, X_{\frac{N}{2}+2}, \ldots, X_{N-1}$ can be obtained by taking the complex conjugate of the previously computed X_r's.

(14.12)
$$X_{N-r} = \overline{X}_r, \quad r = 1, 2, \ldots, N/2 - 1.$$

Using equations (14.11) and (14.12), all the X_r's can be obtained except for $X_{\frac{N}{2}}$. To compute $X_{r+\frac{N}{2}} = F_{r+\frac{N}{2}} + \omega_N^{r+\frac{N}{2}} G_{r+\frac{N}{2}}$, recall that $F_{r+\frac{N}{2}} = F_r$, $G_{r+\frac{N}{2}} = G_r$, and

$\omega_N^{r+\frac{N}{2}} = -\omega_N^r$. Using these properties with $r = 0$, $X_{r+\frac{N}{2}} = X_{\frac{N}{2}}$ can now be computed by

(14.13) $$X_{\frac{N}{2}} = F_0 - G_0 \,.$$

In total, N extra complex additions/subtractions are needed to recover the F_r's and G_r's after one complex FFT is performed on $N/2$ complex numbers, and an additional $N/2$ complex multiplications and $N/2+1$ complex additions/subtractions are needed to compute the X_r's using equations (14.11), (14.12), and (14.13). Therefore, for large N, almost half of the arithmetic operations can be saved by performing the FFT on $N/2$ complex numbers instead of treating the real-valued series as consisting of N complex numbers.

14.3   Notes and References

According to Bergland [4], there are two basic approaches to the evaluation of real-valued time series. The approach which makes use of the conventional complex FFT algorithm and depends on forming an artificial $N/2$-term complex record from each N-term real record was due to Cooley, Lewis, and Welch [31]; an alternative approach was proposed in [4]. The former approach was used by Brigham [17] and Walker [106]. The algorithms developed in this chapter are based on Walker's approach [106]. Other algorithms for computing real-valued series may be found in [4, 8, 88]. Implementation of split-radix FFT algorithms for real and real-symmetric data is described in [38].

Chapter 15

FFTs for Composite N

In order to compute the DFT of a time series of length $N = 2^n = (2^s)^k$, one has the choice of using various radix-2^s FFT algorithms, which require $\Theta\left(N \log_2 N\right)$ arithmetic operations. Specific algorithms covered in previous chapters include radix-2, radix-4, and split-radix FFTs. If $N \neq 2^n$, Bluestein's algorithm, described in Chapter 13, can be used. This algorithm requires $\Theta\left(M \log_2 M\right)$ arithmetic operations, where $2N - 2 \leq M \leq 4N$.

In this chapter, other methods for computing the DFT when $N \neq 2^n$ are presented, i.e., when N is a product of arbitrary integers. It will be shown that for $N = F_1 \times F_2 \times \cdots \times F_m$, the arithmetic operations required are (proportional to) $\mathcal{A} = N \times (F_1 + F_2 + \cdots + F_m)$. Accordingly, $N = 2^m$ is simply a special case when all m factors of N are equal to 2.

It is of historical interest to note that this approach was suggested by Cooley and Tukey in 1965. In their initial article [33], they derived the decimation-in-time FFT for $N = P \times Q$, and stated that a fast Fourier transform with arbitrary m factors could be obtained by successive applications of the two-factor algorithm. However, because the only m-factor formula derived in [33] was for $N = 2^m$, and the manipulation required to extend the formula to many factors is not intuitively obvious, the potential generality of Cooley and Tukey's initial algorithm was often overlooked. In 1980, de Boor [34] provided the full treatment of the FFT for a general N by deriving it in terms of nested multiplication. The derivation in this chapter closely follows his work.

The derivation and efficient implementation of the FFTs for composite N require several important mathematical tools. These are reviewed in Sections 15.1, 15.2, and 15.3 before the two-factor FFT is derived in Section 15.4; more algorithmic tools are provided in Sections 15.5 and 15.6 before the many-factor FFT for general N is derived in Section 15.7.

15.1 Nested-Multiplication as a Computational Tool

15.1.1 Evaluating a polynomial by nested-multiplication

Consider a degree-3 polynomial defined by

$$P(z) = \sum_{\ell=0}^{\ell=3} a_\ell z^\ell = a_3 z^3 + a_2 z^2 + a_1 z + a_0,$$

where z is a complex variable, and the a_i's, $0 \le i \le 3$, are complex coefficients. To evaluate $P(z)$ at $z = \beta$, one may compute $P(\beta)$ by Horner's nested-multiplication rule as shown below.

> Horner's rule:
> $$P(\beta) = ((a_3\beta + a_2)\,\beta + a_1)\,\beta + a_0$$

Observe that each multiplication is accompanied by an addition. To this point, arithmetic complexity has been measured in flops—the number of real multiplicative and additive operations. Since computations often consist of (mainly) sequences of multiply-add pairs, it is also common in the literature to report the number of real or complex "multiply-add" pairs as the arithmetic cost of a computation. At the risk of inconsistency with the rest of the book, arithmetic cost is measured in this chapter by the number of "multiply-add" pairs.

In general, a polynomial of degree K is defined by

$$P_K(z) = \sum_{\ell=0}^{\ell=K} a_\ell z^\ell.$$

To evaluate $P_K(z)$ at $z = \beta$, the following implementation of Horner's rule can be executed, where the coefficients are assumed to be available in the array a. Observe that the **while** loop is executed $K+1$ times; thus, exactly $K+1$ complex multiply-add operations are required to evaluate a degree-K polynomial using this scheme.

Algorithm 15.1 Horner's rule in pseudo-code for computing $P_K(z)$ at $z = \beta$.

begin
    $z := \beta$
    $\ell := K$ Compute *PolyResult* $= P_K(z)$ at $z = \beta$
    PolyResult $:= 0$
    while $\ell \ge 0$ **do**
        PolyResult $:=$ *PolyResult* $* z + a[\ell]$
        $\ell := \ell - 1$
    end while
end

15.1.2   Computing a DFT by nested-multiplication

Recall from Chapter 3 that given a complex series x_ℓ, $0 \leq \ell \leq N - 1$, the discrete Fourier transform (DFT) of x is defined by

$$X_r = \sum_{\ell=0}^{N-1} x_\ell \omega_N^{r\ell}, \quad r = 0, 1, \ldots, N - 1,$$

where ω_N is the Nth primitive root of unity, i.e., $\omega_N \equiv e^{-j2\pi/N}$ with $j \equiv \sqrt{-1}$.

To express the DFT as the evaluation of a polynomial at the N primitive roots, define $P(z)$ of degree $N - 1$ as

$$P_{N-1}(z) = \sum_{\ell=0}^{N-1} x_\ell z^\ell.$$

One immediately obtains

$$X_r = \sum_{\ell=0}^{N-1} x_\ell \left(\omega_N^r\right)^\ell = P_{N-1}\left(\omega_N^r\right), \quad r = 0, 1, \ldots, N - 1.$$

Therefore, each X_r can be computed by evaluating $P_{N-1}(z)$ at $z = \omega_N^r$. The following algorithm computes the DFT by Horner's rule, where $a[\ell] = x_\ell$ on input, and $b[r] = X_r$ on output. Note also that this algorithm incorporates the computation of $z = \omega_N^r$, $0 \leq r \leq N - 1$.

Algorithm 15.2 Computing a DFT of length N by Horner's rule.

begin
  $\theta := 2\pi/N$; $W := exp(-j\theta)$ Compute $W = \omega_N$
  $z := 1$ $z = \omega_N^0 = 1$
  for $r := 0$ **to** $N - 1$ **do**
    $\ell := N - 1$ Compute $X_r = PolyResult = P_{N-1}(z)$
    $PolyResult := 0$
    while $\ell \geq 0$ **do**
        $PolyResult := PolyResult * z + a[\ell]$ $z = \omega_N^r$
        $\ell := \ell - 1$
    end while
    $b[r] := PolyResult$ $b[r] = X_r$ on output
    $z := z * W$ Update $z = \omega_N^{r+1}$
  end

Since N multiply-add operations are required to compute each X_r for $0 \leq r \leq N-1$, the total cost for computing the DFT amounts to N^2 multiply-add operations. (The N multiplications required to update ω_N^r to ω_N^{r+1} are commonly ignored in reporting the cost, because it is negligible compared to the N^2 term for large N.) Thus, on the surface, this approach does not appear to have any advantage over the naive matrix multiplication, and is vastly inferior to the fast algorithms discussed in previous chapters.

However, given that much more efficient algorithms do exist for computing the DFT, it appears likely that Horner's scheme can be manipulated in such a way to reduce the complexity for these specially selected evaluation points. This is indeed the case, but before proceeding with the development, some additional mathematical tools are needed. These are provided in Sections 15.2, 15.3 and 15.5. Sections 15.4 and 15.7 describe how nested multiplication and these tools are used together to develop FFTs for composite N.

15.2 A 2D Array as a Basic Programming Tool

It is a common practice to store a vector in a one-dimensional array, and store a matrix in a two-dimensional array. For example, to process a 2×3 matrix

$$A = \begin{bmatrix} a_{0,0} & a_{0,1} & a_{0,2} \\ a_{1,0} & a_{1,1} & a_{1,2} \end{bmatrix},$$

it is natural to store the matrix in a 2D array A of the same row and column dimensions so that array location $A[\rho, q]$ contains $a_{\rho,q}$, $0 \le \rho \le 1$, and $0 \le q \le 2$.

Individual elements in A can be accessed randomly given the values of p and q. However, if the algorithm can be organized so that memory access proceeds consecutively through its addresses, efficient use of memory on most modern computers is achieved. In order to do this, it is necessary to know whether the programming language used stores arrays row-by-row or column-by-column in physical memory. For example, the C language uses the row-major format, and the Fortran language uses the column-major format. The two formats are depicted in Figure 15.1 for $A[Pdim, Qdim]$, $Pdim = 2$, and $Qdim = 3$. Note from Figure 15.1 that $A[0, 0]$ and $A[1, 0]$ are adjacent in memory

---

Figure 15.1 Row-major and column-major storage formats for A of dimensions 2×3.

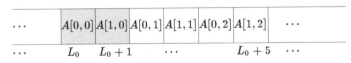

---

if A is stored column by column, but they are $Qdim = 3$ locations apart if A is stored row by row. In general, given ρ and q, the memory address ℓ satisfying $mem[\ell] = A[\rho, q]$ is computed according to the formula depicted in Figure 15.2 below. Note that the row-major scheme dictates that

$$mem[\ell] = A[\rho, q] \quad \text{iff} \quad \ell = L_0 + (\rho \times Qdim + q),$$

and the column-major scheme dictates that

$$mem[\ell] = A[\rho, q] \quad \text{iff} \quad \ell = L_0 + (q \times Pdim + \rho).$$

Figure 15.2 Memory address translation for the two storage schemes.

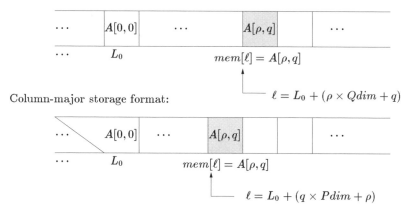

Row-major storage format:

$mem[\ell] = A[\rho, q]$

$\ell = L_0 + (\rho \times Qdim + q)$

Column-major storage format:

$mem[\ell] = A[\rho, q]$

$\ell = L_0 + (q \times Pdim + \rho)$

15.2.1 Row-oriented and column-oriented code templates

If the 2D array $A[Pdim, Qdim]$ is stored in *row-major* format, the following code templates all access A's elements in consecutive order. Recall that $A[\rho, q] = mem[\ell]$, where $\ell = L_0 + (\rho \times Qdim + q)$.

```
for ρ := 0 to Pdim − 1 do
    for q := 0 to Qdim − 1 do
        access   A[ρ, q]
    end for
end for
```

```
ℓ := L₀
for ρ := 0 to Pdim − 1 do
    for q := 0 to Qdim − 1 do
        access   mem[ℓ]
        ℓ := ℓ + 1
    end for
end for
```

```
Ldim := Pdim * Qdim
for ℓ := L₀ to Ldim − 1
    access   mem[ℓ]
end for
```

If the 2D array $A[Pdim, Qdim]$ is stored in *column-major* format, the following code templates all access the array elements in consecutive order. Recall that $A[\rho, q] = mem[\ell]$, where $\ell = L_0 + (q \times Pdim + \rho)$.

```
for q := 0 to Qdim − 1 do
    for ρ := 0 to Pdim − 1 do
        access   A[ρ, q]
    end for
end for
```

```
ℓ := L₀
for q := 0 to Qdim − 1 do
    for ρ := 0 to Pdim − 1 do
        access   mem[ℓ]
        ℓ := ℓ + 1
    end for
end for
```

```
Ldim := Pdim ∗ Qdim
for ℓ := L₀ to Ldim − 1
    access   mem[ℓ]
end for
```

15.3 A 2D Array as an Algorithmic Tool

15.3.1 Storing a vector in a 2D array

With the previous section in mind, it is obvious that one can proceed in the other direction. That is, the code templates can be used to store a one-dimensional array x of size $N = Pdim \times Qdim$ in a two-dimensional array $A[Pdim, Qdim]$ either "row by row" or "column by column" by adapting the code template introduced in the previous subsection. The utility of this idea will emerge in subsequent sections.

```
L₀ := 0;    ℓ := L₀
for ρ := 0 to Pdim − 1 do
    for q := 0 to Qdim − 1 do
        A[ρ, q] := xₗ
        ℓ := ℓ + 1
    end for
end for
```

The previous results on address translation for the row-major storage scheme can thus be applied with $L_0 = 0$, and one obtains

$$A[\rho, q] = x_\ell \quad \text{iff} \quad \ell = \rho \times Qdim + q.$$

In a similar fashion, x can be stored in $A[Pdim, Qdim]$ "column by column" as shown below.

```
L₀ := 0;    ℓ := L₀
for q := 0 to Qdim − 1 do
    for ρ := 0 to Pdim − 1 do
        A[ρ, q] := xₗ
        ℓ := ℓ + 1
    end for
end for
```

With $L_0 = 0$, one obtains

$$A[\rho, q] = x_\ell \quad \text{iff} \quad \ell = q \times Pdim + \rho.$$

15.3.2 Use of 2D arrays in computing the DFT

If the vector x of length $N = P \times Q$ is stored in a two-dimensional array $A[P, Q]$ by columns, then $x_\ell = A[\rho, q]$ when $\ell = q \times P + \rho$. Then the DFT of the vector x can be written as

$$(15.1) \qquad X_r = P_{N-1}(z) = \sum_{\ell=0}^{N-1} x_\ell \, z^\ell = \sum_{\rho=0}^{P-1} \sum_{q=0}^{Q-1} A[\rho, q] \, z^{(q \times P + \rho)},$$

$$\text{where} \quad z = \omega_N^r, \quad \text{for} \quad r = 0, 1, \ldots, N-1.$$

Recall that the **while** loop of Algorithm 15.2 computes $X_r = P_{N-1}(z)$ using the numerical value of $z = \omega_N^r$ and the coefficients $a[\ell] = x_\ell$, $0 \leq \ell \leq N-1$. Given below is a *preliminary* form of the same DFT of x in terms of $A[\rho, q] = x_\ell$.

```
begin
    θ := 2π/N;  W := exp(−jθ)
    z := 1
    for r := 0 to N − 1 do
        X_r := Σ_{ρ=0}^{P−1} Σ_{q=0}^{Q−1} A[ρ, q] * z^(q*P+ρ)
        z := z * W
    end for
end
```

The next step is to simplify the term $z^{(q \times P + \rho)}$. Since $N = P \times Q$, $\omega_N^P = \omega_{N/P} = \omega_Q$, one has

$$z^P = \left(\omega_N^r\right)^P = \left(\omega_N^P\right)^r = \omega_Q^r.$$

Equation (15.1) may then be written as

$$X_r = \sum_{\rho=0}^{P-1} \sum_{q=0}^{Q-1} A[\rho, q] \, z^{(q \times P + \rho)}, \qquad r = 0, 1, \ldots, N-1,$$

$$(15.2) \qquad = \sum_{\rho=0}^{P-1} \left(\sum_{q=0}^{Q-1} A[\rho, q] \, \left(z^P\right)^q \right) z^\rho$$

$$= \sum_{\rho=0}^{P-1} \left(\sum_{q=0}^{Q-1} A[\rho, q] \, \left(\omega_Q^r\right)^q \right) \left(\omega_N^r\right)^\rho,$$

and the corresponding DFT algorithm takes the following form, *which is not in final efficient form yet.* It is still an $\Theta\left(N^2\right)$ algorithm; further changes are required to make it efficient.

Algorithm 15.3 Use of 2D arrays in computing the DFT.

begin

$\quad \theta := 2\pi/N; \quad \theta_y := 2\pi/Q$

$\quad WN := exp(-j\theta); \quad WQ := exp(-j\theta_y)$ $\qquad\qquad$ Compute $WN = w_N$, $WQ = w_Q$

$\quad z := 1; \quad y := 1$ $\qquad\qquad\qquad\qquad\qquad\qquad$ Initialize $z = w_N^0 = 1$, $y = w_Q^0 = 1$

$\quad$ **for** $r := 0$ **to** $N-1$ **do**

$\qquad$ **for** $\rho := 0$ **to** $P-1$ **do**

$\qquad\qquad$ Compute $Y_\rho := \sum_{q=0}^{Q-1} A[\rho, q] * y^q$ by Horner's rule $\qquad$ $y = w_Q^r$

$\qquad$ **end for**

$\qquad$ Compute $X_r := \sum_{\rho=0}^{P-1} Y_\rho * z^\rho$ by Horner's rule $\qquad\qquad$ $z = w_N^r$

$\qquad$ $z := z * WN; \quad y := y * WQ$ $\qquad\qquad\qquad$ Update $z = w_N^{r+1}$, $y = w_Q^{r+1}$

$\quad$ **end for**

end

Algorithm 15.3 requires Q multiply-adds to compute each Y_ρ and P multiply-adds to compute each X_r after $Y_0, Y_1, \ldots Y_{P-1}$ are available. Thus, a total of $N(P \times Q + P) = N^2 + P \times N$ multiply-adds are required to compute all X_r's, $r = 0, 1, \ldots, N-1$. As in the previous section, the $2N$ multiplications for updating z and y are ignored.

Since the 1D array implementation of the DFT in Algorithm 15.2 incurs N^2 multiply-adds, the more complicated implementation of the DFT in Algorithm 15.3 costs more, and as presented so far, has no apparent advantages. In the next section, further modifications are presented which turn this algorithm into an efficient fast Fourier transform (FFT) algorithm.

15.4   An Efficient FFT for $N = P \times Q$

The keys to turning the inefficient DFT Algorithm 15.3 into an efficient fast Fourier transform (FFT) algorithm may be viewed as the consequence of storing the output vector X of length $N = P \times Q$ in a two-dimensional array $B[Q, P]$ column by column. The relevant identities are listed below.

(15.3) $$X_r = B[\hat{q}, \hat{\rho}]$$

if and only if

$$r = \hat{\rho} \times Q + \hat{q}, \quad w_Q^Q = 1, \quad \text{and} \quad w_Q^r = w_Q^{\hat{\rho} \times Q + \hat{q}} = w_Q^{\hat{q}}.$$

The reason for storing X in a $Q \times P$ array requires some explanation. Observe that the key identity is $r = \hat{\rho} \times Q + \hat{q}$, because Q is the period of w_Q. To be able to express $r = \hat{\rho} \times Q + \hat{q}$, X must be stored column by column in $B[Q, P]$ or stored row by row in $C[P, Q]$. Since the input x is stored column by column in $A[P, Q]$, it is consistent to store X in the same manner in $B[Q, P]$ for now.

Substituting $r = \hat{\rho} \times Q + \hat{q}$ and $\omega_Q^r = \omega_Q^{\hat{q}}$ into equation (15.2), one obtains

$$X_{\hat{\rho} \times Q + \hat{q}} = X_r, \qquad r = 0, 1, \ldots, N - 1,$$

(15.4)
$$= \sum_{\rho=0}^{P-1} \left(\sum_{q=0}^{Q-1} A[\rho, q] \, (\omega_Q^r)^q \right) (\omega_N^r)^\rho,$$

$$= \sum_{\rho=0}^{P-1} \left(\sum_{q=0}^{Q-1} A[\rho, q] \, (\omega_Q^{\hat{q}})^q \right) (\omega_N^{\hat{\rho} \times Q + \hat{q}})^\rho,$$

where $0 \le \hat{q} \le Q - 1$, and $0 \le \hat{\rho} \le P - 1$. Observe that the inner sum varies with the indices $\hat{q}$ and ρ, but it is independent of the index $\hat{\rho}$. Therefore, to save operations, the inner sum should be pre-computed for all $0 \le \rho \le P - 1$, and $0 \le \hat{q} \le Q - 1$. At this point, consider that, in general, the resulting $P \times Q$ values may be stored in either $Y[P, Q]$ or its transpose, $Z[Q, P]$, and both options are listed side by side below. Note that Y is computed column by column, array Z is computed row by row, and $y = 1, \omega_Q, \omega_Q^2, \ldots, \omega_Q^{Q-1}$ are updated in order for efficiency.

$y := 1$ $y = \omega_Q^0 = 1$
 for $\hat{q} := 0$ **to** $Q - 1$
     for $\rho := 0$ **to** $P - 1$
        $Y[\rho, \hat{q}] := \sum_{q=0}^{Q-1} A[\rho, q] * y^q$
     end for
     $y := y * WQ$ update $y = \omega_Q^{\hat{q}+1}$
 end for
	$y := 1$ $y = \omega_Q^0 = 1$
	 for $\hat{q} := 0$ **to** $Q - 1$
	     for $\rho := 0$ **to** $P - 1$
	        $Z[\hat{q}, \rho] := \sum_{q=0}^{Q-1} A[\rho, q] * y^q$
	     end for
	     $y := y * WQ$ update $y = \omega_Q^{\hat{q}+1}$
	 end for

Assuming that $Y[P, Q]$ stores the intermediate results, a corresponding FFT algorithm is given below.

Algorithm 15.4 Efficient FFT for $N = P \times Q$—a column-oriented approach.

---

begin
   $\theta := 2\pi/N; \quad \theta_y := 2\pi/Q$
   $WN := \exp(-j\theta); \quad WQ := \exp(-j\theta_y)$ Compute $WN = \omega_N$, $WQ = \omega_Q$
   $z := 1; \quad y := 1$ Initialize $z = \omega_N^0 = 1$, $y = \omega_Q^0 = 1$
   for $\hat{q} := 0$ **to** $Q - 1$
      for $\rho := 0$ **to** $P - 1$
         Compute $Y[\rho, \hat{q}] = \sum_{q=0}^{Q-1} A[\rho, q] * y^q$ by Horner's rule $y = \omega_Q^{\hat{q}}$
      end for
      $y := y * WQ$ Update $y = \omega_Q^{\hat{q}+1}$
   end for
   for $\hat{\rho} := 0$ **to** $P - 1$ **do**
      for $\hat{q} := 0$ **to** $Q - 1$ **do**
         Compute $B[\hat{q}, \hat{\rho}] = X_{\hat{\rho} \times Q + \hat{q}} = \sum_{\rho=0}^{P-1} Y[\rho, \hat{q}] * z^\rho$ by Horner's rule $z = \omega_N^r$
         $z := z * WN$ Update $z = \omega_N^{r+1}$
      end for
   end for
end

Note that because $z = 1, \omega_N, \omega_N^2, \ldots, \omega_N^{N-1}$ are updated and used in the polynomial evaluation in that order, the code in Algorithm 15.4 must also compute $X_0, X_1, \ldots, X_{N-1}$ in that order. If the X_r's are assumed to be stored in array $B[Q, P]$ in a column-major format, then the array $\boldsymbol{B}$ must be computed column by column. Here is a summary of all 2D arrays constructed and accessed in this FFT algorithm for $N = P \times Q$.

- Input x_ℓ's are stored in $A[P, Q]$ column by column.

- Array $A[P, Q]$ is accessed row by row in the computation of $Y[P, Q]$. (Note that how $\boldsymbol{A}$ is accessed is dictated by the polynomial evaluation in the inner loop.)

- Array $Y[P, Q]$ is computed column by column.

- Array $Y[P, Q]$ is then accessed column by column in the computation of array $B[Q, P]$.

- Array $B[Q, P]$ is computed column by column.

- Output X_r's are stored in $B[Q, P]$ column by column.

Thus, this version of the FFT algorithm for $N = P \times Q$ is suitable for implementation when the programming language supports column-major storage scheme for 2D arrays.

If one needs to use a programming language which supports row-major storage scheme, it is now a simple exercise to convert Algorithm 15.4 into using row-oriented computation as much as possible.

To determine the arithmetic cost, observe that Q multiply-adds are required to compute each $Y[\rho, \hat{q}]$, and P multiply-adds are required to compute each $B[\hat{q}, \hat{\rho}]$. The total cost is thus $N(P + Q)$ complex multiply-add operations.

15.5   Multi-Dimensional Array as an Algorithmic Tool

15.5.1   Storing a 1D array into a multi-dimensional array

To store a one-dimensional array $\boldsymbol{x}$ of length $N = N_0 \times N_1 \times \cdots \times N_{\nu-1}$ in a ν-dimensional array $A[N_0, N_1, \ldots, N_{\nu-1}]$, one may extend the row-oriented 2D code template from Section 15.3 as shown below.

```
L₀ := 0;    ℓ := L₀
for n₀ := 0 to N₀ − 1 do
    for n₁ := 0 to N₁ − 1 do
                ⋮
            for nᵥ₋₁ := 0 to Nᵥ₋₁ − 1 do
                A[n₀, n₁, ..., nᵥ₋₁] := xℓ
                ℓ := ℓ + 1
            end for   nᵥ₋₁
                ⋮
        end for   n₁
end for   n₀
```

Thus, the following identity holds using the row-major storage scheme:

(15.5) $$A[n_0, n_1, \ldots, n_{\nu-1}] = x_\ell$$

if and only if

$$\ell = n_0 \prod_{k=1}^{\nu-1} N_k + n_1 \prod_{k=2}^{\nu-1} N_k + \cdots + n_{\nu-2} N_{\nu-1} + n_{\nu-1}.$$

Note that when $\nu = 2$, the identity above is reduced to $A[n_0, n_1] = x_\ell$ iff $\ell = n_0 \times N_1 + n_1$, which is exactly the result expected from storing x in a 2D array $A[N_0, N_1]$ row by row.

Similarly, one can also extend the column-oriented 2D code template from Section 15.3 as shown below.

```
L_0 := 0;    ℓ := L_0
for n_{ν-1} := 0 to N_{ν-1} - 1 do

        ⋮

        for n_1 := 0 to N_1 - 1 do
            for n_0 := 0 to N_0 - 1 do
                A[n_0, n_1, ..., n_{ν-1}] := x_ℓ
                ℓ := ℓ + 1
            end for  n_0
        end for  n_1

        ⋮

end for  n_{ν-1}
```

The following identity holds using the column-major storage scheme:

(15.6) $$A[n_0, n_1, \ldots, n_{\nu-1}] = x_\ell$$

if and only if

$$\ell = n_{\nu-1} \prod_{k=0}^{\nu-2} N_k + n_{\nu-2} \prod_{k=0}^{\nu-3} N_k + \cdots + n_1 N_0 + n_0.$$

When $\nu = 2$, the identity above is reduced to $A[n_0, n_1] = x_\ell$ iff $\ell = n_1 \times N_0 + n_0$, which is as expected from storing x in a 2D array $A[N_0, N_1]$ column by column.

15.5.2 Row-oriented interpretation of ν-D arrays as 2D arrays

If one defines $P_j = N_0 \times N_1 \times \cdots \times N_{j-1}$, $Q_j = N_j \times N_{j+1} \times \cdots N_{\nu-1}$, then the ν-dimensional array $A[N_0, N_1, \ldots, N_{\nu-1}]$ may be interpreted as a 2D array $B[P_j, Q_j]$. That is, $A[n_0, \ldots, n_{j-1}, n_j, \ldots, n_{\nu-1}] = B[\rho, q]$, where $0 \le \rho \le P_j - 1$, and $0 \le q \le Q_j - 1$.

To determine the value of ρ, apply identity (15.5) to $P_j = N_0 \times N_1 \times \cdots \times N_{j-1}$ instead of $N = N_0 \times N_1 \times \cdots \times N_{\nu-1}$. To determine the value of q, apply identity

(15.5) to $Q_j = N_j \times N_{j+1} \times \cdots \times N_{\nu-1}$ instead of $N = N_0 \times N_1 \times \cdots \times N_{\nu-1}$. The following identity results.

(15.7) $A[n_0, \ldots, n_{j-1}, n_j, \ldots, n_{\nu-1}] = B[\rho, q]$

if and only if

$$\rho = n_0 \prod_{k=1}^{j-1} N_k + n_1 \prod_{k=2}^{j-1} N_k + \cdots + n_{j-2}N_{j-1} + n_{j-1},$$

$$\text{and} \quad q = n_j \prod_{k=j+1}^{\nu-1} N_k + n_{j+1} \prod_{k=j+2}^{\nu-1} N_k + \cdots + n_{\nu-2}N_{\nu-1} + n_{\nu-1}.$$

With ℓ defined in identity (15.5), together with ρ, q, P_j and Q_j defined above, it can be verified that

(15.8) $x_\ell = B[\rho, q] = A[n_0, \ldots, n_{j-1}, n_j, \ldots, n_{\nu-1}]$ iff $\ell = \rho \times Q_j + q$.

15.5.3   Column-oriented interpretation of ν-D arrays as 2D arrays

Instead of using identity (15.5), identity (15.6) is now applied to P_j and Q_j separately, and one obtains the following result if the column-major storage scheme is used for storing A and B.

(15.9) $A[n_0, \ldots, n_{j-1}, n_j, \ldots, n_{\nu-1}] = B[\rho, q]$

if and only if

$$\rho = n_{j-1} \prod_{k=0}^{j-2} N_k + n_{j-2} \prod_{k=0}^{j-3} N_k + \cdots + n_1N_0 + n_0,$$

$$\text{and} \quad q = n_{\nu-1} \prod_{k=j}^{\nu-2} N_k + n_{\nu-2} \prod_{k=j}^{\nu-3} N_k + \cdots + n_{j+1}N_j + n_j.$$

With ℓ defined in identity (15.6), together with ρ, q, P_j and Q_j defined above, it can be shown that

(15.10) $x_\ell = B[\rho, q] = A[n_0, \ldots, n_{n-1}, n_j, \ldots, n_{\nu-1}]$ iff $\ell = q \times P_j + \rho$.

15.5.4   Row-oriented interpretation of ν-D arrays as 3D arrays

To interpret a ν-D array $A[N_0, N_1, \ldots, N_{\nu-1}]$ as a 3D array $C[B_j, Q_j, F_j]$, define, for example, $B_j = N_0 \times N_1 \times \cdots N_{j-1}$, $Q_j = N_j$, and $F_j = N_{j+1} \times N_{j+2} \times \cdots N_{\nu-1}$. The following identity results if both arrays A and C are stored in row-major format. Observe that in determining the values of b, q and f, identity (15.5) is applied to the three dimensions B_j, Q_j, and F_j separately.

(15.11) $A[n_0, \ldots, n_{j-1}, n_j, n_{j+1} \ldots, n_{\nu-1}] = C[b, q, f]$

if and only if

$$b = n_0 \prod_{k=1}^{j-1} N_k + n_1 \prod_{k=2}^{j-1} N_k + \cdots + n_{j-2}N_{j-1} + n_{j-1},$$

$$q = n_j, \quad \text{and} \quad f = n_{j+1} \prod_{k=j+2}^{\nu-1} N_k + \cdots + n_{\nu-2}N_{\nu-1} + n_{\nu-1}.$$

Furthermore, it can be verified that

(15.12) $x_\ell = C[b, q, f] = A[n_0, n_1, \ldots, n_{\nu-1}]$ iff $\ell = b\,(Q_j \times F_j) + q \times F_j + f.$

15.5.5   Column-oriented interpretation of ν-D arrays as 3D arrays

The following identity holds if both arrays A and C are stored in column-major format. Observe that, instead of identity (15.5), identity (15.6) is now applied to dimensions B_j, Q_j, and F_j separately to determine the values of b, q, and f.

(15.13) $$A[n_0, \ldots, n_{j-1}, n_j, n_{j+1} \ldots, n_{\nu-1}] = C[b, q, f]$$

if and only if

$$b = n_{j-1} \prod_{k=0}^{j-2} N_k + n_{j-2} \prod_{k=2}^{j-3} N_k + \cdots + n_1 N_0 + n_0,$$

$$q = n_j, \quad \text{and} \quad f = n_{\nu-1} \prod_{k=j+1}^{\nu-2} N_k + \cdots + n_{j+2}N_{j+1} + n_{j+1}.$$

Furthermore, it can be shown that

(15.14) $x_\ell = C[b, q, f] = A[n_0, n_1, \ldots, n_{\nu-1}]$ iff $\ell = f\,(B_j \times Q_j) + q \times B_j + b.$

15.6   Programming Different ν-D Arrays From a Single Array

It has been assumed, either implicitly or explicitly, in previous sections (and in general) that when a ν-D array is suitable for developing and implementing an algorithm, the value of ν is fixed and a ν-D array with desired dimensions is declared and used throughout the program.

For example, the expression $A[Pdim, Qdim]$ is used to declare a two-dimensional array with $Pdim$ rows and $Qdim$ columns. For a problem with $N = P \times Q = 11 \times 21$, after initializing $Pdim := P = 11$ and $Qdim := Q = 21$, $A[Pdim, Qdim]$ declares a 11-by-21 2D array, and A is used as a 11-by-21 array throughout the program. If one also needs a 33-by-11 array in at some point during program execution, a second 2D array B with 33 rows and 11 columns will normally be declared in the program. If, in addition to the two 2D arrays A and B, a 3D array with dimensions $N_0 = 11$, $N_1 = 3$, and $N_2 = 7$ is needed at some point during program execution, then a third array $C[N_0, N_1, N_2]$ can be declared for use in the program.

There are several potentially undesirable consequences from using the standard approach described above. First, to interpret (and access) a 1D array as a ν-D array, one may be required to copy the data from the 1D array to the ν-D array, which incurs extra time and storage.

Second, the 3D array $C[11, 3, 7]$ cannot overwrite the 2D array $A[11, 21]$ even though the data in A may be no longer needed when C is constructed. It is thus possible that storage is wasted when a large number of data arrays cannot be re-used in the program.

Third, if the problem size N is composite, and if the number of arrays with different dimensions to be used in the program depends in some way on the number of factors in N, then the number of arrays declared changes from problem to problem. When this happens, one can no longer use static storage allocation, but instead must resort to dynamic storage allocation.

Fortunately, these unwelcome possibilities can be avoided if different ν-D arrays with different dimensions can be allocated and accessed from a single one-dimensional array by passing the latter through subroutines.

15.6.1   Support from the FORTRAN programming language

In FORTRAN, if a 1D array A of size $N = \prod_{k=0}^{\nu-1} N_k$ is declared in the main program, the same 1D array A may be "passed" and actually used as an m-D array of dimensions $P_0 \times P_1 \times \cdots \times P_{m-1}$ inside a subroutine, where $N = \prod_{k=0}^{m-1} P_k$, and the P_k's are the formal parameters of the subroutine. Note that P_k is not necessarily equal to N_k, and P_k may be the product of several N_k's.

Since $A[P_0, P_1, \ldots, P_{m-1}]$ are declared as so inside a subroutine with P_k's being the formal parameters, the parameter P_k may take on different actual values each time the subroutine is called (as long as $N = \prod_{k=0}^{m-1} P_k$ is satisfied.)

If $A[\ell] = x_\ell$, $0 \le \ell \le N - 1$, in the main program, regardless of the value of m and the values of the P_k's, the column-major storage scheme will guarantee that the x_ℓ's can be accessed in consecutive order if the m-D array $A[P_0, P_1, \ldots, P_{m-1}]$ is accessed column-by-column inside the subroutine.

For example, suppose $A[Ndim]$ with $Ndim = 11 \times 21 = 231$ is declared and initialized in the main program. The following subroutines can be called to use A as a 2D array with different dimensions during program execution if one so desires.

```
Subroutine   UseAs2D(A, Pdim, Qdim, . . .)
Complex   A[Pdim, Qdim]
       ⋮
```

Note that the subroutine above may be invoked with $Pdim = 11$ and $Qdim = 21$ at one point, and it may be invoked with $Pdim = 3$ and $Qdim = 77$ at another point during program execution. Therefore, it affords great flexibility in developing the algorithm and programming it efficiently.

The next subroutine declaration allows the use of arrays A and C as two different 3D arrays inside the subroutine. Again the three dimensions can take on different actual values each time the subroutine is invoked. Furthermore, the roles of arrays

A and C may be easily switched by exchanging their positions in the list of actual parameters.

Subroutine $Use3dA3dC(A, C, Pdim, Qdim, Fdim, ...)$
Complex $A[Pdim, Qdim, Fdim]$, $C[Pdim, Fdim, Qdim]$

⋮

15.6.2   Further adaptation

This programming technique can be further adapted as follows. Suppose $N = \prod_{k=0}^{\nu-1} N_k$, the main program can form different $Pdim$, $Qdim$, and $Fdim$ from the N_k's to invoke *Subroutine Use3dA3dC* during each iteration as shown in the skeleton program below.

---

begin   main program begins

⋮

Complex $A[N]$, $C[N]$ declare A and C in main program

⋮

$P := 1$; $F := N_{\nu-1}$; $Q := N/F$ $N = P \times Q \times F = \prod_{k=0}^{\nu-1} N_k$
$InputA := true$
for $j := \nu - 1$ downto 0 **do**
   if $InputA$ array A contains input data
      Call $Use3dA3dC(A, C, P, Q, F, ...)$
      $InputA := false$
   else   array C contains input data
      Call $Use3dA3dC(C, A, P, Q, F, ...)$
      $InputA := true$
   end if
   $P := P * F$ $P = N_{\nu-1} \times \cdots \times N_j$ for next iteration
   $nextF := j - 1$
   if $nextF \geq 0$
      $F := N_{nextF}$ $F = N_{j-1}$ for next iteration
      $Q := Q/F$ maintain $N = P \times Q \times F$
   end if
end for

⋮

end   main program ends

For example, if $N = \prod_{k=0}^{3} N_k = 3 \times 5 \times 11 \times 17$, input $A[N]$ and output $C[N]$ are to be used as $A[1, 3 \times 5 \times 11, 17]$ and $C[1, 17, 3 \times 5 \times 11]$ in the first call to *Subroutine Use3dA3dC*. In the second call, the roles of C and A are switched: the new content in $C[1 \times 17, 3 \times 5, 11]$ represents the input, and $A[1 \times 17, 11, 3 \times 5]$ is to be overwritten by the output. In the third call, the roles of A and C are again switched and they are used as input $A[17 \times 11, 3, 5]$ and output $C[17 \times 11, 5, 3]$. Lastly, $C[17 \times 11 \times 5, 1, 3]$ serves as the input array, and $A[17 \times 11 \times 5, 3, 1]$ contains the final output as you will see in a concrete example in the next section. This turns out to be a very useful technique in developing and programming efficient FFTs for composite $N = N_0 \times N_1 \times \cdots \times N_{\nu-1}$.

15.7   An Efficient FFT for $N = N_0 \times N_1 \times \cdots \times N_{\nu-1}$

If the vector $\boldsymbol{x}$ of length $N = \prod_{k=0}^{\nu-1} N_k$ is stored in a ν-D array $A[N_0, N_1, \ldots, N_{\nu-1}]$ in column-major format, the DFT of $\boldsymbol{x}$ in terms of $A[n_0, n_1, \ldots, n_{\nu-1}] = x_\ell$, with ℓ defined by identity (15.6), can be interpreted using ν-D, 3-D and 2-D arrays. To begin, express $N = B \times Q \times F$ with $B = 1$, $Q = \prod_{k=0}^{\nu-2} N_k$, and $F = N_{\nu-1}$, and apply the identities from previous sections in an obvious way. (Note that it is not an oversight that A is accessed by rows in the formulation. It turns out that if A is stored in a row-major format, then it would need to be accessed by columns in the FFT formulation. The penalty is the same either way.)

$$X_r = P_{N-1}(z) = \sum_{\ell=0}^{N-1} x_\ell z^\ell, \qquad \text{where} \quad z = \omega_N^r, \quad r = 0, 1, \ldots, N-1,$$

$$= \sum_{n_0=0}^{N_0-1} \sum_{n_1=0}^{N_1-1} \cdots \sum_{n_{\nu-1}=0}^{N_{\nu-1}-1} A[n_0, n_1, \ldots, n_{\nu-1}] \times z^{n_{\nu-1} \prod_{k=0}^{\nu-2} N_k + \cdots + n_1 N_0 + n_0}$$

(15.15)
$$= \sum_{q=0}^{Q-1} \sum_{f=0}^{F-1} C[0, q, f] \, z^{f \times Q + q}, \qquad (\text{Note:} \quad B = 1, \quad B - 1 = 0),$$

$$= \sum_{q=0}^{Q-1} \left(\sum_{f=0}^{F-1} C[0, q, f] \times \left(z^Q \right)^f \right) z^q,$$

$$= \sum_{q=0}^{Q-1} \left(\sum_{f=0}^{F-1} C[0, q, f] \times \left(\omega_{B \times F}^r \right)^f \right) z^q, \qquad \omega_N^Q = \omega_{N/Q} = \omega_{B \times F}.$$

Observe that $\omega_{B \times F}^{B \times F} = 1$. To take advantage of the periodic property of $\omega_{B \times F}$, store output X in $Z[B, F, Q]$ (although dimension $B = 1$ for now, 3D-format is used because $B > 1$ after the initial step) using column-major format so that $r = \left(\hat{q} \times F + \hat{f} \right) B + \hat{b}$ according to identity (15.14). Thus,

(15.16)
$$\omega_{B \times F}^r = \omega_{B \times F}^{\left(\hat{q} \times F + \hat{f} \right) B + \hat{b}} = \omega_{B \times F}^{\hat{f} \times B + \hat{b}},$$

and equation (15.15) can be rewritten as

(15.17)
$$X_r = X_{\left(\hat{q} \times F + \hat{f} \right) B + \hat{b}} = \sum_{q=0}^{Q-1} \left(\sum_{f=0}^{F-1} C[\hat{b}, q, f] \times \left(\omega_{B \times F}^{\hat{f} \times B + \hat{b}} \right)^f \right) z^q,$$

$$\text{where} \quad \hat{b} = b = 0, \quad \text{and} \quad z = \omega_N^r.$$

Therefore, the bracketed inner sum only needs to be computed for $0 \le \hat{f} \le F - 1$, $0 \le \hat{b} \le B - 1$, and $0 \le q \le Q - 1$. The corresponding pseudo-code segment is shown below.

$$
\boxed{
\begin{array}{ll}
B := 1; \ F := N_{\nu-1}; \ Q := N/(B * F) & \text{Initial dimensions for Step 1} \\
\theta := 2\pi/(B * F); \ WBF := exp(-j\theta) & \text{Compute } WBF = \omega_{B \times F} \\
z := 1 & z = \omega_{B \times F}^{0} = 1 \\
\underline{\textbf{for}} \ \hat{f} := 0 \ \underline{\textbf{to}} \ F - 1 & \\
\quad \underline{\textbf{for}} \ \hat{b} := 0 \ \underline{\textbf{to}} \ B - 1 & B - 1 = 0 \\
\quad\quad \underline{\textbf{for}} \ q := 0 \ \underline{\textbf{to}} \ Q - 1 & \\
\quad\quad\quad C''[\hat{b}, \hat{f}, q] := \sum_{f=0}^{F-1} C[\hat{b}, q, f] * z^{f} & z = \omega_{B \times F}^{\hat{f} \times B + \hat{b}} \\
\quad\quad \underline{\textbf{end for}} & \\
\quad\quad z := z * WBF & \text{Update } z = \omega_{B \times F}^{\hat{f} \times B + \hat{b} + 1} \\
\quad \underline{\textbf{end for}} & \\
\underline{\textbf{end for}} &
\end{array}
}
$$

Note that $F \times B \times Q \times F = N \times N_{\nu-1}$ multiply-adds are incurred. After $C''[B, F, Q]$ is available, equation (15.17) is simplified to

$$
(15.18) \qquad X_r = X_{(\hat{q} \times F + \hat{f})B + \hat{b}} = Z[\hat{b}, \hat{f}, \hat{q}] = \sum_{q=0}^{Q-1} C''[\hat{b}, \hat{f}, q] \, z^q,
$$

$$
\text{where} \quad \hat{b} = 0, \quad Q = \prod_{k=0}^{\nu-2} N_k, \quad z = \omega_N^r, \quad \text{for} \quad r = 0, 1, \ldots, N - 1.
$$

If efficiency is not a concern for composite Q, one immediately obtains the complete algorithm as shown below. (The computation of C'' is repeated for easy reference.)

$$
\boxed{
\begin{array}{ll}
B := 1; \ F := N_{\nu-1}; \ Q := N/(B * F) & \text{Initial dimensions for Step 1} \\
\theta := 2\pi/(B * F); \ WBF := exp(-j\theta) & \text{Compute } WBF = \omega_{B \times F} \\
z := 1 & z = \omega_{B \times F}^{0} = 1 \\
\underline{\textbf{for}} \ \hat{f} := 0 \ \underline{\textbf{to}} \ F - 1 & \\
\quad \underline{\textbf{for}} \ \hat{b} := 0 \ \underline{\textbf{to}} \ B - 1 & B - 1 = 0 \\
\quad\quad \underline{\textbf{for}} \ q := 0 \ \underline{\textbf{to}} \ Q - 1 & \\
\quad\quad\quad C''[\hat{b}, \hat{f}, q] := \sum_{f=0}^{F-1} C[\hat{b}, q, f] * z^{f} & z = \omega_{B \times F}^{\hat{f} \times B + \hat{b}} \\
\quad\quad \underline{\textbf{end for}} & \\
\quad\quad z := z * WBF & \text{Update } z = \omega_{B \times F}^{\hat{f} \times B + \hat{b} + 1} \\
\quad \underline{\textbf{end for}} & \\
\underline{\textbf{end for}} & \\
\quad\quad\quad \text{Final part: an inefficient way to compute } X_r\text{'s for composite } Q & \\
\theta := 2\pi/N; \ WN := exp(-j\theta) & \text{Compute } WN = \omega_N \\
z := 1 & z = \omega_N^{0} = 1 \\
\underline{\textbf{for}} \ \hat{q} := 0 \ \underline{\textbf{to}} \ Q - 1 \ \underline{\textbf{do}} & \\
\quad \underline{\textbf{for}} \ \hat{f} := 0 \ \underline{\textbf{to}} \ F - 1 \ \underline{\textbf{do}} & \\
\quad\quad \underline{\textbf{for}} \ \hat{b} := 0 \ \underline{\textbf{to}} \ B - 1 \ \underline{\textbf{do}} & B - 1 = 0 \\
\quad\quad\quad Z[\hat{b}, \hat{f}, \hat{q}] := \sum_{q=0}^{Q-1} C''[\hat{b}, \hat{f}, q] * z^{q} & z = \omega_N^{r} \\
\quad\quad\quad z := z * WN & \text{Update } z = \omega_N^{r+1} \\
\quad\quad \underline{\textbf{end for}} & \\
\quad \underline{\textbf{end for}} & \\
\underline{\textbf{end for}} &
\end{array}
}
$$

On output, consecutive X_r's are stored in $Z[B, F, Q]$ in column-major format. This version of the algorithm is inefficient, because it requires, in total, $N \times (F + Q) = N^2/N_{\nu-1}+O(N)$ complex multiply-adds. (Of course, since $B = 1$ and $N = B \times Q \times F = Q \times F$, the code above simply expresses the two-factor FFT algorithm presented in Section 15.4 for $N = Q \times F$ in a pseudo 3-D format.)

Obviously, in order to improve efficiency in the computation of X_r's, one should make use of the fact that Q is the product of the remaining $\nu - 1$ prime factors.

Now, suppose one *interprets* and *actually uses* the computed $C''[B, F, Q]$ as $C[B', Q', F']$ (as suggested in Section 15.6), with $B' = B \times F$, $F' = N_{\nu-2}$, and $Q' = Q/F'$. Thus $Q = Q' \times F'$, $q = f' \times Q' + q'$ (in column-major format), and one may rewrite the equation (15.18) as

$$
X_r = X_{(\hat{q}' \times F' + \hat{f}')B' + \hat{b}'}
$$

$$
= Z[\hat{b}', \hat{f}', \hat{q}']
$$

$$
= \sum_{q'=0}^{Q'-1} \sum_{f'=0}^{F'-1} C[\hat{b}', q', f'] z^{f' \times Q' + q'}, \quad z = w_N^r, \ r = 0, 1, \ldots, N-1.
$$

(15.19)
$$
= \sum_{q'=0}^{Q'-1} \left(\sum_{f'=0}^{F'-1} C[\hat{b}', q', f'] \times \left(z^{Q'} \right)^{f'} \right) z^{q'},
$$

$$
= \sum_{q'=0}^{Q'-1} \left(\sum_{f'=0}^{F'-1} C[\hat{b}', q', f'] \times \left(w_{B' \times F'}^{r} \right)^{f'} \right) z^{q'},
$$

$$
= \sum_{q'=0}^{Q'-1} \left(\sum_{f'=0}^{F'-1} C[\hat{b}', q', f'] \times \left(w_{B' \times F'}^{\hat{f}' \times B' + \hat{b}'} \right)^{f'} \right) z^{q'}.
$$

Some useful identities are summarized below.

- $N = B \times Q \times F = 1 \times \left(\prod_{k=0}^{\nu-2} N_k \right) \times N_{\nu-1}$.

- $B' = B \times F = 1 \times N_{\nu-1} = N_{\nu-1}$.

- $F' = N_{\nu-2}$.

- $Q' = Q/F' = \prod_{k=0}^{\nu-3} N_k$.

- $N = B' \times Q' \times F'$, and $Q' \times F' = Q$.

- $0 \le r = (\hat{q}' \times F' + \hat{f}')B' + \hat{b}' \le N - 1$.

- $w_{B' \times F'}^{B' \times F'} = 1$.

Now, rename the dimensions $B := B'$, $Q := Q'$, $F := F'$ in equation (15.19) and obtain

(15.20)
$$
\boxed{
\begin{aligned}
X_r &= X_{(\hat{q} \times F + \hat{f})B + \hat{b}} \\
&= Z[\hat{b}, \hat{f}, \hat{q}] \\
&= \sum_{q=0}^{Q-1} \left(\sum_{f=0}^{F-1} C[\hat{b}, q, f] \times \left(w_{B \times F}^{\hat{f} \times B + \hat{b}} \right)^{f} \right) z^{q}, \quad z = w_N^r.
\end{aligned}
}
$$

Since equation (15.20) is identical to equation (15.17), the bracketed inner sum can now be computed using exactly the same code from the initial step with updated $B = N_{\nu-1}$, $F = N_{\nu-2}$, and $Q = \prod_{k=0}^{\nu-3} N_k$. The code-segment is repeated below.

$B := N_{\nu-1}$; $F := N_{\nu-2}$; $Q := N/(B * F)$	Updated dimensions for Step 2
$\theta := 2\pi/(B * F)$; $WBF := exp(-j\theta)$	Compute $WBF = \omega_{B \times F}$
$z := 1$	$z = \omega_{B \times F}^0 = 1$
for $\hat{f} := 0$ **to** $F - 1$	$F = N_{\nu-2}$
$\quad$ **for** $\hat{b} := 0$ **to** $B - 1$	$B = N_{\nu-1}$
$\quad\quad$ **for** $q := 0$ **to** $Q - 1$	
$\quad\quad\quad C''[\hat{b}, \hat{f}, q] := \sum_{f=0}^{F-1} C[\hat{b}, q, f] * z^f$	$z = \omega_{B \times F}^{\hat{f} \times B + \hat{b}}$
$\quad\quad$ **end for**	
$\quad\quad z := z * WBF$	Update $z = \omega_{B \times F}^{\hat{f} \times B + \hat{b} + 1}$
$\quad$ **end for**	
end for	

The remaining computation can again be expressed as

$$(15.21) \qquad X_r = X_{(\hat{q} \times F + \hat{f})B + \hat{b}} = Z[\hat{b}, \hat{f}, \hat{q}] = \sum_{q=0}^{Q-1} C''[\hat{b}, \hat{f}, q] \, z^q,$$

$$\text{where} \quad z = \omega_N^r, \quad \text{for} \quad r = 0, 1, \ldots, N-1.$$

Since equation (15.21) is identical to equation (15.18) except for the new values taken by the dimensions, the process for deriving the second step can be repeated using $B' := B \times F = N_{\nu-1} \times N_{\nu-2}$, $F' := N_{\nu-3}$, and $Q' := Q/F$.

Finally, the νth step repeats the same process using $B := B' = N_{\nu-1} \times N_{\nu-2} \times \cdots \times N_1$, $Q := Q' = 1$, and $F := F' = N_0$, and the values of X_r's are available in $C''[B, F, Q]$ because one now has

$$(15.22) \qquad X_r = X_{(\hat{q} \times F + \hat{f})B + \hat{b}} = \sum_{q=0}^{Q-1} C''[\hat{b}, \hat{f}, q] \, z^q = C''[\hat{b}, \hat{f}, \hat{q}],$$

$$\text{for} \quad Q - 1 = 0, \quad q = 0, \quad \text{and} \quad z^q = 1 .$$

Therefore, consecutive X_r's are available in a column-major format in array $C''[B, F, Q]$, which can be interpreted as $C''[N_{\nu-1}, \ldots, N_1, N_0, 1]$. Observe that $Q = 1$ is a dummy dimension, so the output array C'' can also be interpreted as the *transpose* of the ν-dimensional input array $A[N_0, N_1, \ldots, N_{\nu-1}]$.

Since $N = \prod_{k=0}^{\nu-1} N_k$ and $N \times N_k$ multiply-adds are needed in each of the ν steps, the total cost amounts to $N \times (N_0 + N_1 + \cdots + N_{\nu-1})$ multiply-adds in this ν-factor FFT algorithm.

15.8   Notes and References

Since a non-prime factor is composite, it is logical to assume that N is expressed as the product of prime factors when applying the Cooley-Tukey FFT algorithm to general N. Note, however, that the Cooley-Tukey algorithm does not impose any restriction on the factors, and it should not be confused with the so-called "Prime Factor" Algorithm (PFA), which is only applicable to a very restricted set of N values. For the PFA to work, the factors of N are required to be "coprime", or, in other words, *pairwise* prime. For example, if $N = N_0 \times N_1 \times N_2 = 5 \times 7 \times 16$, then any two factors of N are relatively prime, and one can apply the PFA. However, if the same N is expressed as $N = N_0 \times N_1 \times N_2 \times N_3 = 5 \times 7 \times 2 \times 8$, then because factor $N_2 = 2$ and factor $N_3 = 8$ are not relatively prime, one cannot apply the PFA. When N contains only coprime factors, the PFA provides an alternative which achieves lower operation count than the Cooley-Tukey algorithm.

The PFA was initially proposed by Good [50] in 1960, and its theory was first presented by Kolba and Parks [62] in 1977. A PFA variant with further reduced operation count was developed by Winograd [109] in 1978. An excellent reference for the PFA, Winograd's algorithm (WPFA), and other PFA variants is Nussbaumer's monograph [70, Chapter 5], which also contains the short DFT algorithms that are most frequently used with the PFA or WPFA for factors $N_k = 2, 3, 4, 5, 7, 8, 9$, and 16 [70, pages 144–150]. Johnson and Burrus [57] developed large DFT modules: 11, 13, 17, 19, and 25. More recently, the PFA was adapted for vector-processing and generalized further by Temperton [100, 101, 102].

A modified in-place and in-order form of the PFA was implemented in FORTRAN by Burrus and Eschenbacher [20] in 1981. The program allows coprime factors chosen from the set of 2, 3, 4, 5, 7, 8, 9, and 16. Observe that the maximum number of relatively prime factors is four, and the maximum length is $N = 5 \times 7 \times 9 \times 16 = 5040$. Unfortunately, as pointed out by Burrus and Eschenbacher, the required modification is different for different N, so the program must either be complicated by added control statements or recompiled for a new N. Burrus' in-place, in-order, fixed size transform was modified by Rothweiler [81] in 1982 to produce a general-purpose, in-place, and in-order PFA for variable transform size. The class of in-order (also called self-sorting) in-place algorithms has recently been extended to include radix-2, radix-3, radix-4, radix-5, and mixed-radix FFTs [58, 101, 102].

For $N = 3^m$ and $N = 2^p 3^q 5^r$, specialized radix-3 and mixed radix FFTs exist [37, 102]. An FFT algorithm of radix 3, 6, and 12 was proposed in [92]. Split-radix algorithms for length p^m DFTs were proposed in [105].

Chapter 16

Selected FFT Applications

Chapter 1 contains a development showing how the discrete Fourier transform (DFT) arises naturally in trigonometric interpolation. The FFT algorithms presented in the preceding chapters provide fast implementations of the DFT. Since the DFT arises in a wide variety of applications, it is appropriate to select a few more examples to demonstrate the practical usefulness of the various FFT algorithms.

16.1 Fast Polynomial Multiplication

Consider two polynomials of degree N defined by

$$(16.1) \qquad A_N(z) = \sum_{\ell=0}^{N} a_\ell z^\ell = a_N z^N + \cdots + a_1 z + a_0$$

and

$$(16.2) \qquad B_N(z) = \sum_{\ell=0}^{N} b_\ell z^\ell = b_N z^N + \cdots + b_1 z + b_0,$$

where, in general, z is a complex variable, and a_ℓ and b_ℓ, $0 \le \ell \le N$, are complex coefficients. The product of polynomials $A_N(z)$ and $B_N(z)$ is a polynomial of degree $2N$ defined by

$$(16.3) \qquad C_{2N}(z) = A_N(z) B_N(z) = \sum_{\ell=0}^{2N} c_\ell z^\ell = c_{2N} z^{2N} + \cdots + c_1 z + c_0.$$

Defining $a_k = b_k = 0$ for $N < k \le 2N$, the coefficients c_ℓ are given by

$$(16.4) \qquad c_\ell = \sum_{k=0}^{\ell} a_k b_{\ell-k}, \quad \ell = 0, 1, \ldots, 2N.$$

The arithmetic cost of computing the c_ℓ is measured by the $\Theta\left(N^2\right)$ complex multiplications needed to compute all of the $a_k \times b_{\ell-k}$ terms in equation (16.4). Given below is an example showing the product of two polynomials of degree 3 with real coefficients.

Example 16.1

$$C_6(z) = A_3(z)B_3(z)$$
$$= \left(5z^3 + 2z^2 + 3z + 7\right)\left(4z^3 + 9z^2 + 5z + 2\right)$$
$$= 20z^6 + 53z^5 + 55z^4 + 75z^3 + 82z^2 + 41z + 14$$

Note that by adding zero coefficients, equation (16.4) also defines the product of two polynomials of different degrees. Given below is such an example.

Example 16.2

$$P_4(z) = B_3(z)D_1(z)$$
$$= \left(5z^3 + 2z^2 + 3z + 7\right)(5z + 2)$$
$$= \left(5z^3 + 2z^2 + 3z + 7\right)\left(0 \times z^3 + 0 \times z^2 + 5z + 2\right)$$
$$= 0 \times z^6 + 0 \times z^5 + 25z^4 + 20z^3 + 19z^2 + 41z + 14$$
$$= 25z^4 + 20z^3 + 19z^2 + 41z + 14$$

This seemingly unnecessary act of adding zero coefficients turns out to serve a useful purpose in deriving a fast $\Theta\left(N\log_2 N\right)$ polynomial multiplication algorithm based on FFT computation. The algorithm for computing $C_{2N}(z) = A_N(z)B_N(z)$ consists of the following steps.

Step 1. (Preprocessing) Add N leading zero coefficients to the polynomial $A_N(z)$ of degree N to obtain $\tilde{A}_{2N}(z)$, for which $a_\ell = 0$, $N < \ell \le 2N$.

Step 2. (Preprocessing) Add N leading zero coefficients to the polynomial $B_N(z)$ of degree N to obtain $\tilde{B}_{2N}(z)$, for which $b_\ell = 0$, $N < \ell \le 2N$.

Step 3. (Fast polynomial evaluation) Evaluate $\tilde{A}_{2N}(z)$ at the $2N+1$ primitive roots of unity: $1, \omega_M, \omega_M^2, \ldots, \omega_M^{2N}$, where $M = 2N + 1$. That is, one computes

$$(16.5) \qquad A_r = A_N\left(\omega_M^r\right) = \tilde{A}_{2N}\left(\omega_M^r\right) = \sum_{\ell=0}^{M-1} a_\ell \omega_M^{r\ell}, \quad r = 0, 1, \ldots, M - 1.$$

Equation (16.5) above defines a DFT of length M.

Step 4. (Fast polynomial evaluation) Evaluate $\tilde{B}_{2N}(z)$ at the $2N+1$ primitive roots of unity: $1, \omega_M, \omega_M^2, \ldots, \omega_M^{2N}$, where $M = 2N + 1$. That is, one computes

$$(16.6) \qquad B_r = B_N\left(\omega_M^r\right) = \tilde{B}_{2N}\left(\omega_M^r\right) = \sum_{\ell=0}^{M-1} b_\ell \omega_M^{r\ell}, \quad r = 0, 1, \ldots, M - 1.$$

Equation (16.6) above defines a DFT of length M.

Step 5. Compute $C_r = A_r B_r$, for $r = 0, \ldots, M - 1$. This provides the values of the polynomial $C_{2N}(z) = A_N(z)B_N(z)$ at $M = 2N + 1$ <u>distinct</u> points, uniquely defining $C_{2N}(z)$. The next step provides the coefficients of the polynomial from these values.

Step 6. (Fast polynomial interpolation) Compute the coefficients c_ℓ, $0 \le \ell \le 2N$, by interpolating at the $M = 2N + 1$ distinct values of $C_r = C_{2N}(\omega_M^r)$ computed in Step 5. That is, one needs to solve the following linear system of equations to obtain the coefficients $c_0, c_1, \ldots, c_{2N}$.

$$(16.7) \qquad C_r = \sum_{\ell=0}^{M-1} c_\ell \omega_M^{r\ell}, \quad r = 0, 1, \ldots, M - 1.$$

Since equation (16.7) represents the DFT of the unknown c_ℓ, it can be rewritten as the inverse DFT of the known C_r as shown in Chapter 1, i.e.,

$$(16.8) \qquad c_\ell = \frac{1}{M} \sum_{r=0}^{M-1} C_r \omega_M^{-r\ell}, \quad \ell = 0, 1, \ldots, M - 1.$$

To obtain the c_ℓ, one computes an inverse DFT of length M defined by equation (16.8).

The DFT of length M in Steps 3 and 4, and the inverse DFT (IDFT) in Step 6 may each be computed by Bluestein's FFT and inverse FFT (IFFT) for arbitrary M in $\Theta\left(\widetilde{M} \log_2 \widetilde{M}\right)$ time, where $2M - 2 \le \widetilde{M} < 4M$ as shown in Chapter 13. Since $M = 2N + 1$, $4N \le \widetilde{M} < 8N + 4$, the arithmetic cost of computing the product of two polynomials of degree N has been reduced from $\Theta\left(N^2\right)$ using equation (16.4) to $\Theta\left(N \log_2 N\right)$ as desired.

Alternatively, one may add zero coefficients and use the $\widetilde{M} = 2^k$ primitive roots of unity: $1, \omega_{\widetilde{M}}, \omega_{\widetilde{M}}^2, \ldots, \omega_{\widetilde{M}}^{\widetilde{M}-1}$, to increase the length of the DFT and IDFT in Steps 3, 4, and 6 from M to the nearest $\widetilde{M} = 2^k$. Then one can apply a radix-2 FFT and IFFT to compute the respective DFT and IDFT. The arithmetic cost remains $\Theta\left(N \log_2 N\right)$ as desired.

16.2 Fast Convolution and Deconvolution

The convolution of two vectors a and b, denoted by $a \otimes b$, is mathematically equivalent to polynomial multiplication if the contents of the vectors a and b are interpreted as the coefficients of two polynomials as shown below.

Definition 16.1 Consider two complex vectors $a = [a_0, a_1, \ldots, a_N]$ and $b = [b_0, b_1, \ldots, b_N]$. Then the convolution of a and b is defined by

$$c = a \otimes b$$
$$= [a_0, a_1, \ldots, a_N] \otimes [b_0, b_1, \ldots, b_N]$$
$$= [c_0, c_1, c_2, \ldots, c_{2N}, c_{2N+1}]$$

where $a_k = b_k = 0$, $N < k \le 2N$, $c_{2N+1} = 0$, and

$$(16.9) \qquad c_\ell = \sum_{k=0}^{\ell} a_k b_{\ell-k}, \quad \ell = 0, 1, \ldots, 2N.$$

Given below is an example showing the convolution of two real vectors a and b.

Example 16.3

$$\begin{aligned}
c &= [\,a_0, a_1, a_2, a_3\,] \otimes [\,b_0, b_1, b_2, b_3\,] \\
&= [\,7, 3, 2, 5\,] \otimes [\,2, 5, 9, 4\,] \\
&= [\,14, 41, 82, 75, 55, 53, 20, 0\,] \\
&= [\,c_0, c_1, c_2, c_3, c_4, c_5, c_6, c_7 \equiv 0\,]
\end{aligned}$$

Note that by adding zero components, the same definition applies to compute the convolution of two vectors of different dimensions. Given below is such an example.

Example 16.4

$$\begin{aligned}
c &= [\,a_0, a_1, a_2, a_3\,] \otimes [\,b_0, b_1\,] \\
&= [\,7, 3, 2, 5\,] \otimes [\,2, 5, 0, 0\,] \\
&= [\,14, 41, 19, 20, 25, 0, 0, 0\,] \\
&= [\,14, 41, 19, 20, 25, 0\,] \\
&= [\,c_0, c_1, c_2, c_3, c_4, c_5 \equiv 0\,]
\end{aligned}$$

Therefore, except for adding $c_{2N+1} = 0$ (which must be present to preserve the symmetry), equation (16.9) in the definition of convolution is essentially identical to equation (16.4) which defines the product of two polynomials. Accordingly, the fast $\Theta\,(N \log_2 N)$ FFT-based polynomial multiplication algorithm developed in the last section may be used to compute the convolution of two vectors of length $N+1$. In addition, the polynomial multiplication algorithm provides a constructive proof of the following convolution theorem.

Theorem 16.2 Let $\hat{a} = [\,a_0, a_1, \ldots, a_N, 0, \ldots, 0\,]$ and $\hat{b} = [\,b_0, b_1, \ldots, b_N, 0, \ldots, 0\,]$ be vectors of length $2N+1$ (after a and b are each padded with N zeros). Let $\text{DFT}(\hat{a}) = [\,A_0, A_1, \ldots, A_{2N}\,]$ and $\text{DFT}(\hat{b}) = [\,B_0, B_1, \ldots, B_{2N}\,]$ be the results of applying discrete Fourier transforms to the two sequences of complex numbers. Then the convolution of the vector a of length $N+1$ and the vector b of length $N+1$ is a vector c of length $2N+2$ as given below. (Note that the last element $c_{2N+1} = 0$ need not be computed, and it is added afterwards to preserve symmetry.)

$$a \otimes b = [\,c_0, c_1, \ldots, c_{2N}, 0\,]$$

where

$$[\,c_0, c_1, \ldots, c_{2N}\,] = \text{IDFT}\,(\,[\,A_0 B_0, A_1 B_1, \ldots, A_{2N} B_{2N}\,]\,).$$

For very large N, it may be beneficial to avoid padding a and b with N more zeros. When such a need arises, one may compute $c = a \otimes b$ through one positive and one negative "wrapped" convolutions as suggested in [1]. Each wrapped convolution involves computing two DFTs and one IDFT of length $N+1$ (instead of $2N+1$ in the unwrapped convolution). Because two wrapped convolutions are required, the arithmetic cost of computing $c = a \otimes b$ will not be reduced, although the storage for additional zeros can be saved. Interested readers are referred to [1] for further reading on wrapped convolutions.

"Deconvolution" is introduced in [77] as "the process of *undoing* the smearing in a data set (represented by the vector a) that has occurred under the influence of a known response function (with its discrete values represented by the vector b), for example, because of the known effect of a less-than-perfect measuring apparatus." Mathematically, the objective is to recover the unknown original vector a, assuming that the contents of vectors b and $c = a \otimes b$ are given. It is a simple exercise to *reverse* the convolution process. Assuming that the vector $\hat{b}$ contains the vector b of length $N+1$ and an additional N zeros as before, recall that in deriving the fast polynomial multiplication algorithm it was shown that the vector $C = \mathrm{DFT}(c)$ is the component-wise product of the vector $A = \mathrm{DFT}(\hat{a})$ and the vector $B = \mathrm{DFT}(\hat{b})$, so the currently unknown $\mathrm{DFT}(\hat{a}) = [\, C_0/B_0, C_1/B_1, \ldots, C_{2N}/B_{2N} \,]$, assuming that $B_k \neq 0$, $0 \leq k \leq 2N$. This immediately leads to

$$(16.10) \qquad \hat{a} = \mathrm{IDFT}\left(\, [\, C_0/B_0, C_1/B_1, \ldots, C_{2N}/B_{2N} \,] \,\right),$$

where c and b are given, $C = \mathrm{DFT}(c)$, and $B = \mathrm{DFT}(\hat{b})$. The "deconvoluted" vector a can now be recovered from the first $N+1$ elements in $\hat{a}$. The two DFTs and one IDFT of length $2N+1$ can again be computed using FFT algorithms. Clearly, the arithmetic cost of deconvolution is the same as that of convolution.

16.3 Computing a Toeplitz Matrix-Vector Product

Readers are assumed to be familiar with the definition and the properties of Toeplitz matrices covered in Chapter 13. Numerical computations involving Toeplitz matrices arise frequently in signal processing [55]. The explicit relationship between a convolution and the product of a Toeplitz matrix and a vector was pointed out in [55]. In short, the convolution of vectors a of length $2N+1$ and b of length $N+1$ can be written as the product of a $(3N+1) \times (N+1)$ matrix A and the vector b. Since the square $(N+1) \times (N+1)$ Toeplitz matrix T formed by the $2N+1$ elements from a corresponds to $N+1$ rows in A, the product Tb is a subvector of $\hat{c} = Ab$. An example is given below to demonstrate how to form a Toeplitz matrix-vector product by convoluting two vectors.

Example 16.5 The convolution of a vector a of length $2N+1=7$ and a vector b of length $N+1=4$ defined by

$$
\begin{aligned}
c &= [\, a_0, a_1, a_2, a_3, a_4, a_5, a_6 \,] \otimes [\, b_0, b_1, b_2, b_3 \,] \\
&= [\, 7, 3, 2, 5, 1, 1, 1 \,] \otimes [\, 2, 5, 9, 4 \,] \\
&= [\, 14, 41, 82, 75, 57, 60, 36, 18, 13, 4, 0 \,] \\
&= [\, c_0, c_1, c_2, c_3, c_4, c_5, c_6, c_7, c_8, c_9, c_{10} \equiv 0 \,]
\end{aligned}
$$

can be written as the product of a 10×4 Toeplitz matrix and a vector of length 4 as shown below.

$$
\hat{c} =
\begin{bmatrix}
\hat{c}_9 \\ \hat{c}_8 \\ \hat{c}_7 \\ \hat{c}_6 \\ \hat{c}_5 \\ \hat{c}_4 \\ \hat{c}_3 \\ \hat{c}_2 \\ \hat{c}_1 \\ \hat{c}_0
\end{bmatrix}
=
\begin{bmatrix}
a_6 & & & \\
a_5 & a_6 & & \\
a_4 & a_5 & a_6 & \\
a_3 & a_4 & a_5 & a_6 \\
a_2 & a_3 & a_4 & a_5 \\
a_1 & a_2 & a_3 & a_4 \\
a_0 & a_1 & a_2 & a_3 \\
 & a_0 & a_1 & a_2 \\
 & & a_0 & a_1 \\
 & & & a_0
\end{bmatrix}
\begin{bmatrix}
b_3 \\ b_2 \\ b_1 \\ b_0
\end{bmatrix}
=
\begin{bmatrix}
1 & & & \\
1 & 1 & & \\
1 & 1 & 1 & \\
5 & 1 & 1 & 1 \\
2 & 5 & 1 & 1 \\
3 & 2 & 5 & 1 \\
7 & 3 & 2 & 5 \\
 & 7 & 3 & 2 \\
 & & 7 & 3 \\
 & & & 7
\end{bmatrix}
\begin{bmatrix}
4 \\ 9 \\ 5 \\ 2
\end{bmatrix}
=
\begin{bmatrix}
4 \\ 13 \\ 18 \\ 36 \\ 60 \\ 57 \\ 75 \\ 82 \\ 41 \\ 14
\end{bmatrix}
= c.
$$

Observe that each diagonal of the matrix A is formed entirely by one element from the vector a. The result $\hat{c}_\ell = c_\ell$, $0 \le \ell \le 3N$, is expected as it can be easily verified from the matrix equation $\hat{c} = Ab$ that $\hat{c}_\ell = c_\ell = \sum_{k=0}^{\ell} a_k b_{\ell-k}$, $\ell = 0, 1, \ldots, 3N$, where $a_k = 0$ for $2N < k \le 3N$ and $b_k = 0$ for $N < k \le 3N$.

Since an $(N+1) \times (N+1)$ Toeplitz matrix T formed by elements $a_0, a_1, \ldots, a_{2N}$ consists of $N+1$ rows from the matrix A, the result of Tb is contained in $\hat{c} = Ab$ as shown below.

$$
\begin{bmatrix}
\hat{c}_6 \\ \hat{c}_5 \\ \hat{c}_4 \\ \hat{c}_3
\end{bmatrix}
=
\begin{bmatrix}
a_3 & a_4 & a_5 & a_6 \\
a_2 & a_3 & a_4 & a_5 \\
a_1 & a_2 & a_3 & a_4 \\
a_0 & a_1 & a_2 & a_3
\end{bmatrix}
\begin{bmatrix}
b_3 \\ b_2 \\ b_1 \\ b_0
\end{bmatrix}
=
\begin{bmatrix}
5 & 1 & 1 & 1 \\
2 & 5 & 1 & 1 \\
3 & 2 & 5 & 1 \\
7 & 3 & 2 & 3
\end{bmatrix}
\begin{bmatrix}
4 \\ 9 \\ 5 \\ 2
\end{bmatrix}
=
\begin{bmatrix}
36 \\ 60 \\ 57 \\ 75
\end{bmatrix}
=
\begin{bmatrix}
c_6 \\ c_5 \\ c_4 \\ c_3
\end{bmatrix}.
$$

Therefore, the product of an $(N+1) \times (N+1)$ Toeplitz matrix T (with its diagonals formed by $a_0, a_1, \ldots, a_{2N}$) and the vector b may be recovered from the convolution of $a = [a_0, a_1, \ldots, a_{2N}]$ and $b = [b_0, b_1, \ldots, b_N]$ using the fast FFT-based algorithm introduced in the previous section.

Since a Hankel matrix is obtained from reversing the rows of a Toeplitz matrix, the product of a Hankel matrix and a vector can also be computed by convolution. These FFT-based fast algorithms for matrix-vector multiplication may then be used to obtain the fast implementation of several Toeplitz (or Hankel)-based iterative solvers for linear systems and least squares problems [9, 71, 73].

16.4   Computing a Circulant Matrix-Vector Product

Since the multiplication of a circulant matrix and a vector is one step in Bluesteins's FFT for arbitrary N (explained in Chapter 13), readers are referred to Section 13.1.5 for the definition of circulant matrices, as well as the details of a fast algorithm which diagonalizes a circulant coefficient matrix by a DFT matrix.

As shown in Section 13.1.4, a symmetric Toeplitz matrix may be embedded in an enlarged circulant matrix, so the fast algorithm in Section 13.1.5 also computes the product of a symmetric Toeplitz matrix and a vector.

Since circulant preconditioners may be constructed for Toeplitz least squares problems [67], the FFT-based fast algorithm in Section 13.1.5 can be used to yield a fast implementation of certain special preconditioned conjugate gradient (PCGLS) methods [9, 67].

16.5 Solving a Large Circulant Linear System

It was shown in Section 13.1.5 that the circulant coefficient matrix H in a matrix-vector product $z = Hy$ can be diagonalized by a DFT matrix Ω so that

$$(16.11) \qquad \Omega z = \Omega(Hy) = (\Omega H)y = (D\Omega)y = D(\Omega y),$$

where D is diagonal with its elements computed by a radix-2 FFT, and Ω is a radix-2 DFT matrix. That is, the fast algorithm in Section 13.1.5 computes

$$(16.12) \qquad z = \Omega^{-1}(D(\Omega y)).$$

If, instead of computing the matrix-vector product, one is required to solve for the unknown y when the circulant matrix H and the (right-hand side) vector z are given, one may diagonalize H so that $\Omega H = D\Omega$ as before, and rewrite (16.11) as

$$(16.13) \qquad y = \Omega^{-1}\left(D^{-1}(\Omega z)\right),$$

which is in exactly the same form as (16.12). Therefore, the same fast algorithm for computing the circulant matrix-vector product may be used to solve a circulant system of equations.

16.6 Fast Discrete Sine Transforms

Since the various fast algorithms for computing discrete trigonometric transforms (DTT's) make use of the fast Fourier transforms, the FFT also plays an important role in the wide ranging applications of trigonometric transforms, although its contribution may not always be visible to the end user. The DFT-related transforms considered in this and the following sections include the discrete sine transform (DST) defined by

$$(16.14) \qquad S_r = \sum_{\ell=1}^{N-1} x_\ell \sin\left(\frac{\pi r \ell}{N}\right), \qquad r = 1, 2, \ldots, N-1,$$

and the discrete cosine transform (DCT) defined by

$$(16.15) \qquad C_r = \frac{x_0 + (-1)^r x_N}{2} + \sum_{\ell=1}^{N-1} x_\ell \cos\left(\frac{\pi r \ell}{N}\right), \qquad r = 0, 1, \ldots, N.$$

These two trigonometric transforms (of real data) are mathematically related to the discrete Fourier transform (DFT) because each can be derived from a real (symmetric odd or even) DFT as shown below.

One can "embed" the discrete sine transform (DST) of length $N-1$ defined by equation (16.14) in a DFT, as outlined below.

Step 1. Define a real (symmetric odd) DFT of length $2N$ using the given $N-1$ real-valued data set $\boldsymbol{x} = [\, x_1, x_2, \ldots, x_{N-1} \,]$, i.e.,

$$(16.16) \qquad X_r = \sum_{\ell=0}^{2N-1} x_\ell \omega_{2N}^{r\ell}, \quad r = 0, 1, \ldots, 2N-1,$$

where $\omega_{2N} = e^{-\frac{2j\pi}{2N}} = \cos\left(\frac{\pi}{N}\right) - j\sin\left(\frac{\pi}{N}\right)$, $x_0 \equiv x_N \equiv 0$ and the added x_{N+1}, $x_{N+2}, \ldots, x_{2N-1}$ are chosen to be symmetric odd about x_N, i.e.,

$$(16.17) \qquad x_{2N-\ell} \equiv -x_\ell, \quad \ell = 1, \ldots, N.$$

Step 2. Derive the DST of length $N-1$ embedded inside the DFT of length $2N$ in (16.16) as shown below. Observe that $\boldsymbol{X} = \mathrm{DFT}(\boldsymbol{x})$ contains only pure imaginary numbers if the data sequence in $\boldsymbol{x}$ is real-valued and symmetric odd, with $X_0 = X_N = 0$.

(16.18)

$$
\begin{aligned}
X_r &= \sum_{\ell=0}^{2N-1} x_\ell \omega_{2N}^{r\ell}, \quad r = 0, 1, \ldots, 2N-1, \\[1mm]
&= \sum_{\ell=0}^{N-1} x_\ell \omega_{2N}^{r\ell} + \sum_{\ell=N}^{2N-1} x_\ell \omega_{2N}^{r\ell} \\[1mm]
&= \sum_{\ell=0}^{N-1} x_\ell \omega_{2N}^{r\ell} + \sum_{\ell=1}^{N} x_{2N-\ell} \omega_{2N}^{r(2N-\ell)} \\[1mm]
&= \sum_{\ell=0}^{N-1} x_\ell \omega_{2N}^{r\ell} + \omega_{2N}^{r(2N)} \sum_{\ell=1}^{N} (-x_\ell)\, \omega_{2N}^{-r\ell} \quad (\because x_{2N-\ell} = -x_\ell) \\[1mm]
&= \sum_{\ell=0}^{N-1} x_\ell \omega_{2N}^{r\ell} - \sum_{\ell=1}^{N-1} x_\ell \omega_{2N}^{-r\ell} \quad (\because \omega_{2N}^{2N} = 1, \text{ and } x_N \equiv 0 \text{ was removed}) \\[1mm]
&= \sum_{\ell=1}^{N-1} x_\ell \omega_{2N}^{r\ell} - \sum_{\ell=1}^{N-1} x_\ell \omega_{2N}^{-r\ell} \quad (\because x_0 \equiv 0 \text{ was removed}) \\[1mm]
&= \sum_{\ell=1}^{N-1} x_\ell \left(\omega_{2N}^{r\ell} - \omega_{2N}^{-r\ell} \right) \\[1mm]
&= -2j \sum_{\ell=1}^{N-1} x_\ell \sin\left(\frac{\pi r\ell}{N}\right) \quad (\because \sin(0) = \sin(\ell\pi) = 0, \therefore X_0 = X_N = 0) \\[1mm]
&= -2j \times S_r. \quad (\because S_r \text{ is real-valued}, \therefore X_r \text{ is pure imaginary})
\end{aligned}
$$

Following (16.18), if $[\, S_1, S_2, \ldots, S_{N-1} \,] = \mathrm{DST}\left([\, x_1, x_2, \ldots, x_{N-1} \,] \right)$ denotes the desired results of a discrete sine transform, and $[\, X_0, X_1, \ldots, X_{2N-1} \,] = \mathrm{DFT}\left([\, x_0, x_1, \ldots, x_{2N-1} \,] \right)$ denotes the results of the discrete Fourier transform defined in Step 1, then

$$
\begin{aligned}
[\, S_1, S_2, \ldots, S_{N-1} \,] &= \frac{j}{2}\, [\, X_1, X_2, \ldots, X_{N-1} \,] \\[1mm]
(16.19) \qquad &= \frac{j}{2}\, [\, j\mathrm{Im}\,(X_1), j\mathrm{Im}\,(X_2), \ldots, j\mathrm{Im}\,(X_{N-1}) \,] \\[1mm]
&= -\frac{1}{2}\, [\, \mathrm{Im}\,(X_1), \mathrm{Im}\,(X_2), \ldots, \mathrm{Im}\,(X_{N-1}) \,].
\end{aligned}
$$

Note that because X is pure imaginary, the real-valued S is obtained by dividing $\text{Im}(X)$ by -2 in the actual computation. Hence, S is immediately available after X is computed using a fast algorithm to implement the real symmetric odd DFT defined in Step 1.

To summarize, the task of computing a DST of $N{-}1$ real-valued data items can be accomplished by computing a "real" DFT of length $2N$, which can be implemented by the FFT algorithm specifically tailored to real-valued data described in Chapter 14.

Finally, the same fast algorithm also computes the inverse discrete sine transform (IDST), because the sine transform is its own inverse. The precise relationship

$$\boxed{\;x = \text{IDST}(S) = \tfrac{2}{N}\text{DST}(S) \text{ if and only if } S = \text{DST}(x)\;}$$

can be easily derived through applying the inverse discrete Fourier transform (IDFT) to $X = -2jS$, and noting that $X_{N-r} = X_r^{*}$ (see Chapter 14) implies that X is pure imaginary and $X_{N-r} = -X_r$. (In other words, $\text{Im}(X)$ is a real symmetric odd sequence.) This is shown below.

(16.20)

$$x_\ell = \frac{1}{2N}\sum_{r=0}^{2N-1} X_r \omega_{2N}^{-r\ell}, \quad \ell = 0,1,\ldots,2N-1,$$

$$= \frac{1}{2N}\left(\sum_{r=0}^{N-1} X_r \omega_{2N}^{-r\ell} + \sum_{r=N}^{2N-1} X_r \omega_{2N}^{-r\ell}\right)$$

$$= \frac{1}{2N}\left(\sum_{r=0}^{N-1} X_r \omega_{2N}^{-r\ell} + \sum_{r=1}^{N} X_{2N-r}\omega_{2N}^{-\ell(2N-r)}\right)$$

$$= \frac{1}{2N}\left(\sum_{r=0}^{N-1} X_r \omega_{2N}^{-r\ell} + \omega_{2N}^{-\ell(2N)}\sum_{r=1}^{N}(-X_r)\omega_{2N}^{r\ell}\right) \quad (\because X_{2N-r} = -X_r)$$

$$= \frac{1}{2N}\left(\sum_{r=0}^{N-1} X_r \omega_{2N}^{-r\ell} - \sum_{r=1}^{N-1} X_r \omega_{2N}^{r\ell}\right) \quad (\because \omega_{2N}^{2N} = 1, \text{ and } X_N = 0 \text{ was removed})$$

$$= \frac{j}{N}\sum_{r=1}^{N-1} S_r\left(\omega_{2N}^{r\ell} - \omega_{2N}^{-r\ell}\right) \quad (\because X_0 = 0 \text{ was removed, and } X_r = -2jS_r)$$

$$= \frac{2}{N}\sum_{r=1}^{N-1} S_r \sin\left(\frac{\pi r\ell}{N}\right) \quad (\because S_r\text{'s are real-valued, } \therefore x_\ell \text{ is real.})$$

Thus $x = \tfrac{2}{N}\text{DST}(S)$ if and only if $S = \text{DST}(x)$ as desired. The definition of the DST in (16.14) may also be viewed as multiplying the data sequence $x = [\,x_1, x_2, \ldots, x_{N-1}\,]$ by an $(N{-}1)\times(N{-}1)$ DST coefficient matrix $M_{\sin}$, as shown below for $N = 5$.

$$M_{\sin}x = \begin{bmatrix} \sin\theta & \sin 2\theta & \sin 3\theta & \sin 4\theta \\ \sin 2\theta & \sin 4\theta & \sin 6\theta & \sin 8\theta \\ \sin 3\theta & \sin 6\theta & \sin 9\theta & \sin 12\theta \\ \sin 4\theta & \sin 8\theta & \sin 12\theta & \sin 16\theta \end{bmatrix} \begin{bmatrix} x_1 \\ x_2 \\ x_3 \\ x_4 \end{bmatrix} = \begin{bmatrix} S_1 \\ S_2 \\ S_3 \\ S_4 \end{bmatrix}, \quad \theta \equiv \frac{\pi}{N} = \frac{\pi}{5}.$$

Therefore, the result $\text{DST}(S) = \tfrac{N}{2}x$ also establishes the orthogonality of the columns of $M_{\sin}$ as shown below.

$$M_{\sin} S = M_{\sin} (M_{\sin} x)$$
$$= (M_{\sin} M_{\sin}) x$$
(16.21)
$$= \frac{N}{2} x,$$

which implies that

(16.22) $$M_{\sin} M_{\sin} = \frac{N}{2} I,$$

where $M_{\sin}$ is the $(N-1) \times (N-1)$ DST coefficient matrix, and I is the $(N-1) \times (N-1)$ identity matrix.

16.7   Fast Discrete Cosine Transform

One can also "embed" the discrete cosine transform (DCT) of length $N+1$ defined by (16.15) in a DFT.

Step 1. Define a real (symmetric even) DFT of length $2N$ using the given $N+1$ real-valued data items in $x = [x_0, x_1, \ldots, x_N]$, i.e.,

(16.23) $$X_r = \sum_{\ell=0}^{2N-1} x_\ell \omega_{2N}^{r\ell}, \quad r = 0, 1, \ldots, 2N - 1,$$

where $\omega_{2N} = e^{-\frac{2j\pi}{2N}} = \cos\left(\frac{\pi}{N}\right) - j\sin\left(\frac{\pi}{N}\right)$, and the added $x_{N+1}, x_{N+2}, \ldots, x_{2N-1}$ are chosen to be symmetric even about x_N, i.e.,

(16.24) $$x_{2N-\ell} \equiv x_\ell, \quad \ell = 1, \ldots, N.$$

Step 2. Derive the DCT of length $N+1$ embedded inside the DFT of length $2N$ in (16.23) as shown below. Observe also that $X = \text{DFT}(x)$ is shown to be real-valued if the data sequence in $\boldsymbol{x}$ is real-valued and symmetric even.

$$
\begin{aligned}
X_r &= \sum_{\ell=0}^{2N-1} x_\ell \omega_{2N}^{r\ell}, \quad r = 0, 1, \ldots, 2N-1, \\
&= \sum_{\ell=0}^{N-1} x_\ell \omega_{2N}^{r\ell} + \sum_{\ell=N}^{2N-1} x_\ell \omega_{2N}^{r\ell} \\
&= \sum_{\ell=0}^{N-1} x_\ell \omega_{2N}^{r\ell} + \sum_{\ell=1}^{N} x_{2N-\ell} \omega_{2N}^{r(2N-\ell)} \\
&= \sum_{\ell=0}^{N-1} x_\ell \omega_{2N}^{r\ell} + \omega_{2N}^{r(2N)} \sum_{\ell=1}^{N} x_\ell \omega_{2N}^{-r\ell} \quad (\because x_{2N-\ell} = x_\ell) \\
&= \sum_{\ell=0}^{N-1} x_\ell \omega_{2N}^{r\ell} + \sum_{\ell=1}^{N} x_\ell \omega_{2N}^{-r\ell} \quad (\because \omega_{2N}^{2N} = 1.) \\
&= x_0 + \left(\sum_{\ell=1}^{N-1} x_\ell \omega_{2N}^{r\ell} + \sum_{\ell=1}^{N-1} x_\ell \omega_{2N}^{-r\ell} \right) + x_N \omega_{2N}^{-rN} \\
&= x_0 + (-1)^r x_N + \sum_{\ell=1}^{N-1} x_\ell \left(\omega_{2N}^{r\ell} + \omega_{2N}^{-r\ell} \right) \quad (\because \omega_{2N}^{N} = -1) \\
&= 2 \left(\frac{x_0 + (-1)^r x_N}{2} + \sum_{\ell=1}^{N-1} x_\ell \cos\left(\frac{\pi r \ell}{N} \right) \right) \\
&= 2C_r. \quad \text{(Note: } C_r \text{ and } X_r \text{ are real-valued.)}
\end{aligned}
$$

(16.25)

Following (16.25), if $[C_0, C_1, \ldots, C_N] = \text{DCT}([x_0, x_1, \ldots, x_N])$ denotes the desired results of a discrete cosine transform, and $[X_0, X_1, \ldots, X_{2N-1}] = \text{DFT}([x_0, x_1, \ldots, x_{2N-1}])$ denotes the results of the discrete Fourier transform defined in Step 1, then

(16.26)
$$
[C_0, C_1, \ldots, C_N] = \frac{1}{2}[X_0, X_1, \ldots, X_N].
$$

Similar to the computation of the DST, the task of computing a DCT of $N+1$ data items can be accomplished by computing a real DFT of length $2N$, which can be implemented by the $\Theta(N \log_2 N)$ FFT algorithm specifically tailored to real data described in Chapter 14.

Since the X_r and X_{N-r} are complex conjugate pairs and $\boldsymbol{X}$ is real, $X_r = X_{N-r}$ and one can conclude that $\boldsymbol{X}$ contains a real symmetric even sequence. Similar to the derivation of the IDST, one can apply an IDFT to $\boldsymbol{X} = 2\boldsymbol{C}$ to derive the inverse discrete cosine transform (IDCT), which satisfies the relationship

$$
\boxed{\boldsymbol{x} = \text{IDCT}(\boldsymbol{C}) = \frac{2}{N}\text{DCT}(\boldsymbol{C}) \text{ if and only if } \boldsymbol{C} = \text{DCT}(\boldsymbol{x})} \quad .
$$

Therefore, the same fast algorithm may be used to compute the IDCT because a DCT is also its own inverse. Again, the definition of the DCT in (16.15) may also be viewed as multiplying the data sequence $\boldsymbol{x} = [x_0, x_1, \ldots, x_N]$ by a $(N+1)\times(N+1)$ DCT coefficient matrix $M_{\cos}$, as shown below for $N = 5$.

$$
\begin{bmatrix}
C_0 \\
C_1 \\
C_2 \\
C_3 \\
C_4 \\
C_5
\end{bmatrix}
=
\begin{bmatrix}
1/2 & 1 & 1 & 1 & 1 & 1/2 \\
1/2 & \cos\theta & \cos 2\theta & \cos 3\theta & \cos 4\theta & -1/2 \\
1/2 & \cos 2\theta & \cos 4\theta & \cos 6\theta & \cos 8\theta & 1/2 \\
1/2 & \cos 3\theta & \cos 6\theta & \cos 9\theta & \cos 12\theta & -1/2 \\
1/2 & \cos 4\theta & \cos 8\theta & \cos 12\theta & \cos 16\theta & 1/2 \\
1/2 & -1 & 1 & -1 & 1 & -1/2
\end{bmatrix}
\begin{bmatrix}
x_0 \\
x_1 \\
x_2 \\
x_3 \\
x_4 \\
x_5
\end{bmatrix}
, \quad \theta \equiv \frac{\pi}{N} = \frac{\pi}{5}.
$$

Therefore, the result $\mathrm{DCT}(C) = \frac{N}{2}x$ also establishes the orthogonality of the columns of $M_{\cos}$, i.e.,

$$
(16.27) \qquad\qquad M_{\cos} M_{\cos} = \frac{N}{2}I,
$$

where $M_{\cos}$ is the $(N+1)\times(N+1)$ DCT coefficient matrix, and I is the $(N+1)\times(N+1)$ identity matrix.

The columns of the matrix $M_{\cos}$ are also called the basis vectors of the standard discrete cosine transform of Type I (DCT-I) in [78, 91, 108]. An alternative and insightful way to prove their orthogonality is to show them to be the eigenvectors of a symmetric second-difference matrix [91]. By varying the boundary conditions of the difference equations, Strang shows in [91] that the complete set of DCT-Types I to VIII (discovered by Wang and Hunt in 1985 in [108]) can be derived from the eigenvectors of eight second-difference matrices. Readers are referred to [78, 91, 108] for further discussion on deriving other types of DCTs and their application in image processing.

16.8   Fast Discrete Hartley Transform

The real-valued discrete Hartley transform (DHT) and its inverse (IDHT) are defined for N real-valued data items $x = [x_0, x_1, \ldots, x_{N-1}]$ as

$$
(16.28) \qquad\qquad H_r = \sum_{\ell=0}^{N-1} x_\ell \, \mathrm{cas}\left(\frac{2\pi r\ell}{N}\right), \quad r = 0, 1, \ldots, N-1,
$$

and

$$
(16.29) \qquad\qquad x_\ell = \frac{1}{N} \sum_{r=0}^{N-1} H_r \, \mathrm{cas}\left(\frac{2\pi r\ell}{N}\right), \quad \ell = 0, 1, \ldots, N-1,
$$

where $\mathrm{cas}\left(\frac{2\pi r\ell}{N}\right) \equiv \cos\left(\frac{2\pi r\ell}{N}\right) + \sin\left(\frac{2\pi r\ell}{N}\right)$. The real-valued Hartley transform is its own inverse as highlighted below, and the same fast algorithm may be used to compute either the forward or the inverse Hartley transforms. Since the IDHT can be easily derived through the use of IDFT once the relationship between the DHT and the DFT is established, the former task is left as an exercise for the reader.

$$
\boxed{x = \mathrm{IDHT}(H) = \tfrac{1}{N}\mathrm{DHT}(H) \text{ if and only if } H = \mathrm{DHT}(x)}.
$$

In exactly the same manner as the forward and inverse DST (or DCT) are related to the orthogonality of the columns of DST (or DCT) coefficient matrices, the columns of M_{DHT} from (16.28) can be shown to be orthogonal, and satisfy

$$
(16.30) \qquad\qquad M_{\mathrm{DHT}} M_{\mathrm{DHT}} = N I,
$$

where M_{DHT} is the $N \times N$ DHT coefficient matrix, and I is the $N \times N$ identity matrix.

Comparing the H_r from equation (16.28) to the X_r obtained from transforming the same real-valued data set by the DFT shown below,

(16.31)
$$X_r = \sum_{\ell=0}^{N-1} x_\ell \omega_N^{r\ell}, \quad r = 0, 1, \ldots, N-1,$$
$$= \sum_{\ell=0}^{N-1} x_\ell \left(\cos\left(\frac{2\pi r\ell}{N}\right) - j \sin\left(\frac{2\pi r\ell}{N}\right) \right)$$
$$= \sum_{\ell=0}^{N-1} x_\ell \cos\left(\frac{2\pi r\ell}{N}\right) - j \sum_{\ell=0}^{N-1} x_\ell \sin\left(\frac{2\pi r\ell}{N}\right)$$
$$= \text{Re}\,(X_r) + j\text{Im}\,(X_r),$$

one obtains the following identity which relates the real-valued H_r to the complex-valued X_r:

(16.32)
$$H_r = \text{Re}\,(X_r) - \text{Im}\,(X_r), \quad r = 0, 1, \ldots, N-1.$$

Using (16.32), H is easily obtained by post-processing X, which can be computed using the FFT algorithm introduced in Chapter 14 for real-valued data of length N.

Since the DHT is a real-valued transform, it is desirable that complex arithmetic be avoided. To this end, specialized fast Hartley transform (FHT) algorithms have been developed that use only real arithmetic. Note that the roles played by the DFT and the DHT on transforming real-valued data are *reversible*. That is, if the real-valued $H = \text{DHT}(x)$ has been computed using a specialized FHT, the complex-valued $X = \text{DFT}(x)$ can be obtained using the identities $X_0 = H_0$, $X_{\frac{N}{2}} = -X_0$,

(16.33)
$$\text{Re}\,(X_r) = \frac{1}{2}\,(H_{N-r} + H_r), \quad r = 1, 2, \ldots, N/2 - 1,$$

and

(16.34)
$$\text{Im}\,(X_r) = \frac{1}{2}\,(H_{N-r} - H_r),$$

which are easily derived using the symmetric property of X.

Recall from Chapter 14 that if $X = \text{DFT}(x)$ for real-valued $x = [x_0, x_1, \ldots, x_{N-1}]$, then X_{N-r} and X_r are complex conjugates, i.e.,

(16.35)
$$X_{N-r} = X_r^*, \quad r = 1, 2, \ldots, N/2 - 1.$$

It follows that

(16.36)
$$\text{Re}\,(X_{N-r}) = \text{Re}\,(X_r) \quad \text{and} \quad \text{Im}\,(X_{N-r}) = -\text{Im}\,(X_r).$$

Replacing r by $N - r$ in (16.32), and applying the symmetric relationship from (16.36) to the real and imaginary parts of X_{N-r} yields

(16.37)
$$H_{N-r} = \text{Re}\,(X_{N-r}) - \text{Im}\,(X_{N-r}), \quad r = 1, 2, \ldots, N/2 - 1,$$
$$= \text{Re}\,(X_r) + \text{Im}\,(X_r).$$

Combining the H_r in (16.32) and the H_{N-r} in (16.37) for $1 \leq r \leq N/2-1$, one obtains the expression for Re (X_r) and Im (X_r) given in (16.33). Next, one obtains the relations $X_0 = H_0$ and $X_{\frac{N}{2}} = -X_0$ by substituting $r = 0$ in the definition of X_r and H_r, and the remaining $X_{\frac{N}{2}+1}$, $X_{\frac{N}{2}+2}$, ..., X_{N-1} can be obtained by applying the symmetric property in (16.36) to the available X_1, X_2, ..., $X_{\frac{N}{2}-1}$.

Additional information about the Hartley transform (continuous or discrete) and its applications in digital filtering can be found in the text by Bracewell [13] and in articles by Bracewell [14], Buneman [19], Sorensen et al. [87], and Duhamel and Vetterli [40]. The use of the Hartley transform in geophysical applications is described in [82].

16.9   Fast Chebyshev Approximation

The fast Fourier transform algorithm is useful in finding the best uniform approximation to a continuous function by Chebyshev polynomials. This is not surprising because the discrete Chebyshev transform is closely related to the discrete cosine transform, and the latter has been shown to be related to the DFT. A brief review of Chebyshev polynomials follows.

The family of Chebyshev polynomials is defined by

$$T_N(x) = \cos N\theta, \quad \text{where } \theta = \arccos(x), \text{ i.e., } x = \cos \theta, \ N = 0, 1, \ldots$$

By noting that $\cos N\theta = \text{Re}\left((\cos \theta + j \sin \theta)^N\right)$ and that

(16.38)
$$(\cos \theta + j \sin \theta)^N = \left(\cos \theta + j\sqrt{1 - \cos^2 \theta}\right)^N, \quad \text{assuming } \sin \theta \geq 0$$
$$= \left(x + j\sqrt{1 - x^2}\right)^N, \quad 0 \leq x^2 = \cos^2 \theta \leq 1,$$

one may obtain the Chebyshev polynomial of degree N in x from the binomial expansion of the right-hand side of (16.38), i.e.,

$$T_N(x) = \cos N(\arccos x)$$

(16.39)
$$= \text{Re}\left(\left(x + j\sqrt{1 - x^2}\right)^N\right)$$
$$= x^N + \binom{N}{2} x^{N-2} (x^2 - 1) + \binom{N}{4} x^{N-4} (x^2 - 1)^2 + \cdots$$

To generate the Chebyshev polynomials in increasing degrees, it is convenient to combine the trigonometric identities

$$T_{N+1}(x) = \cos(N + 1)\theta = \cos N\theta \cos \theta - \sin N\theta \sin \theta = xT_N(x) - \sin N\theta \sin \theta$$

and

$$T_{N-1}(x) = \cos(N - 1)\theta = \cos N\theta \cos \theta + \sin N\theta \sin \theta = xT_N(x) + \sin N\theta \sin \theta$$

to obtain the following three-term recurrence.

(16.40) $T_{N+1}(x) = 2xT_N(x) - T_{N-1}(x), \quad N = 1, 2, \ldots$

Given below are sample Chebyshev polynomials of degrees $0, 1, 2, \ldots, N$. While they all satisfy (16.39), the $T_N(x)$'s are shown to be generated using (16.40) when $N \geq 2$. For $-1 \leq x \leq 1$,

$$T_0(x) = \cos 0 = 1$$
$$T_1(x) = \cos \theta = x$$
$$T_2(x) = 2xT_1(x) - T_0(x) = 2x^2 - 1$$
$$T_3(x) = 2xT_2(x) - T_1(x) = 4x^3 - 3x$$
$$T_4(x) = 2xT_3(x) - T_2(x) = 8x^4 - 8x^2 + 1$$

$$\vdots$$

$$T_N(x) = 2^{N-1}x^N + \text{terms of lower degrees}, \quad N \geq 2.$$

Chebyshev polynomials are an important family of orthogonal polynomials, and they possess numerous interesting properties. A thorough discussion is beyond the scope of this book; the objective here is simply to connect the DFT to the usage of Chebyshev polynomials in approximating continuous functions.

Chebyshev abscissae The Chebyshev polynomial $T_N(x) = 0$ has N distinct roots at

$$x_k = \cos \theta_k = \cos \frac{(2k+1)\pi}{2N}, \quad k = 0, 1, \ldots, N-1.$$

It is a simple task to verify that $T_N(x_k) = \cos N\theta_k = 0$, and $0 \leq \theta_k < 2\pi$. They are commonly referred to as the "Chebyshev abscissae."

Chebyshev extrema Since $T_N(x) = \cos N(\arccos x)$, it has the property that $-1 \leq T_N(x) \leq 1$. Furthermore, $T_N(x)$ reaches its $N+1$ extrema, i.e., $T_N(x_r) = \pm 1$, at

$$x_r = \cos \theta_r = \cos \frac{r\pi}{N}, \quad r = 0, 1, \ldots, N,$$

because $T_N(x_r) = \cos N\theta_r = \cos r\pi = \pm 1$.

The connection between Chebyshev approximation and the DFT can now be developed. Since the Chebyshev polynomials $T_N(x)$ are defined for $x \in [-1, 1]$, in order to approximate a continuous function $f(x)$ for $x \in [a, b]$, a pre-processing step is needed to change $f(x)$ to $g(y)$ by substituting

$$x = \frac{a+b}{2} + \frac{a-b}{2}y$$

so that $x \in [a, b]$ implies $y \in [-1, 1]$, and $f(x_k) = g(y_k)$ when x_k and y_k satisfy the relationship above. Instead of approximating $f(x)$, one now approximates the continuous function $g(y)$ for $y \in [-1, 1]$ by a linear combination of Chebyshev polynomials. For a given N,

$$g(y) \approx \sum_{\ell=0}^{N} c_\ell T_\ell(y) = \sum_{\ell=0}^{N} c_\ell \cos \ell\theta, \quad \theta = \arccos y,$$

where the coefficients c_ℓ are the unknowns to be determined.

One may solve for the $N+1$ unknown coefficients by setting up a linear systems of $N+1$ equations. Evaluating $Y_r = g(y_r)$ at $y_r = \cos(r\pi/N)$, for $r = 0, 1, \ldots, N$, one obtains the discrete cosine transform (DCS) as shown below.

$$Y_r = g\left(y_r\right), \qquad r = 0, 1, \ldots, N$$

$$= g\left(\cos\frac{r\pi}{N}\right)$$

$$= \sum_{\ell=0}^{N} c_\ell \cos \ell\left(\frac{r\pi}{N}\right)$$

$$= \frac{\hat{c}_0 + (-1)^r \hat{c}_N}{2} + \sum_{\ell=1}^{N-1} c_\ell \cos\frac{\pi r \ell}{N},$$

where $\hat{c}_0 = 2c_0$, and $\hat{c}_N = 2c_N$. Hence, $\hat{c}_0, c_1, c_2, \ldots, c_{N-1}, \hat{c}_N$ can be obtained by applying the inverse discrete cosine transform (IDCS) to the $Y_r = g(y_r)$. Recall that a forward or an inverse DCS of length $N+1$ may be embedded in a DFT of length $2N$ and computed by the FFT algorithm.

16.10 Solving Difference Equations

The use of the DST, DCT, and DFT algorithms in solving difference equations in the form of initial value problems or boundary value problems is another application area. A survey of difference equations, some specific applications in which boundary value problems arise, and the connections between certain difference equations and the DFT can be found in [15]. Readers are assumed to have some familiarity with difference equations.

Consider a boundary value problem specified by the following nonhomogeneous second-order difference equation with real constant coefficients from [15]:

$$(16.41) \qquad au_{r+1} = -bu_r - au_{r-1} + f_r, \qquad r = 1, \ldots, N-1.$$

The boundary condition is assumed to be Dirichlet with $u_0 = u_N = 0$.

For $N = 5$ the unknowns are u_1, u_2, u_3, and u_4 in (16.41). By rewriting (16.41) as

$$(16.42) \qquad au_{r-1} + bu_r + au_{r+1} = f_r, \qquad r = 1, \ldots, N-1,$$

the four equations defined by (16.42), together with $u_0 = u_5 = 0$, form the symmetric tridiagonal system as shown below.

$$(16.43) \qquad \begin{bmatrix} b & a & & \\ a & b & a & \\ & a & b & a \\ & & a & b \end{bmatrix} \begin{bmatrix} u_1 \\ u_2 \\ u_3 \\ u_4 \end{bmatrix} = \begin{bmatrix} f_1 \\ f_2 \\ f_3 \\ f_4 \end{bmatrix}.$$

The unknown vector u may be obtained using a direct method such as Gaussian elimination with partial pivoting; alternatively, one may use the Cholesky method if the coefficient matrix is symmetric positive definite. Since the cost of solving a tridiagonal system of N equations is $\Theta(N)$ using a direct method, whereas the cost of using a

FFT-based method introduced below is $\Theta\left(N \log_2 N\right)$, it makes no sense to use this FFT-based method for one-dimensional problems.

However, the FFT-based method is useful for solving multidimensional boundary value problems. For example, block tridiagonal systems of the form

$$(16.44) \qquad \begin{bmatrix} T & -I & & & \\ -I & T & -I & & \\ & \ddots & \ddots & \ddots & \\ & & -I & T & -I \\ & & & -I & T \end{bmatrix} \begin{bmatrix} u_1 \\ u_2 \\ \vdots \\ u_{N-1} \\ u_N \end{bmatrix} = \begin{bmatrix} f_1 \\ f_2 \\ \vdots \\ f_{N-1} \\ f_N \end{bmatrix},$$

where T is an $N \times N$ symmetric tridiagonal matrix (which has the same structure as that of T in the 1D example), I is an $N \times N$ identity matrix, and each vector u_r and f_r is of length N, result from finite difference approximation to a two-dimensional Poisson equation with Dirichlet boundary conditions on an $N \times N$ square grid. The evolution of fast (non-iterative) Poisson solvers has a long history which began with the odd-even reduction method and the Fourier analysis method proposed by Hockney in 1965 [16, 21, 35, 53, 93, 97]. Instead of solving (16.44) using the block LU factorization [54] at the cost of $\Theta\left(N^4\right)$, one may use the Buneman variant of the cyclic odd-even reduction algorithm [18, 21] or the FFT-based matrix decomposition method [21] at a significantly reduced cost of $\Theta\left(N^2 \log_2 N\right)$. Both algorithms have been generalized for arbitrary N [11, 17, 97].

To help explain how the FFT-based method may be used to solve the block tridiagonal system (16.44), the simple tridiagonal system (16.43) is considered first.

Example 16.6 Explaining the FFT-based method by a simple example. The connection between the 1D difference equation (16.41) and the DST was noted and an FFT-based algorithm was developed in [15]. This algorithm is described below.

A trial solution vector u is assumed to be a linear combination of the $N-1$ DST basis vectors, i.e.,

$$u = \sum_{\ell=1}^{N-1} x_\ell u^{(\ell)}, \quad \text{where} \quad u_r^{(\ell)} = \sin\left(\frac{\pi r \ell}{N}\right).$$

Observe that $u_0 = u_N = 0$ because $u_0^{(\ell)} = \sin 0 = 0$, and $u_N^{(\ell)} = \sin \pi \ell = 0$ are satisfied for $1 \le \ell \le N-1$. If this trial solution is substituted into the difference equation, then the x_ℓ are the unknown coefficients to be determined. For $N = 5$, the trial solution vector is

$$\begin{bmatrix} u_1 \\ u_2 \\ u_3 \\ u_4 \end{bmatrix} = \sum_{\ell=1}^{N-1=4} x_\ell \begin{bmatrix} u_1^{(\ell)} \\ u_2^{(\ell)} \\ u_3^{(\ell)} \\ u_4^{(\ell)} \end{bmatrix} = \sum_{\ell=1}^{N-1=4} x_\ell \begin{bmatrix} \sin \ell\theta \\ \sin 2\ell\theta \\ \sin 3\ell\theta \\ \sin 4\ell\theta \end{bmatrix},$$

where $\theta = \frac{\pi}{N} = \frac{\pi}{5}$, and $u_0 = 0$, $u_5 = 0$. The equation above, written out in full, is

$$\begin{bmatrix} u_1 \\ u_2 \\ u_3 \\ u_4 \end{bmatrix} = \begin{bmatrix} \sin \theta & \sin 2\theta & \sin 3\theta & \sin 4\theta \\ \sin 2\theta & \sin 4\theta & \sin 6\theta & \sin 8\theta \\ \sin 3\theta & \sin 6\theta & \sin 9\theta & \sin 12\theta \\ \sin 4\theta & \sin 8\theta & \sin 12\theta & \sin 16\theta \end{bmatrix} \begin{bmatrix} x_1 \\ x_2 \\ x_3 \\ x_4 \end{bmatrix},$$

and the trial solution vector u is related to the vector x by $u = M_{\sin} x$.

The derivation above suggests that one may obtain u by applying the DST to x. Replacing u by $M_{\sin} x$ in the tridiagonal system (16.43), one obtains the following system of equations in x.

(16.45) $$T\left(M_{\sin} x\right) = f,$$

where

$$T = \begin{bmatrix} b & a & & \\ a & b & a & \\ & a & b & a \\ & & a & b \end{bmatrix}, \quad \text{and } M_{\sin} x = \begin{bmatrix} \sum_{\ell=1}^{4} x_{\ell} \sin \ell\theta \\ \sum_{\ell=1}^{4} x_{\ell} \sin 2\ell\theta \\ \sum_{\ell=1}^{4} x_{\ell} \sin 3\ell\theta \\ \sum_{\ell=1}^{4} x_{\ell} \sin 4\ell\theta \end{bmatrix}.$$

Substituting the trial solutions

$$u_{r-1} = \sum_{\ell=1}^{N-1} x_{\ell} u_{r-1}^{(\ell)} = \sum_{\ell=1}^{N-1} x_{\ell} \sin(r-1)\ell\theta, \quad \theta = \frac{\pi}{N},$$

$$u_r = \sum_{\ell=1}^{N-1} x_{\ell} u_r^{(\ell)} = \sum_{\ell=1}^{N-1} x_{\ell} \sin r\ell\theta,$$

$$u_{r+1} = \sum_{\ell=1}^{N-1} x_{\ell} u_{r+1}^{(\ell)} = \sum_{\ell=1}^{N-1} x_{\ell} \sin(r+1)\ell\theta,$$

into the rth difference equation in (16.42) gives

$$a \sum_{\ell=1}^{N-1} x_{\ell} \sin(r-1)\ell\theta + b \sum_{\ell=1}^{N-1} x_{\ell} \sin r\ell\theta + a \sum_{\ell=1}^{N-1} x_{\ell} \sin(r+1)\ell\theta = f_r, \quad 1 \le r \le N-1.$$

Collecting the terms with the same coefficients yields

$$a \sum_{\ell=1}^{N-1} x_{\ell}\left(\sin(r-1)\ell\theta + \sin(r+1)\ell\theta\right) + b \sum_{\ell=1}^{N-1} x_{\ell} \sin r\ell\theta = f_r, \quad 1 \le r \le N-1.$$

The expression $\sin(r-1)\ell\theta + \sin(r+1)\ell\theta$ can be simplified to $2 \sin r\ell\theta \cos \ell\theta$ by using the identity $\sin(\alpha \pm \beta) = \sin \alpha \cos \beta \pm \sin \beta \cos \alpha$, which yields

(16.46) $$\sum_{\ell=1}^{N-1} (2a \cos \ell\theta + b) x_{\ell} \sin r\ell\theta = f_r, \quad 1 \le r \le N-1,$$

which is the rth equation of the following system of four equations.

$$M_{\sin} \begin{bmatrix} (2a \cos \theta + b) x_1 \\ (2a \cos 2\theta + b) x_2 \\ (2a \cos 3\theta + b) x_3 \\ (2a \cos 4\theta + b) x_4 \end{bmatrix} = M_{\sin} \Lambda x = \begin{bmatrix} f_1 \\ f_2 \\ f_3 \\ f_4 \end{bmatrix}, \quad \text{where}$$

$$
(16.47) \qquad \Lambda \equiv \begin{bmatrix} 2a\cos\theta + b & & & \\ & 2a\cos 2\theta + b & & \\ & & 2a\cos 3\theta + b & \\ & & & 2a\cos 4\theta + b \end{bmatrix}.
$$

Next, multiply both sides by $M_{\sin}^{-1}$, yielding

$$
\Lambda x_r = M_{\sin}^{-1} f_r.
$$

Therefore, after the right-hand side $X = M_{\sin}^{-1} f$ is computed by applying an inverse sine transform to f, one obtains x by solving the diagonal system $\Lambda x = X$ to obtain

$$
(16.48) \qquad x_\ell = \frac{X_\ell}{2a\cos \ell\theta + b}, \quad 1 \le \ell \le N - 1.
$$

To summarize, the $\Theta\,(N\log_2 N)$ algorithm for solving the difference equation (16.41) consists of the following three steps:

Step 1. Compute $X = \mathrm{IDST}(f) = \frac{2}{N}\mathrm{DST}(f)$ using an FFT algorithm.

Step 2. Compute

$$
x_\ell = \frac{X_\ell}{2a\cos \ell\theta + b}, \quad \theta = \frac{\pi}{N}, \ 1 \le \ell \le N - 1.
$$

Note that $2a\cos \ell\theta + b$ must be nonzero, which implies that T needs to be nonsingular. Generally the source of the problem will ensure this, or some even stronger condition such as positive definiteness or diagonal dominance.

Step 3. Compute $u = \mathrm{DST}(x)$ using an FFT algorithm.

Of course the various systems of equations need not be explicitly set up in the actual solution process; they were used as tools to explain the derivation of the fast algorithm. In particular, the symmetric tridiagonal system of $N-1$ equations

$$
Tu = f
$$

was first converted to the system

$$
TM_{\sin} x = f,
$$

and multiplying both sides by $M_{\sin}^{-1}$ yields

$$
M_{\sin}^{-1} T M_{\sin} x = M_{\sin}^{-1} f.
$$

The relation $TM_{\sin} x = M_{\sin}\Lambda x$ derived above implies that T is diagonalized by the similarity transformation

$$
(16.49) \qquad M_{\sin}^{-1} T M_{\sin} = \Lambda,
$$

or equivalently,

$$
(16.50) \qquad TM_{\sin} = M_{\sin}\Lambda.
$$

Thus, the columns of $M_{\sin}$ and the diagonal elements of Λ are eigenvectors and corresponding eigenvalues of the finite difference matrix T.

Example 16.7 A fast FFT-based method for solving a 2D Poisson equation on a square. The algorithm is developed to solve the block tridiagonal system (16.44). Guided by the development in the first example, each symmetric tridiagonal block T can be diagonalized by the DST matrix as shown above in (16.49), i.e., $M_{\sin}^{-1}TM_{\sin} = \Lambda$, where Λ was defined in (16.47).

The block tridiagonal system (16.44) contains N subsystems of N equations each, namely,

$$
\begin{aligned}
Tu_1 - u_2 &= f_1, \\
-u_{r-1} + Tu_r - u_{r+1} &= f_r, \quad r = 2,\ldots,N-1, \\
Tu_{N-1} - u_N &= f_N.
\end{aligned}
$$

(16.51)

Substituting $T = M_{\sin}\Lambda M_{\sin}^{-1}$ into the rth subsystem, one obtains

$$
-u_{r-1} + M_{\sin}\Lambda M_{\sin}^{-1}u_r - u_{r+1} = f_r.
$$

Next, multiply both sides by $M_{\sin}^{-1}$, yielding

$$
-M_{\sin}^{-1}u_{r-1} + \Lambda M_{\sin}^{-1}u_r - M_{\sin}^{-1}u_{r+1} = M_{\sin}^{-1}f_r.
$$

By defining $x_r = M_{\sin}^{-1}u_r$ and $X_r = M_{\sin}^{-1}f_r$ for $r = 1,2,\ldots,N$, equation (16.44) becomes

$$
\begin{aligned}
\Lambda x_1 - x_2 &= X_1, \\
-x_{r-1} + \Lambda x_r - x_{r+1} &= X_r, \quad r = 2,3,\ldots,N-1, \\
\Lambda x_{N-1} - x_N &= X_N,
\end{aligned}
$$

(16.52)

$$
\text{where} \quad \Lambda = \begin{bmatrix} d_1 & & & \\ & d_2 & & \\ & & \ddots & \\ & & & d_N \end{bmatrix}, \quad x_r = \begin{bmatrix} x_{1r} \\ x_{2r} \\ \vdots \\ x_{Nr} \end{bmatrix}, \quad \text{and} \quad X_r = \begin{bmatrix} X_{1r} \\ X_{2r} \\ \vdots \\ X_{Nr} \end{bmatrix}.
$$

These equations can be rearranged to yield a tridiagonal system of N^2 equations, as demonstrated below, and can be solved in $\Theta(N^2)$ time. For $N=3$ equation (16.52) is

(16.53)

$$
\begin{bmatrix}
d_1 & & & -1 & & & & & \\
& d_2 & & & -1 & & & & \\
& & d_3 & & & -1 & & & \\
-1 & & & d_1 & & & -1 & & \\
& -1 & & & d_2 & & & -1 & \\
& & -1 & & & d_3 & & & -1 \\
& & & -1 & & & d_1 & & \\
& & & & -1 & & & d_2 & \\
& & & & & -1 & & & d_3
\end{bmatrix}
\begin{bmatrix}
x_{11} \\ x_{21} \\ x_{31} \\ x_{12} \\ x_{22} \\ x_{32} \\ x_{13} \\ x_{23} \\ x_{33}
\end{bmatrix}
=
\begin{bmatrix}
X_{11} \\ X_{21} \\ X_{31} \\ X_{12} \\ X_{22} \\ X_{32} \\ X_{13} \\ X_{23} \\ X_{33}
\end{bmatrix},
$$

from which one obtains the following tridiagonal system by rearranging the unknowns and the equations.

(16.54)

$$
\begin{bmatrix}
d_1 & -1 & & & & & & & \\
-1 & d_1 & -1 & & & & & & \\
 & -1 & d_1 & -1 & & & & & \\
 & & -1 & d_2 & -1 & & & & \\
 & & & -1 & d_2 & -1 & & & \\
 & & & & -1 & d_2 & -1 & & \\
 & & & & & -1 & d_3 & -1 & \\
 & & & & & & -1 & d_3 & -1 \\
 & & & & & & & -1 & d_3
\end{bmatrix}
\begin{bmatrix}
x_{11} \\ x_{12} \\ x_{13} \\ x_{21} \\ x_{22} \\ x_{23} \\ x_{31} \\ x_{32} \\ x_{33}
\end{bmatrix}
=
\begin{bmatrix}
X_{11} \\ X_{12} \\ X_{13} \\ X_{21} \\ X_{22} \\ X_{23} \\ X_{31} \\ X_{32} \\ X_{33}
\end{bmatrix}.
$$

To summarize, the algorithm consists of the following three steps.

Step 1. Compute $X_r = M_{\sin}^{-1} f_r$, $r = 1, 2, \ldots, N$, by applying an inverse sine transform (IDST) to each f_r. The total cost is $\Theta\left(N^2 \log_2 N\right)$.

Step 2. Solve a tridiagonal system of N^2 equations to obtain x_r, $r = 1, 2, \ldots, N$. The total cost is $\Theta\left(N^2\right)$ using Gaussian elimination or Cholesky factorization.

Step 3. Compute $u_r = M_{\sin} x_r$, $r = 1, 2, \ldots, N$, by applying a sine transform (DST) to each x_r obtained in step 2. The total cost is $\Theta\left(N^2 \log_2 N\right)$.

Details of the algorithm vary with different boundary conditions; details about using DFT and other DFT-related transforms to solve difference equation BVPs with other types of boundary conditions can be found in Pickering [76] and Briggs and Henson [15]. The cyclic reduction method [18, 21], which is an alternative for solving block tridiagonal systems, is described in Golub and Van Loan [49]; a version generalized for arbitrary N is described in [97]. In [93], Swarztrauber combined the FFT-based Fourier Analysis (FA) method and the Cyclic Reduction (CR) method to obtain a $\Theta\left(N^2 \log_2 \log_2 N\right)$ hybrid FACR algorithm for solving discrete 2D Poisson equations on a square grid.

Part III

Parallel FFT Algorithms

Chapter 17

Parallelizing the FFTs: Preliminaries on Data Mapping

The discussion in Chapters 4 to 9 has focused on providing a unified algorithmic treatment of the NR, RN and NN variants for implementing the sequential radix-2 FFTs. These variations give the options of providing the input time series, and receiving the output frequencies, in either the "natural ordering" or the "bit-reversed ordering." Regardless of which of the three choices is made, one can choose to use either the "DIF" FFT algorithm or the "DIT" FFT algorithm.

The block diagram in Figure 17.1 depicts these various options.

Figure 17.1 Top level design chart for implementing the sequential FFT.

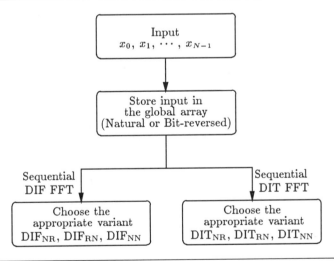

A key step in parallelizing the FFT on multiprocessor computers concerns *the mapping of array addresses to processors.* Figure 17.2 depicts such a process. Recall from the previous chapters that each of the NR, RN, and NN algorithms can be completely specified using the *n*-bit binary address of a representative element. In this chapter, this binary address notation will be used to facilitate the mapping of array locations to multiple processors, to aid in the description and classification of the many known parallel FFT algorithms, and to help in the development of new ones.

Figure 17.2 Top level design chart for implementing the parallel FFT.

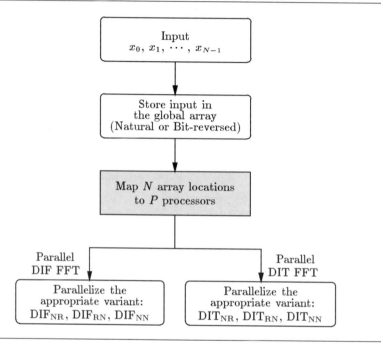

17.1   Mapping Data to Processors

Multiprocessors fall into two general categories: *shared-memory* multiprocessors and *local-memory* (or *distributed-memory*) multiprocessors. As their names imply, they are distinguished by whether each processor can directly access the entire memory available, or whether the memory is partitioned into portions which are private to each processor.

For shared-memory architectures, the main challenge in parallelizing a sequential algorithm is to subdivide the computation among the processor in such a way that the load is balanced, and memory conflicts are kept low. For FFT algorithms, this is a relatively simple task.

In terms of the design of algorithms, local-memory machines impose the additional burden of requiring that the data, as well as the computation, be partitioned. In addition to identifying parallelism in the computation, and assigning computational tasks to individual processors, the data associated with the computation must be distributed among the processors, and communicated among them as necessary. The challenge is to do this in such a way that each processor has the data it needs in its local memory at the time that it needs it, and the amount of communication required among the processors during the computation is kept acceptably low.

A useful way to define different partitionings is to associate each processor with a data item as follows. Since each location in a $N = 2^n$ element array has a n-bit binary address, and $P = 2^d$ processors can each be identified by a unique d-bit binary ID number, a class of partitionings can be specified by designating d *consecutive* bits from the n-bit address as the processor ID number as shown in Figure 17.3 for an example with $N = 32$ and $P = 4$.

This class of mappings is referred to as the generalized "Cyclic Block Mapping" with blocksize $= 2^i$ for $i = 0, 1, \cdots, n - d$. The $n - d + 1$ cyclic block mappings for $n = 5$ and $d = 2$ are illustrated in Figure 17.4, where the array locations mapped to processor P_0 are shaded to highlight the "cyclic" nature with various block sizes.

Figure 17.3 Mapping array locations to processors.

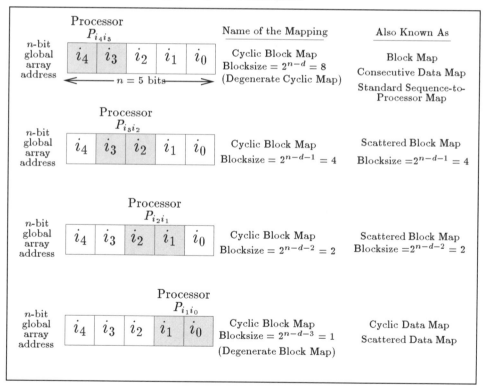

Mapping $N = 2^n = 32$ Array Locations
to $P = 2^d = 4$ Processors

17.2 Properties of Cyclic Block Mappings

The important properties of the class of cyclic block mappings are listed below. It is assumed that $N = 2^n$, $P = 2^d$, and element x_m is stored in $a[m]$, where $0 \leq m \leq N - 1$

Figure 17.4 The $n - d + 1$ cyclic block mappings for $N = 2^n = 32$ and $P = 2^d = 4$.

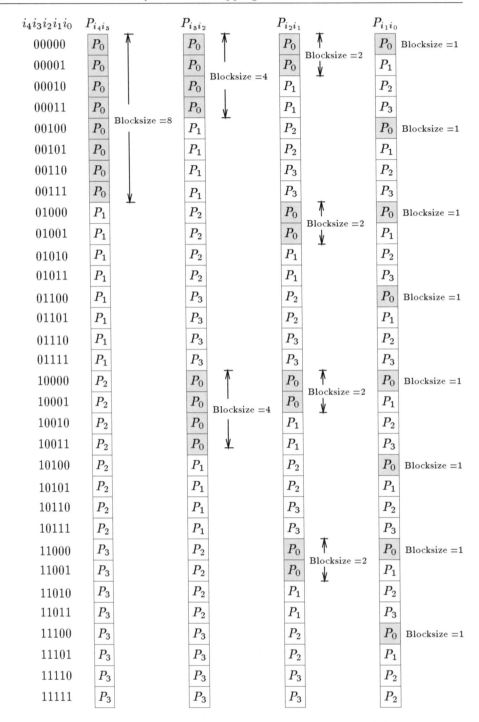

From E. Chu and A. George [28], *Linear Algebra and its Applications*, 284:95–124, 1998. With permission.

and the binary representation of $m = i_{n-1}i_{n-2}\cdots i_0$. The initial n-bit global array address used in the mapping is thus $i_{n-1}i_{n-2}\cdots i_0$. Although a natural ordering is assumed in this chapter so that the concept of data mapping can be introduced in a straightforward manner, the notations are readily adapted to other initial orderings, and those cases will be dealt with when they arise in the following chapters.

◆ **Property 1.** Each cyclic block mapping is defined by designating $i_k i_{k-1}\cdots i_{k-d+1}$ as processor ID number, where $k = n-1,\ n-2,\ \cdots,\ d-1$. There are thus $n-d+1$ different mappings.

◆ **Property 2.** The block size is 2^{k-d+1} for each k defined in property 1. Each mapping in this class can thus be uniquely identified by its block size.

◆ **Property 3.** When the left-most d bits are taken as the processor ID number, the block size is equal to $\frac{N}{P}$, and one has the standard block mapping, which is also known as consecutive data mapping.

◆ **Property 4.** When the right-most d bits are taken as the processor ID number, the block size is equal to one, and one has the standard cyclic mapping.

◆ **Property 5.** Each processor is always assigned $\frac{N}{P}$ locations in total, i.e., this class of mappings ensures even data distribution.

◆ **Property 6.** In parallelizing any one of the four unordered in-place FFTs, each processor can always compute the butterflies involving the $\frac{N}{P}$ local data points independently, because these data correspond to array locations corresponding to the $n-d$ address bits specified by the braces below:

$$\overbrace{i_{n-1}\cdots i_{k+1}}\ \text{designated } d\text{-bit for Processor ID } \overbrace{i_{k-d}\cdots i_0}.$$

◆ **Property 7.** To compute the butterflies involving the address bits used to define the processor ID number, data can always be exchanged between two processors with ID numbers different in exactly one bit, although these exchanges can involve either $\frac{N}{P}$ or $\frac{1}{2}\left(\frac{N}{P}\right)$ data points and they may or may not be pipelined. These design issues will be addressed in the subsequent chapters.

In view of properties 5, 6, and 7, it is not surprising that many mappings used in the literature for parallelizing the *in-place* FFTs belong to the class of Cyclic Block Mappings (CBMs). This class of mappings was also used in parallelizing the ordered FFTs, although in a less straightforward manner.

17.3 Examples of CBM Mappings and Parallel FFTs

CBM mappings have received considerable study in the literature dealing with parallelizing FFTs [23, 36, 46, 56, 59, 90, 95, 104, 107]. These works vary in the choice of the blocksize, whether DIF or DIT transforms are used, and whether the input and/or output is in unordered (reverse-binary) or in natural order, and so on. All these treatments will be brought into a common framework in subsequent chapters.

To give an overview, some examples are cited in Table 17.1. Observe that each CBM mapping is identified by its unique blocksize. The perfect shuffle scheme [90] was already discussed in Section 10.2.2, the other parallel FFT algorithms cited in Table 17.1 are reviewed in the specified sections in Chapter 21, and the underlying techniques can be found in the specified sections in Chapters 19 and 20.

Table 17.1 Examples of cyclic block mappings (CBMs) and parallel FFTs.

Examples of Parallel FFTs using $P = 2^d$ Processors

Input Data: $N = 2^n$ Complex Points

References	CBM Blocksize	FFT Variant	Machine/Implementation
Stone [90], 1971 (Sec. 10.2.2)	$\frac{N}{P} = 1$	radix-2 DIT_{NR}	Perfect Shuffle (with $P = N$)
Jamieson, Mueller & Siegel [56], 1986 (Sec. 20.1.2 & 21.2.2)	2^k for $k = 0$ (cyclic)	radix-2 DIF_{NR}	SIMD System with $P = \frac{N}{2^k}$ for $1 \leq k \leq \log_2 N$
Walton [107],1986 (Sec. 20.2 & 21.2.1)	$\frac{N}{P} = 2^{n-d}$ (consecutive)	radix-2 DIT_{RN}	32-node Ametek Hypercube
Swarztrauber [95], 1987 (Sec. 20.1.2 & 21.2.4)	$\frac{N}{P} = 2^{n-d}$ (consecutive)	radix-2 DIF_{NR} plus intermediate reordering i-cycles	Hypercube (not implemented)
Chamberlain [23], 1988 (Sec. 19.2.1 & 21.1.1)	$\frac{N}{P} = 2^{n-d}$ (consecutive)	radix-2 DIF_{NR} & Its Inverse	64-node Intel iPSC Hypercube & Linear Array (via Reflected-Binary Gray-Codes)
Tong & Swarztrauber [104], 1991 (Sec. 20.1.2 & 21.2.4)	2^k for $k = n - d$ (consecutive) and $k = 0$ (cyclic)	radix-2 DIF_{NR} plus intermediate reordering i-cycles	CM-2 Hypercube (16K 1-bit processors)
Johnsson &Krawitz [59], 1992 (Sec. 19.2.3 & 21.1.4)	$\frac{N}{P} = 2^{n-d}$ (consecutive)	radix-2 DIF_{RN} and DIT_{NR}	2048-processor CM-200 (with Boolean Cube Network)
Dubey, Zubair & Grosch [36], 1994 (Sec. 20.1.2 & 21.2.3)	2^k for $k = 0, 1, \cdots, n - d$.	radix-2 DIF_{NR} plus an ad-hoc rearrangement phase	64-node Intel iPSC/860 Hypercube
Fabbretti et. al. [46], 1996 (Sec. 19.2.3 & 21.1.3)	$\frac{N}{P} = 2^{n-d} = 4^k$ (consecutive)	radix-4 DIT_{NR} plus local split-radix	128-node nCUBE2 Hypercube

Chapter 18

Computing and Communications on Distributed-Memory Multiprocessors

The development of efficient parallel FFT algorithms is the focus of Part III of this book. As revealed by our discussion on mapping data to processors in Chapter 17, the designs of parallel FFT algorithms are mainly driven by data distribution, not by the way the processors are physically connected (through shared memory or by way of a communication network.) This is a feature not shared by parallel numerical algorithms in general.

In general, shared-memory architectures are relatively easy to program, since only the computation and not the data needs to distributed among the processors. In many applications, including the FFTs, implementing software to utilize multiple processors on these architectures is a simple task. However, the drawback for these shared-memory machines is that most are bus-based; each processor communicates with the single shared memory via a shared bus, and such architectures do not scale up well.

On the other hand, regardless of the interconnection topology, implementing software on distributed-memory machines is generally a more challenging task; the computation must be suitably distributed over the processors in a manner that balances the load and keep the processors gainfully employed for as large a fraction of the overall execution time as possible, and the data must be distributed and conveyed among the processors so that they are available in the processors when needed. It is fair to say that programming in this mode requires a relatively high level of sophistication.

There is large body of literature on algorithms specifically designed to exploit distributed-memory computers having a hypercube interconnection network topology. Indeed, for a considerable period of time, this was the primary architecture for distributed-memory multiprocessors. More recently, distributed-memory multiprocessor machines have adopted a mesh topology. Apparently, the hypercube multiprocessor was used by many FFT researchers for the following reasons: (1) for all data mappings introduced, any two processors who are required to share data are directly connected on the hypercube—this perfect match is inherent in the FFT computation, not by design; and this match permits all communications to occur on disjoint paths and eliminates traffic congestion; and (2) the hypercube is not a single topology—many other topologies may be embedded in it: a ring embedded in a hypercube is a ring, a mesh embedded in a hypercube is a mesh, etc. Therefore, it accommodates the FFT on a ring and FFT on a mesh, if they are so desired, without compromise.

Recently, equipment vendors, recognizing the desirability of presenting a shared-memory programming environment to their users, but at the same time wishing to provide scalable architectures, have developed hybrid architectures, with numerous shared-memory modules all connected via a hypercube network. The memory in each of the modules in the aggregate form a uniform address space. Each processor can address the full address space, but some of the addresses are in its local memory, and some are in the local memory of some other processor(s). (Hence the name "non-uniform memory architecture" (NUMA) machines.) The programmer is presented with a single uniform address space, and the underlying hardware and software automatically retrieve and/or migrate data to a processor when it addresses data outside its local memory. Thus, one would expect that algorithms, such as the FFT algorithms in this book, which map naturally onto a hypercube architecture, would perform will on these hybrid machines. Most of the data accesses would be local, and those that involved data in other modules would naturally involve machines that are (topologically) close.

18.1 Distributed-Memory Message-Passing Multiprocessors

Distributed-memory message-passing multiprocessors can have a substantial number of processors, often ranging into several hundred or even a few thousand. Each processor has its own private memory, and can execute its own program and operate on its own data. There is neither globally shared memory nor connection between one processor to another processor's memory modules. Instead, the processors are connected by a network of communication channels, and the processors share data and/or synchronize with each other by sending and receiving messages over the network.

Builders of such machines must make a tradeoff between the cost of equipment and the richness of the connection topology. Networks of various types and topologies are used. They can be of fixed topology, such as a ring or a mesh, or can be packet-switched or circuit-switched networks. For example, the 512 processors of Intel's DELTA Mesh are connected as a 16-by-32 mesh, whereas the 128 processors of Intel's iPSC/860 Hypercube are connected by a hypercube network.

While the hypercube is a fixed topology, it is highly flexible in the sense that it is rich enough to support a variety of other topologies within it. For example, rings, meshes, trees, a torus, etc. can be embedded in a hypercube, so algorithms which are most efficient on any of these topologies can also be developed, evaluated and used on the hypercube.

The hypercube multiprocessor architecture is the subject of Section 18.2. The embedding of a ring by reflected-binary Gray-code was used in some FFT applications, and it is the subject of Section 18.3.

18.2 The d-Dimensional Hypercube Multiprocessors

There are 2^d identical node processors in a hypercube of dimension d. Each processor is uniquely identified by an integer in $\{0, 1, 2, \ldots, 2^d - 1\}$. If each processor ID in the set $\{0, 1, 2, \ldots, 2^d - 1\}$ is represented by a d-bit binary string, $b_{d-1}b_{d-2} \ldots b_0$, then the hypercube network is constructed by physically connecting each pair of processors whose IDs differ in one single bit b_k, $0 \leq k \leq d - 1$. Figure 18.1 illustrates the binary hypercube topologies of dimension $d = 0$, 1, 2, and 3. Each solid line in Figure 18.1 represents a communication channel.

Figure 18.1 Binary hypercubes of dimension 0, 1, 2, and 3.

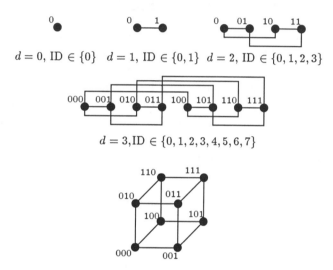

$d = 0$, ID $\in \{0\}$ $d = 1$, ID $\in \{0, 1\}$ $d = 2$, ID $\in \{0, 1, 2, 3\}$

$d = 3$, ID $\in \{0, 1, 2, 3, 4, 5, 6, 7\}$

An equivalent 3D drawing of a hypercube of dimension $d = 3$.

Additional properties of a d-dimensional hypercube are summarized below.

- Each of the $p = 2^d$ processors has exactly d neighbors.

- A d-dimensional hypercube has $p \times (d/2) = d \times 2^{d-1}$ communication channels.

- Two processors are said to be h hops apart if they are connected by a path consisting of h communication channels. Note that $1 \leq h \leq d$ on a hypercube of dimension d.

- A $(d+1)$-dimensional hypercube is constructed by "pairwise" connections between the processors of two d-dimensional hypercubes.

- A hypercube may be shared by multiple users, i.e., disjoint subcubes may be allocated to run different application programs.

- The processor ID is used to specify the receiver of the message and distinguish different actions (in the form of code segments) for different processors.

- A non-blocking **send** and a blocking **receive** are message-passing primitives normally provided by the hypercube operating system; messages that arrive at the destination processor before the execution of the **receive** are placed in a queue until needed.

- Circuit-switched networks are normally used by hypercubes. If a specified receiver is not a neighbor of the sender, a header packet is sent to reserve all of the channels required to build a path. After this "circuit" is established, the message is transmitted, and an end-of-message indicator releases the channels. Overlapped paths cause network congestion.

18.2.1   The subcube-doubling communication algorithm

The hypercube topology provides connectivity that supports a highly efficient *subcube-doubling* communication algorithm. Below, the algorithm is explained in the context of a fictitious application, where each message exchange is followed by local computation. (For a concrete example of parallel **FFT** computation on four processors, see Figures 19.1, 19.2, 19.3, and 19.4 in Chapter 19.) Recall that the hypercube network is constructed by connecting each pair of processors whose IDs differ in one single bit, and it should be understood that the application program is executed *concurrently* by each processor on a hypercube of dimension d.

Algorithm 18.1 Casting subcube-doubling communication in a fictitious application.

$MyID :=$ *System ID of this processor*
$d :=$ *dimension of the hypercube used for this run*
$k := d - 1$ Data exchange begins
while $k \geq 0$ **do**   e.g., See Figures 19.1, 19.2, 19.3, & 19.4.
    Compose (my message) according to the
        application algorithm
    send *(my message) to the processor with ID*
        different from MyID in bit b_k
    receive *a message*   As an exchange for my message
    Perform possible local computation   e.g., One stage of FFT
        according to the application algorithm   butterfly computation
    $k := k - 1$
end while
Perform remaining local computation   e.g., Continue and complete FFT
    according to the application algorithm   computation on each
                                      processor's local data

Using a hypercube of dimension $d = 3$, the d communication steps in the basic subcube-doubling algorithm are illustrated in Figure 18.2. To accomplish the d exchanges, each processor pairs with another processor whose ID is different in bit b_k, $k = d - 1, d - 2, \ldots, 0$. For example, processor P_0 in Figure 18.2 accomplishes the $d = 3$ exchanges by communicating with processors P_4, P_2, and P_1 sequentially. The binary IDs of the latter three processors are 100, 010, and 001, respectively.

Figure 18.2 The d communication steps in the subcube-doubling algorithm ($d = 3$).

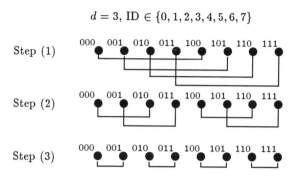

A benefit of using the subcube-doubling communication algorithm is that at each stage of the algorithm, the 2^{d-1} pairs of processors exchange data *concurrently* on 2^{d-1} *disjoint channels*, so there is no traffic congestion. The communication time is no different from that incurred by exchanging a message (of the same length) on a single channel between two processors, and the distance each message travels is always one. These communication advantages are fully exploited by the parallel FFT algorithms developed for hypercubes.

18.2.2 Modeling the arithmetic and communication cost

The following rules are commonly observed in modeling a hypercube system:

1. All processors are assumed to be running at the same speed, and be operational throughout the entire computation.

2. All processors execute the same program concurrently, and they coordinate their asynchronous activities by sending and receiving messages to and from other processors.

3. Each floating-point operation, namely *multiply, divide, subtract,* or *add,* is assumed to take τ units of time.

4. A non-blocking **send** and a blocking **receive** are the only message-passing primitives assumed.

5. The communication time is assumed to be a linear function of the message length (in bytes) and the distance (in hops) defined by

$$t_{comm}(N, h) = \alpha + \beta N + (h - 1)\gamma,$$

where α is the startup time, β is the time required to one byte across a single communication channel, and γ is the per-hop delay.

6. All communication channels are assumed to be full-duplex, hence the bidirectional data (exchange) rates are the same or close to the unidirectional data rates.

7. When all pairs of processors exchange L complex numbers by concurrently execut-
   ing one step of the subcube-doubling algorithm, the bidirectional communication
   time is modeled by unidirectional one-hop communication time of

$$t_{comm}(N, 1) = t_{comm}(16 \times L, 1) = \alpha + 16\beta L.$$

Note that $N = 16 \times L$ bytes because the real and imaginary parts of a complex
number are each an 8-byte double-precision floating-point number.

8. The time for data exchanges is not overlapped by computation. Thus, the total
   elapsed time on a processor is the sum of the arithmetic time and the communication time.

9. If the message length remains a constant, which is the case in the parallel FFT
   algorithms, then the total communication time incurred by the subcube-doubling
   algorithm is modeled by

$$T_{comm}^{FFT} = d \times t_{comm}(16 \times L, 1) = d \times (\alpha + 16\beta L).$$

10. The transmission rate λ per 8-byte (64-bit) floating-point number is derived from
    the time taken to transmit the entire message:

$$\lambda \approx (\alpha + 16\beta L)/(2 \times L).$$

Note that λ includes the contribution from the startup time.

A linear least-squares fit can be used to estimate the startup time α, the per-byte cost
β, and the per-hop penalty γ from experimental data of message-passing times versus
message length. The actual values used for the parameters τ, α, β, γ, and λ in the
model are hardware dependent, and they are appraised on a variety of systems in the
next section.

18.2.3   Hardware characteristics and implications on algorithm design

The configurations of three hypercubes and the DELTA mesh, along with their hardware characteristics, as reported by Dunigan [42, 44], are cited in Table 18.1. The
various benchmark tests which were performed to collect these data are described in
the notes and references section.

Note that while the floating-point arithmetic time τ decreases from $41.5 - 43$ μsec
on the iPSC/1 to 0.1 μsec on the iPSC/860, the speed of communication hardware
improves at a much slower pace. The consequence is that the ratio of τ to the 8-byte
transmission rate λ, which is the Comm./Comp. ratio in Table 1, becomes very large
and remains so to this date: $\tau/\lambda = 26, 59, 1000, 775$ on iPSC/1, iPSC/2, iPSC/860,
and the DELTA, respectively.

Since the "Comm./Comp." ratio has increased 40 fold within a decade, the importance of reducing the communication requirement in the parallel algorithms is crucial,
regardless of whether the communication time is negligible or not on a particular machine or on a particular generation of machines. Therefore, one should always make

Table 18.1 Hardware characteristics of three iPSC hypercubes and the DELTA mesh.

Machine Configurations				
Machine	**DELTA Mesh**	**iPSC/860**	**iPSC/2**	**iPSC/1**
Node CPU	i860	i860	80386/80387	80286/80287
Clock rate	40 Mhz	40 Mhz	16 Mhz	8 Mhz
Memory/node	16 MB	8 MB	4 MB	512 KB
Physical network	2D-Mesh	Hypercube	Hypercube	Hypercube
Peak data rate	22 MB/s	2.8 MB/s	2.8 MB/s	1.25 MB/s
Node OS	NX v3.3	NX v3.2	NX v2.2	v3.0
C compiler	PGI Rel 3.0	GH v1.8.5	C-386 1.8.3.A	Xenix 3.4

Arithmetic Times from Tests in C				
Machine	**DELTA Mesh**	**iPSC/860**	**iPSC/2**	**iPSC/1**
INTEGER*4 +	0.1 μs	0.1 μs	0.6 μs	5.0 μs
INTEGER*4 *	0.3 μs	0.3 μs	1.5 μs	36.5 μs
REAL*8 +	0.1 μs	0.1 μs	6.6 μs	41.5 μs
(Double precision +)	(10 Mflops)	(10 Mflops)	(0.15 Mflops)	(0.024 Mflops)
REAL*8 *	0.1 μs	0.1 μs	7.0 μs	43.0 μs
(Double precision *)	(10 Mflops)	(10 Mflops)	(0.14 Mflops)	(0.023 Mflops)

Coefficients of Communication				
Machine	**DELTA Mesh**	**iPSC/860**	**iPSC/2**	**iPSC/1**
Startup (α)	72 μs	136(75) μs	697(390) μs	862.2 μs
Byte transfer (β)	0.08 μs	0.4 μs	0.4 μs	1.8 μs
Hop penalty (γ)	0.05 μs	33(11) μs	30(10) μ [12]	–

Figures of Merit				
Machine	**DELTA Mesh**	**iPSC/860**	**iPSC/2**	**iPSC/1**
8192-byte transfer speed	11.9 MB/s	2.6 MB/s	2.3 MB/s	0.5 MB/s
REAL*8 * + * + *	18 Mflops	18 Mflops	0.29 Mflops	0.04 Mflops
8-byte transfer time	62 μs	80 μs	390 μs	1120 μs
8-byte multiply time	0.08 μs	0.08 μs	6.6 μs	43.0 μs
Comm./Comp. ratio	775	1000	59	26

Data, unless noted otherwise, are cited from Oak Ridge National Laboratory Technical Reports by T. H. Dunigan [42, 44], with permission. Table is reproduced from E. Chu [25], *The International Journal of High Performance Computing Applications*, 13(2):124–145, 1999, with permission.

an effort to reduce the number of messages, the length of each message, the distance each message travels, and avoid or minimize traffic congestion when designing parallel algorithms.

Note also from Table 18.1 that the message start-up time α remains very high relative to the byte-transfer time β throughout generations of machines. (The α values inside the brackets are for messages shorter than 100 bytes.) When the start-up time is so large, sending a single long message can be much more economical than sending multiple short messages. Therefore, pipelining a large number of short messages is not a viable communicating strategy on these multiprocessors. (Note that when fine-grain pipelined communication is used, the assumption is $\alpha \approx 0$, which was the case on some special-purpose hypercubes.)

Another parameter which stands out in Table 1 is the integer *add* time, which is no longer negligible compared to τ, the floating-point arithmetic time, on newer machines. As reported in Table 1, an integer *add* takes 0.1 μsec on iPSC/860, which is the same as $\tau = 0.1$ μsec on iPSC/860. Since integer *add* is used to compute the addresses of array elements during program execution, they could account for a significant part of the actual execution time. One should keep this in mind in designing data structures for implementing sequential or parallel algorithms—excessive indirect indexing should be avoided.

18.3   Embedding a Ring by Reflected-Binary Gray-Code

Topology embedding is simply mapping a subset of the hypercube communication channels to a different topology. As illustrated in Figure 18.3 below, a ring consisting of $p = 2^d$ processors can be formed by traversing a set of $p = 2^d$ channels selected from the $d\left(2^{d-1}\right)$ channels in a hypercube network. Observe that the three drawings of the embedded 8-processor ring show the same ring, although one figure may be easier to visualize than other forms.

To program a communication algorithm on the embedded ring, each processor needs to "know" the system IDs of processors in consecutive ring positions and store them in a convenient way, such as in a linear array *Ring*, as depicted for $p = 8$ in Figure 18.4 below.

Note that the binary representations of any two processor IDs in $Ring[\ell]$ and $Ring[\ell + 1]$, $0 \le \ell \le p - 2$, differ in exactly one bit, as do the two IDs in $Ring[0]$ and $Ring[p - 1]$. Therefore, the task of embedding a ring in a hypercube amounts to constructing an array *Ring* to contain the *reflected-binary Gray-code*, defined recursively as follows.

If the k-bit Gray-code of length $L = 2^k$ is stored in the array

$$Ring[0 : L - 1] = \{G_0, G_1, \ldots, G_{L-1}\},$$

then the $(k + 1)$-bit Gray-code of length $2 \times L$ is given by

$$Ring[0 : 2 \times L - 1] = \{0G_0, 0G_1, \ldots, 0G_{L-1}, 1G_{L-1} \ldots, 1G_1, 1G_0\}.$$

Note that the recurrence begins with $k = 1$ and $L = 2$. In Figure 18.5, the reflected-binary Gray-code of length $L = 8$ is constructed step by step.

Figure 18.3 Traversing an embedded ring in hypercubes of dimension 0, 1, 2, and 3.

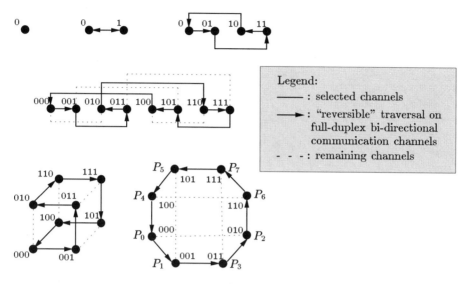

Traversing a ring in hypercubes of dimensions 0, 1, 2, and 3.

Figure 18.4 Storing the embedded 8-processor ring in array $Ring[0:7]$.

Processor named by $Ring$[index]:	P_0	P_1	P_3	P_2	P_6	P_7	P_5	P_4
Array $Ring$:	0	1	3	2	6	7	5	4
index:	0	1	2	3	4	5	6	7

Figure 18.5 Generating reflect-binary Gray-code in $Ring[0:7]$.

Array $Ring$'s content expressed in binary Gray-code, $L = 2$:	0	1						

binary Gray-code, $L = 4$:	00	01	11	10				

binary Gray-code, $L = 8$:	000	001	011	010	110	111	101	100

If one views the Gray-codes as binary representations of numbers, then G_ℓ and $0G_\ell$ represent the same value, and the value of $1G_\ell$ is obtained by adding $L = 2^k$ to G_ℓ if the current length is L, beginning with $L = 2$. The following pseudo-code algorithm constructs the reflected-binary Gray-code. Note that an auxiliary array $PosInRing$ records the ring position for each processor, and the identity $PosInRing[Ring[\ell]] = \ell$ holds.

Algorithm 18.2 Constructing a reflected-binary Gray-code.

$MyID := $ System ID of this processor
$d := $ dimension of the hypercube used for this run
$Nproc := 2 * *d$ length of Gray-code to be computed
$Ring[0] := 0;\quad Ring[1] := 1;\quad L := 2$ initialize Gray-code of length $L = 2$
$PosInRing[0] := 0;\quad PosInRing[1] := 1$
while $L < Nproc$ **do**   reflect the Gray code $d - 1$ times
    $Last := 2 * L - 1$
    for $\ell := 0$ **to** $L - 1$ reflect Gray code of length L
        $ID := Ring[\ell] + L$
        $Ring[Last] := ID$
        $PosInRing[ID] := Last$
        $Last := Last - 1$
    end for
    $L := 2 * L$ double length L after reflection
end while
$MyRingPos := PosInRing[MyID]$

18.4 A Further Twist—Performing Subcube-Doubling Communications on a Ring Embedded in a Hypercube

Consider the simple application which used the subcube-doubling communication algorithm as described in Section 18.2.1. How would that task be accomplished if only the embedded ring is employed? In this case, the same program will work correctly if every reference to P_ℓ is directed to $Ring[\ell]$ at run time. The only effect is that messages are now exchanged between processors whose positions in the embedded ring differ in one single bit. The modified algorithm is given below.

The three subcube-doubling communication steps performed on an embedded 8-processor ring are illustrated in Figure 18.6 below. Observe that the distance between *sender* and *receiver* of each message becomes two in the first and the second step, and the distance becomes one in the last step. This is the direct result of the following theorem, which is proved in [23].

Theorem 18.1 If array $Ring[0 : L - 1] = \{G_0, G_1, \ldots, G_{L-1}\}$ contains the reflected-binary Gray-code of length $L = 2^d$ for $d \geq 2$, then $Ring[\ell] = G_\ell$ and $Ring[m] = G_m$ differ in exactly two bits if ℓ and m differ only in bit b_k for $0 < k < d$.

Algorithm 18.3 Imposing subcube-doubling communication on an embedded ring.

$MyID :=$ *System ID of this processor*
$MyRingPos := PosInRing[MyID]$
$d :=$ *dimension of the hypercube used for this run*
$k := d - 1$
$\qquad\qquad\qquad\qquad\qquad\qquad\qquad$ Data exchange begins
while $k \geq 0$ **do**
$\qquad$ *Compose (my message) according to the*
$\qquad\qquad$ *application algorithm*
$\qquad$ **send** *(my message) to the processor with its RingPos*
$\qquad\qquad$ *different from MyRingPos in bit* b_k
$\qquad$ **receive** *a message* $\qquad\qquad$ As an exchange for my message
$\qquad$ *Perform possible local computation* $\qquad$ e.g., One stage of FFT
$\qquad\qquad$ *according to the application algorithm* $\qquad$ butterfly computation
$\qquad$ $k := k - 1$
end while
Perform remaining local computation $\qquad\qquad$ e.g., Continue and complete FFT
$\qquad$ *according to the application algorithm* $\qquad\qquad$ computation on each
$\qquad\qquad\qquad\qquad\qquad\qquad\qquad\qquad\qquad\qquad$ processor's local data

---

Figure 18.6 Subcube-doubling communications on an embedded ring.

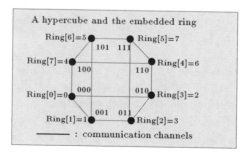

A hypercube and the embedded ring

──── : communication channels

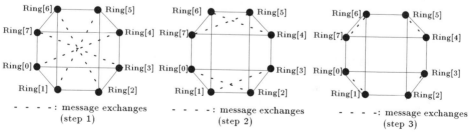

- - - - : message exchanges (step 1)
- - - - : message exchanges (step 2)
- - - - : message exchanges (step 3)

Subcube-Doubling Communications on the Embedded Ring

18.5   Notes and References

The hardware characteristics presented in Table 18.1 are the results of the following benchmark tests.

18.5.1   Arithmetic time benchmarks

The Caltech test suite [61] was selected to benchmark the integer and floating point operations on DELTA, iPSC/860, iPSC/2 and iPSC/1 in [42, 44]. The times illustrate differences in both CPU speed and compiler. The test program measures the time for performing a binary arithmetic operation and an assignment in a loop. For example, the following loops (expressed in C) were timed (on Intel iPSC and Caltech/JPL Mark II hypercubes) in [61]:

Test 1. Loop overhead : for (i=0; i < NumTimes; ++i){   ;}

Test 2. Integer *op* = +, * : for (i=0; i < NumTimes; ++i){j = k *op* l ;}

Test 3. Floating point *op* = +, * : for (i=0; i < NumTimes; ++i){a = b *op* d ;}

Note that to derive the actual time per operation, the loop overhead measured in Test 1 is subtracted from the time measured in Test 2 or Test 3. For integer parameters, the C data types are *short* and *long*; for real numbers, the C data types are *float* and *double*. The corresponding Fortran data types are INTEGER*2, INTEGER*4, REAL*4, REAL*8. Although Fortran types are used in Table 18.1 for clarity following the presentation in [42, 44], all tests in [42, 44, 61] were run in C.

Note also that arithmetic performance, namely time-per-operation, improves if a more complex expression is used in the test because intermediate results can be kept in internal registers. Given below is the loop implementing a five-operation expression which was also timed in [42, 44, 61]

Test 4. Five-operation *+*+* : for (i=0; i < NumTimes; ++i){a=b*c+b*e+d*c;}

Since the Mflops (Million floating-point operations per second) rate is inversely proportional to the arithmetic time per operation, the resulting Mflops rate can be significant higher from benchmarking a compound expression. For the DELTA and iPSC/860, while 10 Mflops rate results from benchmarking Test 3, an 18 Mflops rate results from benchmarking Test 4 [44].

18.5.2   Unidirectional times on circuit-switched networks

When applying the benchmark coefficients in Table 18.1 to model unidirectional communication times, the following should be noted.

1. All of the communication data rates in Table 18.1 have been measured on idle machines [43]. Therefore, additional latency should be accounted for if the communication algorithm employed causes channel contention on a particular network.

2. The DELTA, iPSC/860, and iPSC/2 all use circuit-switching to manage the communication channels. For all three machines, the communication time (in microseconds) for sending an N-byte message h hops away on an idle network is modeled by

$$(18.1) \qquad t_{comm}(N, h) = \alpha + N\beta + (h - 1)\gamma,$$

where α, β, and γ represent the startup time, byte transfer rate, and hop penalty as noted in Table 18.1. When the message is sent to a nearest neighbor, $t_{comm}(N, 1) = \alpha + N\beta$ is commonly referred to as the (one hop) "message passing" time. Note that for the iPSC/2 and iPSC/860, two values of α and two values of γ are given, because such overhead is smaller for short messages of length less than 100 bytes [12, 42].

18.5.3 Bidirectional times on full-duplex channels

Since the subcube-doubling communication incurs message exchanges between all pairs of processors throughout, one needs to address how the measured bidirectional data rates compare with the more widely reported unidirectional data rates. Data rates for "message exchange" between two nearest neighbors on an iPSC/2 were examined by Bomans and Roose in [12]. They show that the "full-duplex" nature of the communication channels can be effectively exploited when both nodes first (and concurrently) send their information and then receive the incoming message. The three exchange methods tested in [12] are described in Table 18.2. Their results show that for mes-

Table 18.2 Exchange methods tested by Bomans et al. in [12].

Exchange Method	Node 0	Node 1
(Concurrent) blocking	csend/crecv	csend/crecv
(Concurrent) non-blocking	irecv/csend/msgwait	irecv/csend/msgwait
(Sequential) send/reply	csend/crecv	crecv/csend

sages shorter than 100 bytes, the (concurrent) blocking method is the fastest; however, for messages of size N with $100 \leq N \leq 384$ bytes, the time for the blocking communication increases faster than the time for the non-blocking method; for longer messages, both concurrent methods show an equal increase in time per byte. Therefore, after the 384-byte threshold is passed, the proportional benefit of the non-blocking method decreases as messages get longer, and it is about 11% for a message of 1 Mbyte [12].

Bomans and Roose further compare the fastest concurrent method (either the blocking or the non-blocking one depending on the message size) with the sequential send/reply method on the iPSC/2. Note that when two messages of the *same length* are exchanged, the sequential send/reply takes as long as a round-trip echo test: namely a test node sends a message to an echo node; the echo node receives the message and sends it back to the test node. The results presented in [12] show that for messages with length ranging from 0 to 1024 bytes, the concurrent exchange time is 62% to 59% of the sequential send/reply time; and for messages with length of 16 Kbytes to 1 Mbytes, the concurrent exchange time is reduced to 52% to 50% of the send/reply time (or the more familiar round-trip echo time.) Since the unidirectional data rates for "message passing" from Node 0 to Node 1 is measured by 50% of the round-trip echo time, these results show that for messages longer than 16 Kbytes, there is virtually no penalty in conducting a concurrent bidirectional message exchange compared to passing a unidirectional message from one node to a nearest neighbor.

Chapter 19

Parallel FFTs without Inter-Processor Permutations

This chapter treats a simple class of parallel FFTs, namely those algorithms which involve no inter-processor data permutations. That is, no part of a processor's initial complement of data migrates to another processor.

For purposes of this chapter, it is assumed that the multiprocessor available has $P = 2^d$ processors, and each processor has its own local memory. That is, the machine in question is a distributed-memory multiprocessor, where each processor is connected to the others via a communication network with a prescribed topology. A common topology is a hypercube, but others such as a regular grid or a ring are also commonly used.

19.1 A Useful Equivalent Notation: | PID |Local M

As discussed in Chapter 17, a key step in parallelizing the FFT on such multiprocessor computers is the mapping of the elements of x to the processors. Assuming that the x elements are stored in the global array a in natural order, i.e., $a[m] = x_m$, $m = i_{n-1}i_{n-2}\cdots i_0$, the array address based notation

$$i_{n-1}\cdots i_{k+1}|i_k \cdots i_{k-d+1}|i_{k-d}\cdots i_0$$

has been used to denote that bits $i_k \cdots i_{k-d+1}$ are chosen to specify the data-to-processor allocation.

In general, since any d bits can be used to form the processor ID number, it is easier to recognize the generic communication pattern if one concatenates the bits representing the ID into one group denoted by "PID," and refers to the remaining $n - d$ bits, which are concatenated to form the local array address, as "Local M."

For the class of cyclic block mappings (CBMs) introduced in Chapter 17, one can use the following equivalent notation, where the leading d bits are always used to identify the processor ID number.

$$|\,\text{PID}\,|\,\text{Local } M = |i_k \cdots i_{k-d+1}|\overbrace{i_{n-1}\cdots i_{k+2}i_{k+1}}\ \overbrace{i_{k-d}\cdots i_1 i_0}\,.$$

In either notation, the d consecutive bits are marked by the symbol " | " at both ends. The two notations are equivalent and both are used in the text.

To fully demonstrate the usage of the " |PID|Local M" notation, Tables 19.1, 19.2, 19.3, and 19.4 below show the local data of each processor after a naturally ordered input series of $N = 32$ elements is divided among $P = 4$ processors using all possible cyclic block mappings.

For each mapping, the content of each local array element $A[M]$ is identified by $a[m]$ in the adjacent column. Observe that $m = i_4 i_3 i_2 i_1 i_0$ can be easily reconstructed for each $A[M]$ in each processor, because the given |PID|Local M specifies exactly which bits can be recovered from the processor ID and which bits can be recovered from the local M.

Table 19.1 Identifying P_{00}'s different local data sets for all cyclic block mappings.

Local data of processor P_{00} expressed in terms of

global array element $a[m]$, $m = i_4 i_3 i_2 i_1 i_0$,

for all possible cyclic block mappings (CBMs)

Blocksize=8 |PID|Local M	$i_4 i_3|i_2 i_1 i_0$	$P_{i_4 i_3}$ = P_{00} $a[m]$	Blocksize=4 |PID|Local M	$i_3 i_2|i_4 i_1 i_0$	$P_{i_3 i_2}$ = P_{00} $a[m]$	Blocksize =2 |PID|Local M	$i_2 i_1|i_4 i_3 i_0$	$P_{i_2 i_1}$ = P_{00} $a[m]$	Blocksize=1 |PID|Local M	$i_1 i_0|i_4 i_3 i_2$	$P_{i_1 i_0}$ = P_{00} $a[m]$
$i_4 i_3|000$	$a[0]$	$i_3 i_2|000$	$a[0]$	$i_2 i_1|000$	$a[0]$	$i_1 i_0|000$	$a[0]$
$i_4 i_3|001$	$a[1]$	$i_3 i_2|001$	$a[1]$	$i_2 i_1|001$	$a[1]$	$i_1 i_0|001$	$a[4]$
$i_4 i_3|010$	$a[2]$	$i_3 i_2|010$	$a[2]$	$i_2 i_1|010$	$a[8]$	$i_1 i_0|010$	$a[8]$
$i_4 i_3|011$	$a[3]$	$i_3 i_2|011$	$a[3]$	$i_2 i_1|011$	$a[9]$	$i_1 i_0|011$	$a[12]$
$i_4 i_3|100$	$a[4]$	$i_3 i_2|100$	$a[16]$	$i_2 i_1|100$	$a[16]$	$i_1 i_0|100$	$a[16]$
$i_4 i_3|101$	$a[5]$	$i_3 i_2|101$	$a[17]$	$i_2 i_1|101$	$a[17]$	$i_1 i_0|101$	$a[20]$
$i_4 i_3|110$	$a[6]$	$i_3 i_2|110$	$a[18]$	$i_2 i_1|110$	$a[24]$	$i_1 i_0|110$	$a[24]$
$i_4 i_3|111$	$a[7]$	$i_3 i_2|111$	$a[19]$	$i_2 i_1|111$	$a[25]$	$i_1 i_0|111$	$a[28]$

Table 19.2 Identifying P_{01}'s different local data sets for all cyclic block mappings.

Local data of processor P_{01} expressed in terms of
global array element $a[m]$, $m = i_4 i_3 i_2 i_1 i_0$,
for all possible cyclic block mappings (CBMs)

Blocksize=8 $\|$PID$\|$Local M $\|i_4 i_3\|i_2 i_1 i_0$	$P_{i_4 i_3}$ $= P_{01}$ $a[m]$	Blocksize=4 $\|$PID$\|$Local M $\|i_3 i_2\|i_4 i_1 i_0$	$P_{i_3 i_2}$ $= P_{01}$ $a[m]$	Blocksize =2 $\|$PID$\|$Local M $\|i_2 i_1\|i_4 i_3 i_0$	$P_{i_2 i_1}$ $= P_{01}$ $a[m]$	Blocksize=1 $\|$PID$\|$Local M $\|i_1 i_0\|i_4 i_3 i_2$	$P_{i_1 i_0}$ $= P_{01}$ $a[m]$
$\|i_4 i_3\|000$	$a[8]$	$\|i_3 i_2\|000$	$a[4]$	$\|i_2 i_1\|000$	$a[2]$	$\|i_1 i_0\|000$	$a[1]$
$\|i_4 i_3\|001$	$a[9]$	$\|i_3 i_2\|001$	$a[5]$	$\|i_2 i_1\|001$	$a[3]$	$\|i_1 i_0\|001$	$a[5]$
$\|i_4 i_3\|010$	$a[10]$	$\|i_3 i_2\|010$	$a[6]$	$\|i_2 i_1\|010$	$a[10]$	$\|i_1 i_0\|010$	$a[9]$
$\|i_4 i_3\|011$	$a[11]$	$\|i_3 i_2\|011$	$a[7]$	$\|i_2 i_1\|011$	$a[11]$	$\|i_1 i_0\|011$	$a[13]$
$\|i_4 i_3\|100$	$a[12]$	$\|i_3 i_2\|100$	$a[20]$	$\|i_2 i_1\|100$	$a[18]$	$\|i_1 i_0\|100$	$a[17]$
$\|i_4 i_3\|101$	$a[13]$	$\|i_3 i_2\|101$	$a[21]$	$\|i_2 i_1\|101$	$a[19]$	$\|i_1 i_0\|101$	$a[21]$
$\|i_4 i_3\|110$	$a[14]$	$\|i_3 i_2\|110$	$a[22]$	$\|i_2 i_1\|110$	$a[26]$	$\|i_1 i_0\|110$	$a[25]$
$\|i_4 i_3\|111$	$a[15]$	$\|i_3 i_2\|111$	$a[23]$	$\|i_2 i_1\|111$	$a[27]$	$\|i_1 i_0\|111$	$a[29]$

Table 19.3 Identifying P_{10}'s different local data sets for all cyclic block mappings.

Local data of processor P_{10} expressed in terms of
global array element $a[m]$, $m = i_4 i_3 i_2 i_1 i_0$,
for all possible cyclic block mappings (CBMs)

Blocksize=8 $\|$PID$\|$Local M $\|i_4 i_3\|i_2 i_1 i_0$	$P_{i_4 i_3}$ $= P_{10}$ $a[m]$	Blocksize=4 $\|$PID$\|$Local M $\|i_3 i_2\|i_4 i_1 i_0$	$P_{i_3 i_2}$ $= P_{10}$ $a[m]$	Blocksize =2 $\|$PID$\|$Local M $\|i_2 i_1\|i_4 i_3 i_0$	$P_{i_2 i_1}$ $= P_{10}$ $a[m]$	Blocksize=1 $\|$PID$\|$Local M $\|i_1 i_0\|i_4 i_3 i_2$	$P_{i_1 i_0}$ $= P_{10}$ $a[m]$
$\|i_4 i_3\|000$	$a[16]$	$\|i_3 i_2\|000$	$a[8]$	$\|i_2 i_1\|000$	$a[4]$	$\|i_1 i_0\|000$	$a[2]$
$\|i_4 i_3\|001$	$a[17]$	$\|i_3 i_2\|001$	$a[9]$	$\|i_2 i_1\|001$	$a[5]$	$\|i_1 i_0\|001$	$a[6]$
$\|i_4 i_3\|010$	$a[18]$	$\|i_3 i_2\|010$	$a[10]$	$\|i_2 i_1\|010$	$a[12]$	$\|i_1 i_0\|010$	$a[10]$
$\|i_4 i_3\|011$	$a[19]$	$\|i_3 i_2\|011$	$a[11]$	$\|i_2 i_1\|011$	$a[13]$	$\|i_1 i_0\|011$	$a[14]$
$\|i_4 i_3\|100$	$a[20]$	$\|i_3 i_2\|100$	$a[24]$	$\|i_2 i_1\|100$	$a[20]$	$\|i_1 i_0\|100$	$a[18]$
$\|i_4 i_3\|101$	$a[21]$	$\|i_3 i_2\|101$	$a[25]$	$\|i_2 i_1\|101$	$a[21]$	$\|i_1 i_0\|101$	$a[22]$
$\|i_4 i_3\|110$	$a[22]$	$\|i_3 i_2\|110$	$a[26]$	$\|i_2 i_1\|110$	$a[28]$	$\|i_1 i_0\|110$	$a[26]$
$\|i_4 i_3\|111$	$a[23]$	$\|i_3 i_2\|111$	$a[27]$	$\|i_2 i_1\|111$	$a[29]$	$\|i_1 i_0\|111$	$a[30]$

Table 19.4 Identifying P_{11}'s different local data sets for all cyclic block mappings.

Local data of processor P_{11} expressed in terms of
global array element $a[m]$, $m = i_4 i_3 i_2 i_1 i_0$,
for all possible cyclic block mappings (CBMs)

Blocksize=8 $\|$PID$\|$Local M $\|i_4 i_3\|i_2 i_1 i_0$	$P_{i_4 i_3}$ $= P_{11}$ $a[m]$	Blocksize=4 $\|$PID$\|$Local M $\|i_3 i_2\|i_4 i_1 i_0$	$P_{i_3 i_2}$ $= P_{11}$ $a[m]$	Blocksize =2 $\|$PID$\|$Local M $\|i_2 i_1\|i_4 i_3 i_0$	$P_{i_2 i_1}$ $= P_{11}$ $a[m]$	Blocksize=1 $\|$PID$\|$Local M $\|i_1 i_0\|i_4 i_3 i_2$	$P_{i_1 i_0}$ $= P_{11}$ $a[m]$
$\|i_4 i_3\|000$	$a[24]$	$\|i_3 i_2\|000$	$a[12]$	$\|i_2 i_1\|000$	$a[6]$	$\|i_1 i_0\|000$	$a[3]$
$\|i_4 i_3\|001$	$a[25]$	$\|i_3 i_2\|001$	$a[13]$	$\|i_2 i_1\|001$	$a[7]$	$\|i_1 i_0\|001$	$a[7]$
$\|i_4 i_3\|010$	$a[26]$	$\|i_3 i_2\|010$	$a[14]$	$\|i_2 i_1\|010$	$a[14]$	$\|i_1 i_0\|010$	$a[11]$
$\|i_4 i_3\|011$	$a[27]$	$\|i_3 i_2\|011$	$a[15]$	$\|i_2 i_1\|011$	$a[15]$	$\|i_1 i_0\|011$	$a[15]$
$\|i_4 i_3\|100$	$a[28]$	$\|i_3 i_2\|100$	$a[28]$	$\|i_2 i_1\|100$	$a[22]$	$\|i_1 i_0\|100$	$a[19]$
$\|i_4 i_3\|101$	$a[29]$	$\|i_3 i_2\|101$	$a[29]$	$\|i_2 i_1\|101$	$a[23]$	$\|i_1 i_0\|101$	$a[23]$
$\|i_4 i_3\|110$	$a[30]$	$\|i_3 i_2\|110$	$a[30]$	$\|i_2 i_1\|110$	$a[30]$	$\|i_1 i_0\|110$	$a[27]$
$\|i_4 i_3\|111$	$a[31]$	$\|i_3 i_2\|111$	$a[31]$	$\|i_2 i_1\|111$	$a[31]$	$\|i_1 i_0\|111$	$a[31]$

19.1.1   Representing data mappings for different orderings

When the input x elements are stored in a in bit-reversed order, i.e., $a[r] = x_m$, where $m = i_{n-1} i_{n-2} \cdots i_0$, and $r = i_0 \cdots i_{n-2} i_{n-1}$, a cyclic block mapping should be denoted by

$$i_0 \cdots i_{k-d} | i_{k-d+1} \cdots i_k | i_{k+1} \cdots i_{n-1} ,$$

or the equivalent

$$\| \text{PID} \| \text{Local } M = |i_{k-d+1} \cdots i_k| \overbrace{i_0 \cdots i_{k-d}} \overbrace{i_{k+1} \cdots i_{n-1}}$$

instead. For example, suppose $N = 32$ and the mapping is denoted by $|i_0 i_1|i_2 i_3 i_4$. To locate $x_m = x_{26}$, one writes down $m = 26_{10} = 11010_2 = i_4 i_3 i_2 i_1 i_0$, from which one knows that x_{26} is stored in $a[r]$, $r = i_0 i_1 i_2 i_3 i_4 = 01011_2 = 11_{10}$, and that $a[11] = x_{26}$ is located in processor $P_{i_0 i_1} = P_{01}$.

It is useful to keep in mind that the bit sequence $i_{n-1} i_{n-2} \cdots i_0$ is always the binary representation of the subscript m of data element x_m or its derivative $x_m^{(k)}$, and the order in which these bits appear in the array address r, when $a[r] = x_m$ or $a[r] = x_m^{(k)}$, refers to permutations that x_m or its derivatives undergo in a initially or during the computation.

This convention is strictly adhered to throughout this text.

To make this absolutely explicit, the mappings demonstrated in Tables 19.1, 19.2, 19.3, 19.4 are repeated for bit-reversed input in Tables 19.5, 19.6, 19.7, and 19.8. Note that the actual distribution of array elements appears unaltered from that in the corresponding table for naturally ordered input, because each cyclic block map determines how to distribute $a[r]$ based on the value of the address r *independent of* the content of $a[r]$.

The reason one must have some way to indicate that $a[r] = x_m$ for bit-reversed input is that the FFT arithmetic operations are performed on the "content" of $a[r]$, and its content matters in specifying how an array element is used and updated in both sequential and parallel FFT algorithms. Recall that a different sequential algorithm must be used for differently ordered input.

Observe that both $r = i_0 i_1 i_2 i_3 i_4$ and $m = i_4 i_3 i_2 i_1 i_0$ can be easily reconstructed for each $A[M]$ in each processor, because, as noted before, the given |PID|Local M specifies exactly which bits can be recovered from the processor ID and which bits can be recovered from the local M.

Table 19.5 Identifying P_{00}'s (bit-reversed) local data sets for all cyclic block mappings.

Local data of processor P_{00} expressed in terms of
global array element $a[r] = x_m$, $r = i_0 i_1 i_2 i_3 i_4$, and $m = i_4 i_3 i_2 i_1 i_0$
for all possible cyclic block mappings (CBMs)

Blocksize=8 |PID|Local M	$P_{i_0 i_1}$ = P_{00}	Blocksize=4 |PID|Local M	$P_{i_1 i_2}$ = P_{00}	Blocksize =2 |PID|Local M	$P_{i_2 i_3}$ = P_{00}	Blocksize=1 |PID|Local M	$P_{i_3 i_4}$ = P_{00}
$|i_0 i_1|i_2 i_3 i_4$	$a[r]$	$|i_1 i_2|i_0 i_3 i_4$	$a[r]$	$|i_2 i_3|i_0 i_1 i_4$	$a[r]$	$|i_3 i_4|i_0 i_1 i_2$	$a[r]$
$|i_0 i_1|000$	$a[0]$	$|i_1 i_2|000$	$a[0]$	$|i_2 i_3|000$	$a[0]$	$|i_3 i_4|000$	$a[0]$
$|i_0 i_1|001$	$a[1]$	$|i_1 i_2|001$	$a[1]$	$|i_2 i_3|001$	$a[1]$	$|i_3 i_4|001$	$a[4]$
$|i_0 i_1|010$	$a[2]$	$|i_1 i_2|010$	$a[2]$	$|i_2 i_3|010$	$a[8]$	$|i_3 i_4|010$	$a[8]$
$|i_0 i_1|011$	$a[3]$	$|i_1 i_2|011$	$a[3]$	$|i_2 i_3|011$	$a[9]$	$|i_3 i_4|011$	$a[12]$
$|i_0 i_1|100$	$a[4]$	$|i_2 i_3|100$	$a[16]$	$|i_2 i_3|100$	$a[16]$	$|i_3 i_4|100$	$a[16]$
$|i_0 i_1|101$	$a[5]$	$|i_1 i_2|101$	$a[17]$	$|i_2 i_3|101$	$a[17]$	$|i_3 i_4|101$	$a[20]$
$|i_0 i_1|110$	$a[6]$	$|i_1 i_2|110$	$a[18]$	$|i_2 i_3|110$	$a[24]$	$|i_3 i_4|110$	$a[24]$
$|i_0 i_1|111$	$a[7]$	$|i_1 i_2|111$	$a[19]$	$|i_2 i_3|111$	$a[25]$	$|i_3 i_4|111$	$a[28]$

Table 19.6 Identifying P_{01}'s (bit-reversed) local data sets for all cyclic block mappings.

Local data of processor P_{01} expressed in terms of
global array element $a[r] = x_m$, $r = i_0 i_1 i_2 i_3 i_4$, and $m = i_4 i_3 i_2 i_1 i_0$
for all possible cyclic block mappings (CBMs)

Blocksize=8 |PID|Local M $\|i_0 i_1\|i_2 i_3 i_4$	$P_{i_0 i_1}$ $= P_{01}$ $a[r]$	Blocksize=4 |PID|Local M $\|i_1 i_2\|i_0 i_3 i_4$	$P_{i_1 i_2}$ $= P_{01}$ $a[r]$	Blocksize =2 |PID|Local M $\|i_2 i_3\|i_0 i_1 i_4$	$P_{i_2 i_3}$ $= P_{01}$ $a[r]$	Blocksize=1 |PID|Local M $\|i_3 i_4\|i_0 i_1 i_2$	$P_{i_3 i_4}$ $= P_{01}$ $a[r]$
$\|i_0 i_1\|000$	$a[8]$	$\|i_1 i_2\|000$	$a[4]$	$\|i_2 i_3\|000$	$a[2]$	$\|i_3 i_4\|000$	$a[1]$
$\|i_0 i_1\|001$	$a[9]$	$\|i_1 i_2\|001$	$a[5]$	$\|i_2 i_3\|001$	$a[3]$	$\|i_3 i_4\|001$	$a[5]$
$\|i_0 i_1\|010$	$a[10]$	$\|i_1 i_2\|010$	$a[6]$	$\|i_2 i_3\|010$	$a[10]$	$\|i_3 i_4\|010$	$a[9]$
$\|i_0 i_1\|011$	$a[11]$	$\|i_1 i_2\|011$	$a[7]$	$\|i_2 i_3\|011$	$a[11]$	$\|i_3 i_4\|011$	$a[13]$
$\|i_0 i_1\|100$	$a[12]$	$\|i_2 i_3\|100$	$a[20]$	$\|i_2 i_3\|100$	$a[18]$	$\|i_3 i_4\|100$	$a[17]$
$\|i_0 i_1\|101$	$a[13]$	$\|i_1 i_2\|101$	$a[21]$	$\|i_2 i_3\|101$	$a[19]$	$\|i_3 i_4\|101$	$a[21]$
$\|i_0 i_1\|110$	$a[14]$	$\|i_1 i_2\|110$	$a[22]$	$\|i_2 i_3\|110$	$a[26]$	$\|i_3 i_4\|110$	$a[25]$
$\|i_0 i_1\|111$	$a[15]$	$\|i_1 i_2\|111$	$a[23]$	$\|i_2 i_3\|111$	$a[27]$	$\|i_3 i_4\|111$	$a[29]$

Table 19.7 Identifying P_{10}'s (bit-reversed) local data sets for all cyclic block mappings.

Local data of processor P_{10} expressed in terms of
global array element $a[r] = x_m$, $r = i_0 i_1 i_2 i_3 i_4$, and $m = i_4 i_3 i_2 i_1 i_0$
for all possible cyclic block mappings (CBMs)

Blocksize=8 |PID|Local M $\|i_0 i_1\|i_2 i_3 i_4$	$P_{i_0 i_1}$ $= P_{10}$ $a[r]$	Blocksize=4 |PID|Local M $\|i_1 i_2\|i_0 i_3 i_4$	$P_{i_1 i_2}$ $= P_{10}$ $a[r]$	Blocksize =2 |PID|Local M $\|i_2 i_3\|i_0 i_1 i_4$	$P_{i_2 i_3}$ $= P_{10}$ $a[r]$	Blocksize=1 |PID|Local M $\|i_3 i_4\|i_0 i_1 i_2$	$P_{i_3 i_4}$ $= P_{10}$ $a[r]$
$\|i_0 i_1\|000$	$a[16]$	$\|i_1 i_2\|000$	$a[8]$	$\|i_2 i_3\|000$	$a[4]$	$\|i_3 i_4\|000$	$a[2]$
$\|i_0 i_1\|001$	$a[17]$	$\|i_1 i_2\|001$	$a[9]$	$\|i_2 i_3\|001$	$a[5]$	$\|i_3 i_4\|001$	$a[6]$
$\|i_0 i_1\|010$	$a[18]$	$\|i_1 i_2\|010$	$a[10]$	$\|i_2 i_3\|010$	$a[12]$	$\|i_3 i_4\|010$	$a[10]$
$\|i_0 i_1\|011$	$a[19]$	$\|i_1 i_2\|011$	$a[11]$	$\|i_2 i_3\|011$	$a[13]$	$\|i_3 i_4\|011$	$a[14]$
$\|i_0 i_1\|100$	$a[20]$	$\|i_2 i_3\|100$	$a[24]$	$\|i_2 i_3\|100$	$a[20]$	$\|i_3 i_4\|100$	$a[18]$
$\|i_0 i_1\|101$	$a[21]$	$\|i_1 i_2\|101$	$a[25]$	$\|i_2 i_3\|101$	$a[21]$	$\|i_3 i_4\|101$	$a[22]$
$\|i_0 i_1\|110$	$a[22]$	$\|i_1 i_2\|110$	$a[26]$	$\|i_2 i_3\|110$	$a[28]$	$\|i_3 i_4\|110$	$a[26]$
$\|i_0 i_1\|111$	$a[23]$	$\|i_1 i_2\|111$	$a[27]$	$\|i_2 i_3\|111$	$a[29]$	$\|i_3 i_4\|111$	$a[30]$

Table 19.8 Identifying P_{11}'s (bit-reversed) local data sets for all cyclic block mappings.

Local data of processor P_{11} expressed in terms of
global array element $a[r] = x_m$, $r = i_0 i_1 i_2 i_3 i_4$, and $m = i_4 i_3 i_2 i_1 i_0$
for all possible cyclic block mappings (CBMs)

Blocksize=8 $\|$PID$\|$Local M $\|i_0 i_1\|i_2 i_3 i_4$	$P_{i_0 i_1}$ $= P_{11}$ $a[r]$	Blocksize=4 $\|$PID$\|$Local M $\|i_1 i_2\|i_0 i_3 i_4$	$P_{i_1 i_2}$ $= P_{11}$ $a[r]$	Blocksize =2 $\|$PID$\|$Local M $\|i_2 i_3\|i_0 i_1 i_4$	$P_{i_2 i_3}$ $= P_{11}$ $a[r]$	Blocksize=1 $\|$PID$\|$Local M $\|i_3 i_4\|i_0 i_1 i_2$	$P_{i_3 i_4}$ $= P_{11}$ $a[r]$
$\|i_0 i_1\|000$	$a[24]$	$\|i_1 i_2\|000$	$a[12]$	$\|i_2 i_3\|000$	$a[6]$	$\|i_3 i_4\|000$	$a[3]$
$\|i_0 i_1\|001$	$a[25]$	$\|i_1 i_2\|001$	$a[13]$	$\|i_2 i_3\|001$	$a[7]$	$\|i_3 i_4\|001$	$a[7]$
$\|i_0 i_1\|010$	$a[26]$	$\|i_1 i_2\|010$	$a[14]$	$\|i_2 i_3\|010$	$a[14]$	$\|i_3 i_4\|010$	$a[11]$
$\|i_0 i_1\|011$	$a[27]$	$\|i_1 i_2\|011$	$a[15]$	$\|i_2 i_3\|011$	$a[15]$	$\|i_3 i_4\|011$	$a[15]$
$\|i_0 i_1\|100$	$a[28]$	$\|i_2 i_3\|100$	$a[28]$	$\|i_2 i_3\|100$	$a[22]$	$\|i_3 i_4\|100$	$a[19]$
$\|i_0 i_1\|101$	$a[29]$	$\|i_1 i_2\|101$	$a[29]$	$\|i_2 i_3\|101$	$a[23]$	$\|i_3 i_4\|101$	$a[23]$
$\|i_0 i_1\|110$	$a[30]$	$\|i_1 i_2\|110$	$a[30]$	$\|i_2 i_3\|110$	$a[22]$	$\|i_3 i_4\|110$	$a[27]$
$\|i_0 i_1\|111$	$a[31]$	$\|i_1 i_2\|111$	$a[31]$	$\|i_2 i_3\|111$	$a[31]$	$\|i_3 i_4\|111$	$a[31]$

19.2 Parallelizing In-Place FFTs Without Inter-Processor Permutations

Computing the butterflies involving the address bits used to define the processor ID number will involve exchange of data between processors whose ID numbers differ in exactly one bit. Of course these processors may or may not be physically adjacent, depending on the network topology.

For example, if the p processors form a hypercube network, data communication between such a pair is always between neighboring processors. If the p processors form a linear array or a two-dimensional grid, such a pair of processors can be physically many hops apart, and the simultaneous data exchange between all pairs can cause traffic congestion in the network.

In general, in the parallel context, if permutations are allowed, it may turn out that part of a processor's complement of data may migrate to another processor. This chapter deals with the case where no permutations are performed.

19.2.1 Parallel DIF$_{NR}$ and DIT$_{NR}$ algorithms

Consider first the sequential FFTs for naturally ordered input data. To parallelize the DIF$_{NR}$ algorithm from Chapter 4, one may use any one of the cyclic block mappings from Chapter 17. For $N = 32$, the computation using the block (or consecutive data) mapping is depicted below.

$$|i_4 i_3|i_2 i_1 i_0 \qquad |i_4 i_3|i_2 i_1 i_0 \qquad |\tau_4 i_3|i_2 i_1 i_0 \qquad |\tau_4 \tau_3|i_2 i_1 i_0 \qquad |\tau_4 \tau_3|\tau_2 i_1 i_0 \qquad |\tau_4 \tau_3|\tau_2 \tau_1 i_0$$

(Initial Map) $\Longleftrightarrow$ $\Longleftrightarrow$

The shorthand notation previously used for sequential FFT is augmented above by two additional symbols. The double-headed arrow $\Longleftrightarrow$ indicates that $\frac{N}{P}$ data elements must be exchanged between processors in advance of the butterfly computation. The symbol $\underset{\triangle}{i_k}$ identifies two things:

- First, it indicates that the incoming data from another processor are the elements whose addresses differ from a processor's own data in bit i_k.

- Second, it indicates that all pairs of processors whose binary ID number differ in bit i_k send each other a copy of their own data.

The required data communications before the first stage of butterfly computation are explicitly depicted in Figures 19.1 and 19.2; the required data communications before the second stage of butterfly computation are depicted in Figures 19.3 and 19.4.

Of course, the other three possible cyclic block mappings may be used, and the corresponding parallel algorithms can be similarly expressed in the shorthand notations below.

$$i_4|i_3 i_2|i_1 i_0 \qquad i_4|i_3 i_2|i_1 i_0 \qquad \tau_4|i_3 i_2|i_1 i_0 \qquad \tau_4|\tau_3 i_2|i_1 i_0 \qquad \tau_4|\tau_3 \tau_2|i_1 i_0 \qquad \tau_4|\tau_3 \tau_2|\tau_1 i_0$$

(Initial Map) $\Longleftrightarrow$ $\Longleftrightarrow$

$$i_4 i_3|i_2 i_1|i_0 \qquad i_4 i_3|i_2 i_1|i_0 \qquad \tau_4 i_3|i_2 i_1|i_0 \qquad \tau_4 \tau_3|i_2 i_1|i_0 \qquad \tau_4 \tau_3|\tau_2 i_1|i_0 \qquad \tau_4 \tau_3|\tau_2 \tau_1|i_0$$

(Initial Map) $\Longleftrightarrow$ $\Longleftrightarrow$

$$i_4 i_3 i_2|i_1 i_0| \qquad i_4 i_3 i_2|i_1 i_0| \qquad \tau_4 i_3 i_2|i_1 i_0| \qquad \tau_4 \tau_3 i_2|i_1 i_0| \qquad \tau_4 \tau_3 \tau_2|i_1 i_0| \qquad \tau_4 \tau_3 \tau_2|\tau_1 i_0|$$

(Initial Map) $\Longleftrightarrow$ $\Longleftrightarrow$

Since the input sequence is in natural order, after *in-place* butterfly computation, the output is known to be in bit-reversed order. Therefore, the processor initially allocated $x_{i_4 i_3 i_2 i_1 i_0}$ will finally have $X_{i_0 i_1 i_2 i_3 i_4}$ if inter-processor permutation is not allowed. In fact, $X_{i_0 i_1 i_2 i_3 i_4}$ is the last derivative which overwrites $x_{i_4 i_3 i_2 i_1 i_0}$ in the same location.

Since the DIT$_{NR}$ algorithm differs from the DIF$_{NR}$ only in the application of twiddle factors, the shorthand notations given above also represent the DIT$_{NR}$ algorithm.

Figure 19.1 Data sent and received by processors P_0 and P_2.

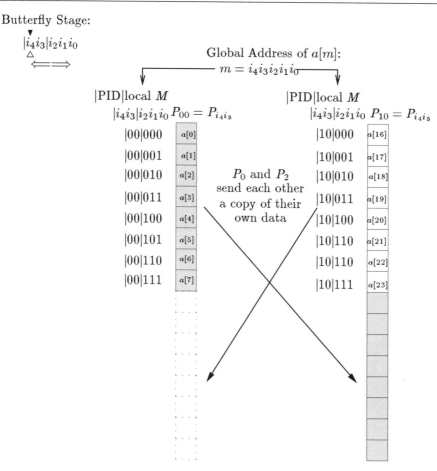

Figure 19.2 Data sent and received by processors P_1 and P_3.

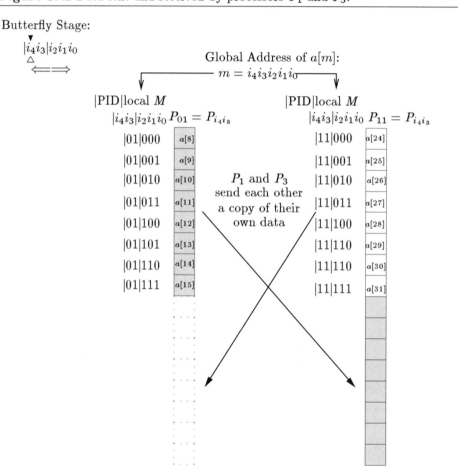

Figure 19.3 Data sent and received by processors P_0 and P_1.

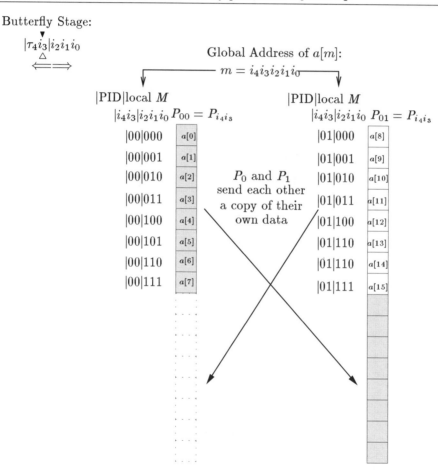

Butterfly Stage:

$|\tau_4 i_3|i_2 i_1 i_0$

Global Address of $a[m]$:

$m = i_4 i_3 i_2 i_1 i_0$

|PID|local M

$|i_4 i_3|i_2 i_1 i_0$ $P_{00} = P_{i_4 i_3}$

$|00|000$	$a[0]$
$|00|001$	$a[1]$
$|00|010$	$a[2]$
$|00|011$	$a[3]$
$|00|100$	$a[4]$
$|00|101$	$a[5]$
$|00|110$	$a[6]$
$|00|111$	$a[7]$

P_0 and P_1 send each other a copy of their own data

|PID|local M

$|i_4 i_3|i_2 i_1 i_0$ $P_{01} = P_{i_4 i_3}$

$|01|000$	$a[8]$
$|01|001$	$a[9]$
$|01|010$	$a[10]$
$|01|011$	$a[11]$
$|01|100$	$a[12]$
$|01|110$	$a[13]$
$|01|110$	$a[14]$
$|01|111$	$a[15]$

Figure 19.4 Data sent and received by processors P_2 and P_3.

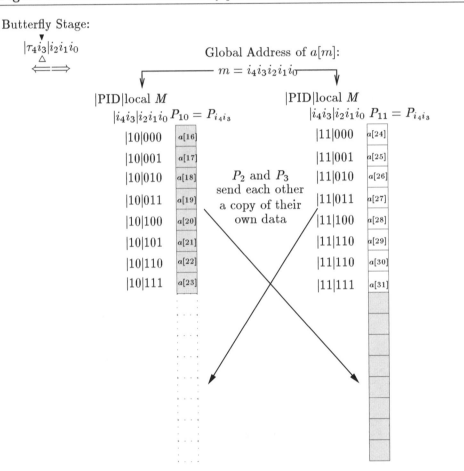

19.2.2 Interpreting the data mapping for bit-reversed output

As a concrete example, suppose the initial mapping is "$|i_4 i_3|i_2 i_1 i_0$." Then processor P_0 initially contains $a[0] = x_0, a[1] = x_1, a[2] = x_2, \cdots, a[7] = x_7$ in their natural order as depicted in Figure 19.1. When the parallel FFT ends, processor P_0 contains $a[0] = X_0, a[1] = X_{16}, a[2] = X_8, a[3] = X_{24}, a[4] = X_4, a[5] = X_{20}, a[6] = X_{12}, a[7] = X_{28}$, which are the first eight elements in the output array in Figure 4.4. In this case, x and X are said to be *comparably mapped* to the processors [64, page 160].

Note that $|i_4 i_3|$ is a subsequence from the subscript of x_m when $m = i_4 i_3 i_2 i_1 i_0$, but $|i_4 i_3|$ is obviously *not* a subsequence of the subscript of X_r when $r = i_0 i_1 i_2 i_3 i_4$. Thus when the data mapping is a CBM of naturally ordered input x, it is not a CBM with respect to the subscript of the bit-reversed output data element X_r.

19.2.3 Parallel DIF_{RN} and DIT_{RN} algorithms

The remaining two in-place sequential FFT variants deal with bit-reversed input data, and they are the DIF_{RN} algorithm from Chapter 5 and the DIT_{RN} algorithm from Chapter 8. For the same example of length $N = 32$, the parallel algorithms corresponding to the four possible cyclic block mappings are represented below.

19.2.4 Interpreting the data mapping for bit-reversed input

Observe that the mapping $|i_0 i_1|i_2 i_3 i_4$ maps $x_{i_4 i_3 i_2 i_1 i_0}$ to processor $P_{i_0 i_1}$. That is, processor P_0 will be allocated $a[0] = x_0$, $a[1] = x_{16}$, $a[2] = x_8$, $a[3] = x_{24}$, $a[4] = x_4$, $a[5] = x_{20}$, $a[6] = x_{12}$, $a[7] = x_{28}$.

Thus, while the initial mapping is a cyclic block mapping with respect to the array address $r = i_0 i_1 i_2 i_3 i_4$, it is obviously *not* a CBM mapping with respect to the subscript of the bit-reversed input data element x_m.

However, since $a[r] = X_r$ on output and $r = i_0 i_1 i_2 i_3 i_4$, the mapping will be a CBM mapping with respect to the subscript of naturally ordered output data element X_r.

19.3   Analysis of Communication Cost

As noted above, butterfly computations will cause communication between processors if the two input elements are stored in different processors. Since both input elements are needed to update each of them, the two processors involved must exchange the $\frac{N}{P}$ data elements for each other to update their local data.

The butterflies in any one of the parallel FFTs introduced in the previous section require data to be exchanged in exactly $d = \log_2 P = 2$ stages, regardless of the blocksize used in the mapping. Algorithms of this type are described in [23, 46, 59]. This is also version 1 of the distributed-memory FFTs in [64, pages 156–162].

19.4   Uneven Distribution of Arithmetic Workload

A possible consequence of this class of schemes is that one half of the processors update their local data according to a formula not involving the twiddle factor. For example, in the parallel DIF$_{NR}$ algorithm, one half of the processors each update $\frac{N}{P}$ elements according to

$$y_\ell = (x_\ell + x_{\ell+N/2}),$$

while the other half of the processors update their local data according to a formula involving the multiplication of a pre-computed twiddle factor, i.e., they each update $\frac{N}{P}$ elements according to

$$z_\ell = (x_\ell - x_{\ell+N/2}) \, w_N^\ell \, .$$

Thus, the arithmetic workload is not evenly divided among all processors unless each processor computes both y_ℓ and z_ℓ. This problem is addressed in the next chapter.

Chapter 20

Parallel FFTs with Inter-Processor Permutations

This chapter treats the class of parallel FFTs which employ inter-processor data permutations. That is, part of a processor's initial complement of data may migrate to another processor to accomplish one or all of the following goals:

- To balance arithmetic workload among the processors.

- To reduce communication cost.

- To have the output data elements arranged in a desired ordering.

This chapter contains a description of a collection of algorithms similar to those in the previous chapter which evenly distribute all butterfly computations among the processors, and also reduce the message lengths from $\frac{N}{P}$ elements to $\frac{1}{2}\frac{N}{P}$ in each of the $\log_2 P + 1$ concurrent message exchanges. The key to achieving this involves data exchanges among processors to effect permutations as well as simply to convey data for purposes of computing butterflies.

20.1 Improved Parallel $\mathrm{DIF_{NR}}$ and $\mathrm{DIT_{NR}}$ Algorithms

It was shown in Chapter 19 that when no inter-processor permutation was allowed in the parallel $\mathrm{DIF_{NR}}$ FFT, the computation of each butterfly was unevenly split between two processors. To avoid this difficulty, an alternative which allows each processor to *replace* one half of its own data with the incoming data is described below. The discussion that follows will focus on the $\mathrm{DIF_{NR}}$ FFT; as will be apparent by the end of the section, the substance of the discussion is the same for the $\mathrm{DIT_{NR}}$ FFT.

20.1.1 The idea and a modified shorthand notation

The idea can be explained using a familiar example: suppose that $N = 32$, and a consecutive data map denoted by "$|i_4i_3|i_2i_1i_0$" is used to distribute data among the four processors. Figures 20.1 and 20.2 show how the data are permuted within each pair of processors *in advance* of the first stage of butterfly computation, and how each processor can then compute exactly the same number of "whole" butterflies—which, of course, implies equal division of arithmetic work.

A shorthand notation must now reflect both the permutation and the computation accomplished in Figures 20.1 and 20.2. A notation which serves these purposes is obtained by modifying the notation for the corresponding parallel FFT (without data permutation between processors) from Chapter 19.

The modified notation begins with the initial map and the first stage of butterfly computation represented by

$$
\begin{array}{cc}
\blacktriangledown \\
|i_4 i_3|i_2 i_1 i_0 \quad |i_2 i_3|i_4 i_1 i_0 \\
\qquad\qquad \vartriangle \; \vartriangle \\
(\text{Initial Map}) \quad \longleftarrow\!\longrightarrow
\end{array}
$$

The symbol $\longleftarrow\!\longrightarrow$ denotes one concurrent message exchange of $\frac{1}{2}\frac{N}{P}$ data elements within all pairs of processors, which occurs in the butterfly stages involving bits which form the processor ID number.

Observe that after data are distributed to individual processors according to the initial mapping $|i_4 i_3|i_2 i_1 i_0$, the element $x_{i_4 i_3 i_2 i_1 i_0}$ in $a[i_4 i_3 i_2 i_1 i_0]$ can be found in $A[i_2 i_1 i_0]$ in processor $P_{i_4 i_3}$. For example, $a[19] = x_{19}$ is shown to be initially in $A[3]$ in P_2 in Figure 20.1, $a[14] = x_{14}$ is shown to be initially in $A[6]$ in P_1 in Figure 20.2, and so on.

When bit i_4 in the PID and bit i_2 in the local M switch their positions in the shorthand notation, the mapping is changed to "$|i_2 i_3|i_4 i_1 i_0$", which means that the data in $a[i_4 i_3 i_2 i_1 i_0]$ can now be found in $A[i_4 i_1 i_0]$ in $P_{i_2 i_3}$. For example, $a[19] = x_{19}$ is relocated to $A[7]$ in P_0 after the inter-processor permutation shown in Figure 20.1, $a[14] = x_{14}$ is relocated to $A[2]$ in P_3 after the inter-processor permutation shown in Figure 20.2, and so on.

To identify the one half of the data each processor must send out, the symbol $\vartriangle$ is used to label two different bits: the bit $\underset{\vartriangle}{i_k}$, which has just been permuted from PID into Local M, and the bit $\underset{\vartriangle}{i_\ell}$, which has just been permuted from Local M to the PID. In the example above, $\underset{\vartriangle}{i_4}$ and $\underset{\vartriangle}{i_2}$ have switched their respective positions in the PID and the Local M.

Because i_k was in PID before the switch, $i_k = 1$ in one processor, and $i_k = 0$ in the other processor. On the other hand, because i_ℓ was in Local M before the switch, $i_\ell = 0$ for half of the data, and $i_\ell = 1$ for another half of the data. Consequently, the value of i_k, the PID bit, is equal to i_ℓ, the local M bit, for half of the data elements in each processor, and the notation which represents the switch of these two bits identifies both the PID of the other processor as well as the data to be sent out or received. To depict exactly what happens, the data exchange between two processors and the butterfly computation represented by

$$
\begin{array}{c}
\blacktriangledown \\
|i_2 i_3|i_4 i_2 i_1 i_0 \\
\;\vartriangle \;\; \vartriangle
\end{array}
$$

is shown in its entirety in Figures 20.1 and 20.2.

Figure 20.1 DIF$_{\rm NR}$ butterfly computation (1st stage) with data migration between P_0 and P_2.

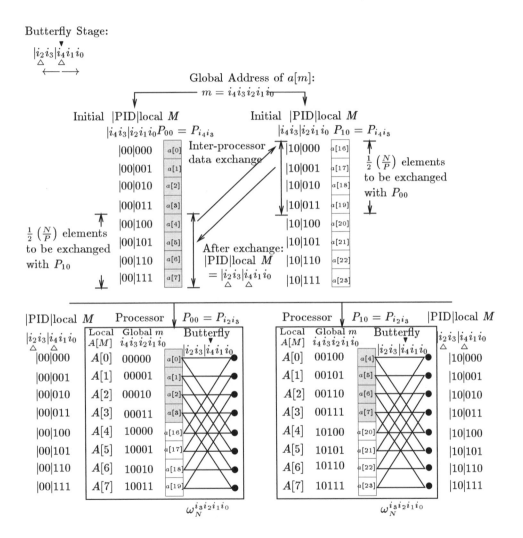

From E. Chu and A. George [28], *Linear Algebra and its Applications*, 284:95–124, 1998. With permission.

Figure 20.2 DIF_{NR} butterfly computation (1st stage) with data migration between P_1 and P_3.

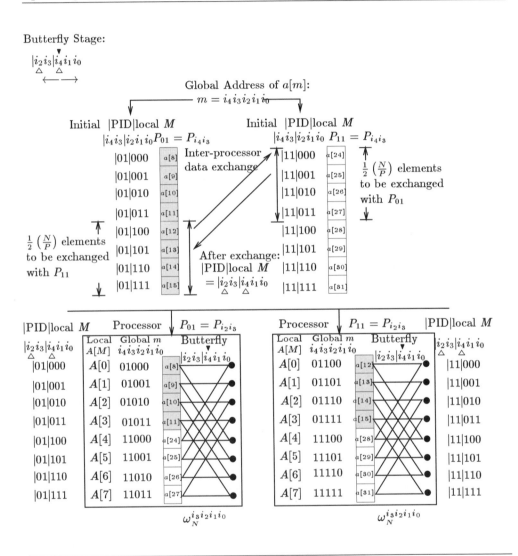

20.1.2   The complete algorithm and output interpretation

Using the shorthand notation developed in the previous section, the complete parallel algorithm corresponding to DIF_{NR} FFT is represented below for the $N = 32$ example.

$$|i_4 i_3|i_2 i_1 i_0 \qquad |i_2 i_3|i_4 i_1 i_0 \quad |i_2 \tau_4|i_3 i_1 i_0 \quad |\tau_3 \tau_4|i_2 i_1 i_0 \quad |\tau_3 \tau_4|\tau_2 i_1 i_0 \quad |\tau_3 \tau_4|\tau_2 \tau_1 i_0$$

$$\text{(Initial Map)} \qquad \longleftarrow \! \longrightarrow \qquad \longleftarrow \! \longrightarrow \qquad \longleftarrow \! \longrightarrow$$

To provide complete information for this example, the second stage of butterfly computations with inter-processor permutation is depicted in Figures 20.3 and 20.4; the third stage of butterfly computations with inter-processor permutation, together with the remaining two stages of local butterfly computations, are depicted in Figures 20.5 and 20.6.

To determine the data mapping for the output elements, observe the following.

- The *in-place* butterfly computation in the DIF_{NR} FFT algorithm ensures that

$$a[i_4 i_3 i_2 i_1 i_0] = x^{(5)}_{i_4 i_3 i_2 i_1 i_0} = X_{i_0 i_1 i_2 i_3 i_4} .$$

- The final mapping $|\tau_3 \tau_4|\tau_2 \tau_1 \tau_0$ indicates that the final content in $a[i_4 i_3 i_2 i_1 i_0]$ is now located in $A[i_2 i_1 i_0]$ in processor $P_{i_3 i_4}$ (instead of the initially assigned processor $P_{i_4 i_3}$).

Accordingly, the output data element $X_{i_0 i_1 i_2 i_3 i_4}$, which overwrites the data in $a[i_4 i_3 i_2 i_1 i_0]$, is finally contained in $A[i_2 i_1 i_0]$ in $P_{i_3 i_4}$.

Figure 20.3 DIF$_{\text{NR}}$ butterfly computation (2nd stage) with data migration between P_0 and P_1.

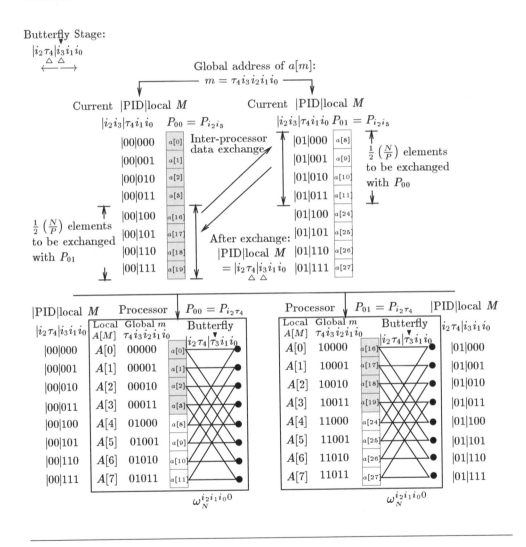

Figure 20.4 DIF$_{NR}$ butterfly computation (2nd stage) with data migration between P_2 and P_3.

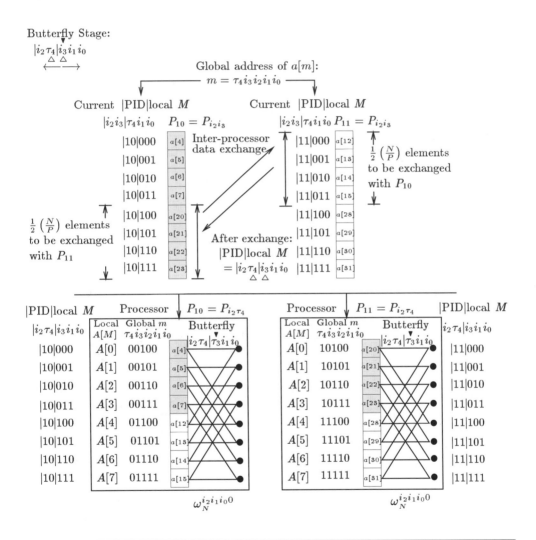

Figure 20.5 $\mathrm{DIF_{NR}}$ butterfly computation (3rd stage) with data migration between P_0 and P_2.

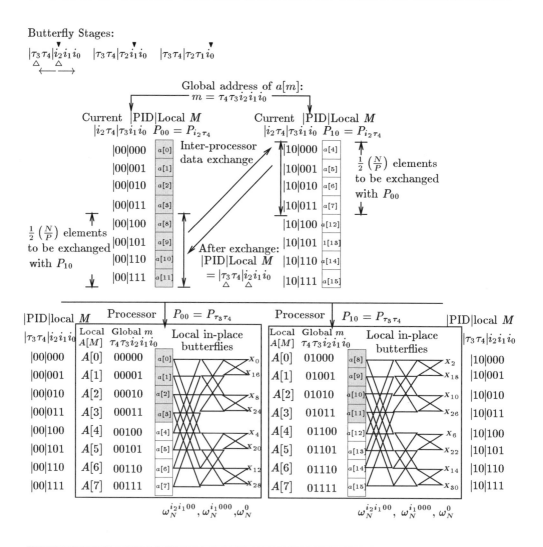

Figure 20.6 DIF$_{\text{NR}}$ butterfly computation (3rd stage) with data migration between P_1 and P_3.

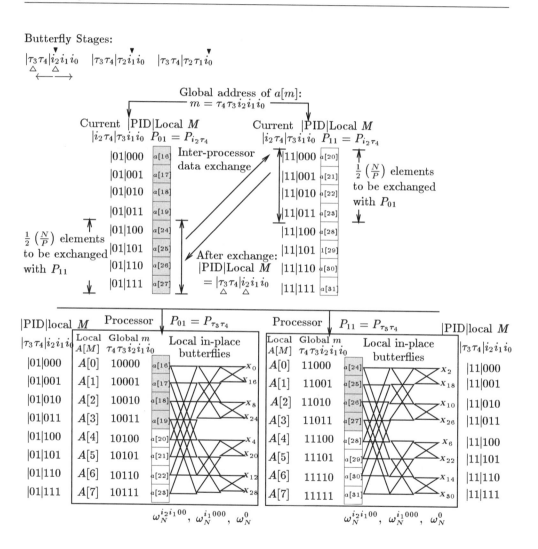

20.1.3 The use of other initial mappings

The parallel algorithms using other initial mappings may be completely specified using the same notations. Given below are the three parallel DIF FFT algorithms corresponding to the three other cyclic block mappings.

$|i_3 i_2|i_4 i_1 i_0$	$|i_3 i_2|i_4 i_1 i_0$	$|\tau_4 i_2|i_3 i_1 i_0$	$|\tau_4 \tau_3|i_2 i_1 i_0$	$|\tau_4 \tau_3|\tau_2 i_1 i_0$	$|\tau_4 \tau_3|\tau_2 \tau_1 i_0$	$|\tau_2 \tau_3|\tau_4 \tau_1 \tau_0$
Initial		$\longleftarrow\longrightarrow$	$\longleftarrow\longrightarrow$			$\longleftarrow\longrightarrow$
$|$PID$|$Local M						(Optional)

$|i_2 i_1|i_4 i_3 i_0$	$|i_2 i_1|i_4 i_3 i_0$	$|i_2 i_1|\tau_4 i_3 i_0$	$|\tau_4 i_1|i_2 \tau_3 i_0$	$|\tau_4 \tau_2|i_1 \tau_3 i_0$	$|\tau_4 \tau_2|\tau_1 \tau_3 i_0$	$|\tau_1 \tau_2|\tau_4 \tau_3 \tau_0$
Initial			$\longleftarrow\longrightarrow$	$\longleftarrow\longrightarrow$		$\longleftarrow\longrightarrow$
$|$PID$|$Local M						(Optional)

$|i_1 i_0|i_4 i_3 i_2$	$|i_1 i_0|i_4 i_3 i_2$	$|i_1 i_0|\tau_4 i_3 i_2$	$|i_1 i_0|\tau_4 \tau_3 i_2$	$|\tau_4 i_0|i_1 \tau_3 \tau_2$	$|\tau_4 \tau_1|i_0 \tau_3 \tau_2$	$|\tau_0 \tau_1|\tau_4 \tau_3 \tau_2$
Initial			$\longleftarrow\longrightarrow$	$\longleftarrow\longrightarrow$		$\longleftarrow\longrightarrow$
$|$PID$|$Local M						(Optional)

Observe that the last permutation is optional because the actual mapping of the output elements can be determined given any mapping, although one mapping may be more convenient than the other if the output data elements are used to continue with a subsequent phase of computation.

It is worth noting that in the examples above, if the optional permutation step is performed, then the array elements which were mapped to one processor will stay together (at the same local address) in a different processor.

Further discussion regarding the optional permutation step and the final mapping is deferred to Section 20.3.

As noted at the beginning of this section, the sequential DIT_{NR} FFT differs from the DIF_{NR} FFT only in the application of twiddle factors. Therefore, the specifications of the various DIT_{NR} versions of the parallel algorithm remain the same as those of the DIF_{NR} versions given above.

20.2 Improved Parallel DIF_{RN} and DIT_{RN} Algorithms

With the explicit understanding that $a[i_0 i_1 i_2 i_3 i_4] = x_{i_4 i_3 i_2 i_1 i_0}$ when input data elements are in bit-reversed order, the parallel FFT (with inter-processor permutation) corresponding to the DIF_{RN} FFT or DIT_{RN} FFT can be specified for the four possible CBM mappings as given below. Note that the final mapping is always specified for $a[i_0 i_1 i_2 i_3 i_4]$, whose initial content of $x_{i_4 i_3 i_2 i_1 i_0}$ is overwritten by the output element $X_{i_0 i_1 i_2 i_3 i_4}$ after the five stages of *in-place* butterfly computation.

$|i_0 i_1|i_2 i_3 i_4$	$|i_0 i_1|i_2 i_3 i_4$	$|i_0 i_1|i_2 i_3 \tau_4$	$|i_0 i_1|i_2 \tau_3 \tau_4$	$|i_0 \tau_2|i_1 \tau_3 \tau_4$	$|\tau_1 \tau_2|i_0 \tau_3 \tau_4$	$|\tau_1 \tau_0|\tau_2 \tau_3 \tau_4$
Initial			$\longleftarrow\longrightarrow$	$\longleftarrow\longrightarrow$		$\longleftarrow\longrightarrow$
$|$PID$|$Local M						(Optional)

$$|i_1 i_2|i_0 i_3 i_4 \quad |i_1 i_2|i_0 i_3 i_4 \quad |i_1 i_2|i_0 i_3 \tau_4 \quad |i_1 \tau_3|i_0 i_2 \tau_4 \quad |\tau_2 \tau_3|i_0 i_1 \tau_4 \quad |\tau_2 \tau_3|i_0 \tau_1 \tau_4 \quad |\tau_2 \tau_1|\tau_0 \tau_3 \tau_4$$

Initial
|PID|Local M (Optional)

$$|i_2 i_3|i_0 i_1 i_4 \quad |i_2 i_3|i_0 i_1 i_4 \quad |i_2 \tau_4|i_0 i_1 i_3 \quad |\tau_3 \tau_4|i_0 i_1 i_2 \quad |\tau_3 \tau_4|i_0 i_1 \tau_2 \quad |\tau_3 \tau_4|i_0 \tau_1 \tau_2 \quad |\tau_3 \tau_2|\tau_0 \tau_1 \tau_4$$

Initial
|PID|Local M (Optional)

$$|i_3 i_4|i_0 i_1 i_2 \quad |i_3 i_0|i_4 i_1 i_2 \quad |\tau_4 i_0|i_3 i_1 i_2 \quad |\tau_4 i_0|\tau_3 i_1 i_2 \quad |\tau_4 i_0|\tau_3 i_1 \tau_2 \quad |\tau_4 \tau_3|i_0 \tau_1 \tau_2$$

Initial
|PID|Local M

As noted earlier, in addition to evenly distributing all butterfly computations among the processors, the message length is reduced from $\frac{N}{P}$ elements to $\frac{1}{2}\frac{N}{P}$ in each of the $\log_2 P + 1$ concurrent message exchanges.

20.3 Further Technical Details and a Generalization

Note that in most examples given in this chapter, the PID bit in question is always exchanged with the leftmost bit in the Local M, which is often referred to as the "pivot" [95, 104]. In these cases, as shown in Figures 20.1, 20.2, 20.3, 20.4, 20.5, and 20.6, the data migrated between processors are consecutively stored in either the top half or the bottom half of each processor's local array A.

However, in one example in Section 20.2, the PID bit in question is always exchanged with the second bit from the right in the Local M; in another example in Section 20.2, the PID bit is always exchanged with the rightmost bit in the Local M. Thus, the pivot could be arbitrarily chosen, if one so desires, from the bits of the Local M.

Since the ID number is formed by consecutive bits when a cyclic block mapping is used, whenever a PID bit is permuted into the local pivot position, it will be exchanged with the next PID bit and occupy the latter's position back in the PID field. After d exchanges, one has the following scenario: the rightmost PID bit is in the Local M, and the pivot i_ρ or τ_ρ from Local M is still in the leftmost position in PID. The example in Section 20.1.2 demonstrates the case involving the i_ρ bit from Local M, and the three examples in Section 20.1.3 demonstrate the case involving the τ_ρ bit from Local M.

Therefore, one more permutation involving these two bits will get i_ρ or τ_ρ back into its original position in the Local M, and the rightmost PID bit would be cyclic-shifted into the leftmost position in the PID as shown below.

$$|\tau_{k-d+1}\tau_k \cdots \tau_{k-d+2}|\tau_{n-1} \cdots \tau_{k+2}\tau_{k+1}\, i_{k-d} \cdots i_{\rho+1}i_\rho i_{\rho-1} \cdots i_1 i_0$$

or

$$|\tau_{k-d+1}\tau_k \cdots \tau_{k-d+2}|\tau_{n-1} \cdots \tau_{\rho+1}\tau_\rho \tau_{\rho-1} \cdots \tau_{k+2}\tau_{k+1}\, i_{k-d} \cdots i_1 i_0$$

Observe that as long as the PID is not formed by the leftmost d bits, there would be at least one τ_ρ bit available when the butterfly computation reaches any PID bit. One thus has the option of using a τ_ρ bit (instead of i_ρ bit) as the pivot. In this case, the τ_ρ bit may stay as the leftmost bit in the PID if the local data is not required to remain together in one processor, and one concurrent message exchange can be saved (by not performing the so-called optional permutation for examples in Section 20.1.3), with the final mapping determined by

$$\left| \tau_\rho \tau_k \cdots \tau_{k-d+2} \right| \overbrace{\tau_{n-1} \cdots \tau_{\rho+1} \tau_{k-d+1} \tau_{\rho-1} \cdots \tau_{k+2} \tau_{k+1}} \overbrace{i_{k-d} \cdots i_1 i_0} .$$

The PID in such a final mapping is no longer formed by consecutive bits. For $N = 2^5$, such a final "$|\text{PID}|\text{Local } M$" is a permutation of the $n = 5$ bits, which still uniquely determines the data mapping of $a[i_4 i_3 i_2 i_1 i_0] = X_{i_0 i_1 i_2 i_3 i_4}$.

In certain contexts, it is important to have the output elements X mapped to the processors in a specific way to facilitate subsequent computations. For example, the final distribution of X (with respect to its bit-reversed subscript) is required to be identical to the initial one for x (with respect to its naturally-ordered subscript) in the solution. This has motivated the development of a number of algorithms which are reviewed in the next chapter.

Chapter 21

A Potpourri of Variations on Parallel FFTs

In this chapter, readers are introduced to a number of parallel FFTs, each of which uses a special "wrinkle" to achieve some objective. To describe these in a single framework, the following conventions and interpretations from previous chapters are adopted, and are exploited in a consistent manner. They are also employed in developing some additional variations that appear to be new.

♦ 1. **(Cyclic Block Maps and Ordered Input)** An initial mapping denoted by

$$|\,\text{PID}\,|\,\text{Local } M = |i_k \cdots i_{k-d+1}|\overbrace{i_{n-1} \cdots i_{k+2} i_{k+1}}\,\overbrace{i_{k-d} \cdots i_1 i_0}$$

indicates a cyclic block map (CBM) of array elements $a[m]$, with the understanding that $a[m] = x_m$, and $m = i_{n-1} \cdots i_1 i_0$.

Since $a[m] = x_m$, the mapping may be conveniently interpreted either with respect to the array address m or with respect to the subscript of naturally ordered x_m.

♦ 2. **(Cyclic Block Maps and Ordered Output)** An initial mapping denoted by

$$|\,\text{PID}\,|\,\text{Local } M = |i_{k-d+1} \cdots i_k|\overbrace{i_0 \cdots i_{k-d}}\,\overbrace{i_{k+1} \cdots i_{n-1}}$$

indicates a CBM map of array elements $a[r]$, with the understanding that $a[r] = x_m$, $r = i_0 i_1 \cdots i_{n-1}$, and $m = i_{n-1} \cdots i_1 i_0$.

If the final mapping remains the same, then because X_r overwrites x_m in $a[r]$ using a RN algorithm, the mapping may be determined either with respect to the array address r or with respect to the subscript of the naturally ordered output element X_r.

♦ 3. **(Equivalent Cyclic Block Maps)** If inter-processor permutations (or local rearrangements of data) are allowed in carrying out the parallel FFT, the final "|PID|Local M" may not be a CBM map of array elements $a[\ell]$, and furthermore, it is possible that $a[\ell] \neq x_\ell$, and $a[\ell] \neq X_\ell$.

For example, the method in Chapter 20 yielded a final mapping of

$$|\,\text{PID}\,|\,\text{Local } M = |\tau_1\tau_0|\tau_2\tau_3\tau_4$$

where $a[i_1i_0i_2i_3i_4] = x^{(5)}_{i_4i_3i_2i_1i_0} = X_{i_0i_1i_2i_3i_4}$; thus $a[\ell] \neq x^{(5)}_\ell$, and $a[\ell] \neq X_\ell$. In such cases, one can determine the mapping directly with respect to the subscript of the output data element X_r, $r = i_0i_1i_2i_3i_4$.

For the example above, the final map $|\tau_1\tau_0|\tau_2\tau_3\tau_4$ dictates that elements X_0 to X_7 are stored in natural order in P_0, elements X_8 to X_{15} are stored in natural order in P_2 (instead of P_1 from a consecutive data map of X_r), elements X_{16} to X_{24} are stored in natural order in P_1 (instead of P_2 from a consecutive data map of X_r), and elements X_{25} to X_{31} are stored in natural order in P_3.

Since processors P_1 and P_2 are indistinguishable from each other (except for their ID numbers), this mapping may be viewed as (and is indeed) an "equivalent" consecutive data map of naturally ordered X_r.

Of course the same definition may be used in describing the mapping for input data.

In general, given a cyclic block map, an equivalent cyclic block map may be obtained by permuting the bits within the PID field.

♦ 4. **(Unordered Output)** If the final "$|$PID$|$Local M" is *neither* a CBM *nor* an equivalent CBM with respect to the subscript of naturally ordered output element X_r, then the output is considered "unordered." (Again, the same may be said about input data under similar circumstances.)

21.1 Parallel FFTs without Inter-Processor Permutations

The basic parallel FFTs from Chapter 19 that do not employ inter-processor permutation form the basis for the work found in [23, 46, 59, 110]. The important ideas in those articles are described in the subsections that follow.

21.1.1 The PID in Gray code

The variant in [23] can be characterized as replacing the PID in a consecutive block map with its Gray code (see Section 18.3 for a discussion of Gray codes). For $N = 32$ and $P = 4$, one uses $p[i_4i_3]$ to replace i_4i_3 in the PID field if $p[i_4i_3]$ contains the Gray-code. Using the reflected-binary Gray-code, $p[i_4i_3] = p[0] = 0$, $p[i_4i_3] = p[1] = 1$, $p[i_4i_3] = p[2] = 3$, and $p[i_4i_3] = p[3] = 2$ are obtained. Note that a Gray code cannot be obtained by reordering the PID bits into i_3i_4.

In general, an initial mapping denoted by

$$|\,\text{PID}\,|\,\text{Local } M = |\overbrace{p[i_k \cdots i_{k-d+1}]}\overbrace{i_{n-1} \cdots i_{k+2}i_{k+1}\ i_{k-d} \cdots i_1i_0}$$

is an "equivalent" cyclic block map (CBM) of naturally ordered input $\boldsymbol{x}$. Of course, the content of $p[i_k \cdots i_{k-d+1}]$ is not limited to Gray code, nor is it limited to permutations of the sequence $i_k \cdots i_{k-d+1}$.

With the understanding that $p[i_4 i_3]$ contains the Gray code, the parallel algorithm in [23] can be completely specified for $N = 32$ and $P = 4$ as shown below.

$$|p[i_4 i_3]|i_2 i_1 i_0 \quad |p[i_4 i_3]|i_2 i_1 i_0 \quad |p[\tau_4 i_3]|i_2 i_1 i_0 \quad |p[\tau_4 \tau_3]|i_2 i_1 i_0 \quad |p[\tau_4 \tau_3]|\tau_2 i_1 i_0 \quad |p[\tau_4 \tau_3]|\tau_2 \tau_1 i_0$$

(Initial Map) $\Longleftarrow\!\!\Longrightarrow$ $\Longleftarrow\!\!\Longrightarrow$

Observe that processor P_{PID} is now identified by $P_{p[i_4 i_3]}$ (instead of $P_{i_4 i_3}$). For example, before the first stage of butterfly computation, P_α and P_β send each other a copy of their own data, where $\alpha = p[0 i_3]$ and $\beta = p[1 i_3]$. Accordingly, for $i_3 = 0$, because $\alpha = p[0] = 0$ and $\beta = p[2] = 3$, data exchange occurs between P_0 and P_3— note that the binary representations of α and β differ in both bits; for $i_3 = 1$, because $\alpha = p[1] = 1$ and $\beta = p[3] = 2$, data exchange occurs between P_1 and P_2—note again that the binary representations of α and β differ in both bits.

According to Theorem 18.1 in Chapter 18, if $P = 2^d$, and $p[k]$ contains the Gray code of k for $0 \leq k \leq P - 1$, then $p[i] = \alpha$ and $p[j] = \beta$ differ in *at most* two bits if i and j differ in exactly one bit.

The implications are:

- The total communication cost of this algorithm is still d concurrent exchanges of $\frac{N}{P}$ elements, but the communicating processors are not always neighbors on a hypercube.

- The two communicating processors are at most two hops away on a hypercube.

While evidentially not as efficient as some other parallel FFT algorithms, this provides the flexibility needed when such a mapping is imposed on the FFT computation. This scenario occurs when a Gray code mapping must be used to implement other parallel algorithms which precede (or follow) the FFT computation.

21.1.2   Using an ordered FFT on local data

The algorithm described in [110] begins with a cyclic map of naturally ordered input, and results in reversely-ordered bits in the Local M in the final mapping. This is accomplished by having each processor apply an ordered (sequential) DIF_{NN} (or DIT_{NN}) FFT to its local data. The algorithm in [110] is specified for $N = 32$ and $P = 4$ below.

$$|i_1 i_0|i_4 i_3 i_2 \quad |i_1 i_0|i_3 i_2 i_4 \quad |i_1 i_0|i_2 i_3 \tau_4 \quad |i_1 i_0|i_2 \tau_3 \tau_4 \quad |i_1 i_0|\tau_2 \tau_3 \tau_4 \quad |\tau_1 i_0|\tau_2 \tau_3 \tau_4$$

(Initial Map) $\Longleftarrow\!\!\Longrightarrow$ $\Longleftarrow\!\!\Longrightarrow$

Since the initial mapping is cyclic with naturally ordered x, the resulting mapping is, in general, given by

$$| d \text{ PID bits } |\overbrace{i_d i_{d+1} \cdots i_{n-2} i_{n-1}},$$

where the PID preserves its initial arrangement: $i_{d-1} i_{d-2} \cdots i_1 i_0$.

Since the Local M is formed by the rightmost $n - d$ bits of X_r's subscript $r = i_0 \cdots i_{d-1} i_d i_{d+1} \cdots i_{n-2} i_{n-1}$, the mapping is equivalent to a consecutive block map with respect to the subscript of naturally ordered X, and the term "block-equivalent" is used when referring to this mapping.

To use this method, one simply distributes naturally ordered x before the computation using a cyclic map. Application of the algorithm yields consecutively ordered X_r's within each processor.

Note that the input data must be distributed among the processors one way or the other, and the communication cost for data distribution is the same regardless of the initial mapping. Therefore, the cyclic initial map used by this method does not cause extra communication, and this is one way to obtain the block-equivalent mapping of naturally ordered X.

It will be shown in Chapter 22 that the communication cost can be halved in an improved algorithm which achieves equivalent results.

21.1.3   Using radix-4 and split-radix FFTs

Sequential radix-4 FFTs were introduced in Chapter 11, and split-radix FFTs were introduced in Chapter 12. A quick review of these two chapters reveals that either one of these two algorithms pairs up the same elements as required in the corresponding radix-2 FFT, and that they each differ from the radix-2 FFT only in how the arithmetic operations are performed on the elements.

The algorithm proposed by Fabbretti [46] begins with a consecutive data map of naturally ordered input. For $N = 2^n$ and $P = 2^d$, each processor performs radix-4 computation after each data exchange in the first d stages of butterfly computations, and each processor performs sequential split-radix FFT on its local data in each of the remaining $n - d$ stages of butterfly computations.

For $N = 32$ and $P = 4$, the parallel algorithm in [46] can thus be specified as shown below.

The communication cost of this algorithm remains d concurrent exchanges of $\frac{N}{P}$ elements.

21.1.4   FFTs for Connection Machines

Parallel FFTs on the Connection Machine (CM) are considered by Johnsson in [59]. Since the CM processors are connected by a hypercube network which allows *concurrent* communications on *all* channels, it is proposed in [59] that data elements exchanged between processors (via different channels) be pipelined across butterfly stages as explained below. (Note that one cannot pipeline data exchanges without pipelining butterfly computations.)

With regard to data mapping, the algorithms in [59] feature a consecutive data map, because all CM compilers use this scheme to distribute input data.

To reduce both arithmetic operations and the need for memory bandwidth, radix-4 or radix-8 FFTs are also recommended in [59] for computing local butterflies.

By using the symbol $\underset{\text{pipelined}}{\Longleftrightarrow}$ to indicate that both data exchanges and butterfly computations are "pipelined" element by element on the CM, the parallel algorithm in [59] can be described for $N = 32$ and $P = 2^d = 8$ as shown below.

$\blacktriangledown$	$\blacktriangledown$	$\blacktriangledown$	$\blacktriangledown$	$\blacktriangledown$	$\blacktriangledown$
$\lvert i_4 i_3 i_2 \lvert i_1 i_0$	$i_4 i_3 i_2 \lvert i_1 i_0$	$\lvert \tau_4 i_3 i_2 \lvert i_1 i_0$	$\lvert \tau_4 \tau_3 i_2 \lvert i_1 i_0$	$\lvert \tau_4 \tau_3 \tau_2 \lvert i_1 i_0$	$\lvert \tau_4 \tau_3 \tau_2 \lvert \tau_1 i_0$
	$\triangle$	$\triangle$	$\triangle$		
(Initial Map)	$\underset{\text{pipelined}}{\Longleftrightarrow}$	$\underset{\text{pipelined}}{\Longleftrightarrow}$	$\underset{\text{pipelined}}{\Longleftrightarrow}$		
	radix-2	radix-2	radix-2	radix-4	radix-4

Note that both communications and butterfly computations are pipelined *element by element*. For the example above, P_0 will begin such fine-grain pipelined communications by exchanging its first element with that of P_4; after this element is updated, P_0 will exchange this updated first element with that of P_2 (along a different channel) at exactly the same time as P_0 exchanges its second element with that of P_4; after these two elements are updated, P_0 can exchange each of its first three elements with the respective element in P_1, P_2, and P_4 concurrently via three different channels. From this point on, P_0 can continue to *concurrently* exchange three consecutive elements (beginning with its second element, then its third element, and so on) each with the respective element in P_1, P_2, and P_4 until its last element has been exchanged with that in P_1. At this point, every element of P_0 has been exchanged with its corresponding element in P_4, P_2, and P_1, and every element has been updated three times in the desired order.

By symmetry, every other processor pipelines its data exchanges and butterfly computations element by element along all its $d = 3$ channels in a symmetric manner.

In total, there are K concurrent exchanges of a single element along all channels, where $K = (d - 1) + \frac{N}{P}$. The communication cost is thus

$$K \times (\alpha + \hat{\beta}) \approx \left(d + \frac{N}{P} \right) \alpha + \left(d + \frac{N}{P} \right) \hat{\beta},$$

where α is the start-up time per message, and $\hat{\beta}$ is the transfer time per complex number. Note that α is negligible on the Connection Machine.

On the other hand, the cost of d (non-pipelined) concurrent exchanges of $\frac{N}{P}$ elements is measured by

$$d \times \left(\alpha + \frac{N}{P} \hat{\beta} \right) = d\alpha + \left(d \times \frac{N}{P} \right) \hat{\beta}.$$

Thus, a fine-grain pipelined method is a good choice when α is negligible, and that is why the pipelined method is used for CM FFT algorithms. In that case, the pipelined communication time is approximately $1/d$ of the non-pipelined time.

However, the non-pipelined scheme becomes more efficient when $\alpha \gg \hat{\beta}$, which is the case on Intel's iPSC hypercubes and other distributed-memory multiprocessors such as the Delta and Paragon meshes, so the non-pipelined concurrent exchanges are commonly used on these multiprocessors.

21.2 Parallel FFTs with Inter-Processor Permutations

The FFT algorithms in [36, 56, 95, 104, 107] all employ interprocessor-permutations which, as introduced in Chapter 20, may be viewed as switching a PID bit with an address bit, the pivot. These algorithms are described in the subsections that follow.

21.2.1 Restoring the initial map at every stage

The parallel algorithm in [107] restores its initial map at the end of every stage by having each processor return the other processor's updated data. The message exchanges are thus doubled as shown below for an example with $N=32$ and $P=2^d=4$.

$$
\begin{array}{lllll}
|i_4 i_3|i_2 i_1 i_0 & |i_2 i_3|i_4 i_1 i_0 & |\tau_4 i_3|i_2 i_1 i_0 & |\tau_4 i_2|i_3 i_1 i_0 & |\tau_4 \tau_3|i_2 i_1 i_0 \\
 & & \text{restore} & & \text{restore} \\
\text{(Initial Map)} & \longleftrightarrow & \xrightarrow{\text{map}} & \longleftrightarrow & \xrightarrow{\text{map}} \\
\end{array}
$$

$$
\begin{array}{lll}
|\tau_4 \tau_3|i_2 i_1 i_0 & |\tau_3 \tau_4|\tau_2 i_1 i_0 & |\tau_3 \tau_4|\tau_2 \tau_1 i_0 \\
\end{array}
$$

The communication cost of this variant is $2d$ concurrent message exchanges of $\frac{1}{2}\frac{N}{P}$ elements. Since the initial map is fully restored, the inter-processor permutations are not reflected in the final mapping at all.

21.2.2 Pivoting on the right-most bit in local M

The first example in [56] begins with a cyclic map of naturally ordered input assuming that $\frac{N}{P} = 2$. The algorithm is described below for $N = 32$ and $P = 2^d = 16$.

$$
\begin{array}{lllll}
|i_3 i_2 i_1 i_0|i_4 & |i_3 i_2 i_1 i_0|i_4 & |\tau_4 i_2 i_1 i_0|i_3 & |\tau_4 \tau_3 i_1 i_0|i_2 & |\tau_4 \tau_3 \tau_2 i_0|i_1 & |\tau_4 \tau_3 \tau_2 \tau_1|i_0 \\
\end{array}
$$

Initial
|PID|Local M
(cyclic)   (unordered)

To ensure that the stride for data in butterfly operations is always one (for good locality) when $\frac{N}{P} > 2$, it was proposed in [56] that inter-processor or local permutations should be employed so that the right-most bit in the local M is always the pivot.

For example, if $N = 32$ and $P = 4$, the parallel algorithm in [56] begins with an initial map of "$|i_1 i_0|i_3 i_2 i_4$." Compared with a cyclic map denoted by "$|i_1 i_0|i_4 i_3 i_2$," the bits in the local M has been cyclic shifted one bit to the left, which ensures that the first local butterfly involves only neighboring elements.

$$
\begin{array}{llllll}
|i_1 i_0|i_3 i_2 i_4 & |i_1 i_0|i_3 i_2 i_4 & |i_1 i_0|i_2 \tau_4 i_3 & |i_1 i_0|\tau_4 \tau_3 i_2 & |\tau_2 i_0|\tau_4 \tau_3 i_1 & |\tau_2 \tau_1|\tau_4 \tau_3 i_0 \\
\end{array}
$$

Initial
|PID|Local M
(cyclic-variant)   (unordered)

Since the initial map is cyclic or its variant (with bits shifted in the local M), the pivot permuted into the PID is always a τ_ρ bit instead of an i_ρ bit, and the communication cost is always d concurrent exchanges of $\frac{1}{2}\frac{N}{P}$ elements.

However, the final mapping is neither a CBM nor its equivalent with respect to the subscript of the output element X_r, where $r = i_0 i_1 i_2 i_3 i_4$. The output is thus "unordered."

21.2.3 All-to-all inter-processor communications

The work by Dubey et al. [36] deals with all possible initial CBM mappings for given N and P on a hypercube. The initial CBM for x and the final CBM for X are required to be identical. Under this condition, the authors show that in addition to $d+1$ (or d) concurrent exchanges between all pairs of processors with IDs different in one bit, their generalized subroutine could, in the worst case, require each processor to send data to all the other processors. This requirement may cause severe data contention, depending on the network topology.

The algorithms in [36] can be described by adding the *all-to-all* communication step as shown below. Using a consecutive block map to distribute naturally ordered input elements x_m, $m = i_4 i_3 i_2 i_1 i_0$, the parallel algorithm for $N = 32$ and $P = 4$ is represented by

$$
\begin{array}{|cccccccc|}
\hline
|i_4 i_3|i_2 i_1 i_0 & |i_2 i_3|i_4 i_1 i_0 & |i_2 \tau_4|i_3 i_1 i_0 & |\tau_3 \tau_4|i_2 i_1 i_0 & |\tau_3 \tau_4|\tau_2 i_1 i_0 & |\tau_3 \tau_4|\tau_2 \tau_1 i_0 & |\tau_0 \tau_1|\tau_2 \tau_3 \tau_4 \\
 & & & & & & \text{possibly} \\
\text{(Initial Map)} & \longleftrightarrow & \longleftrightarrow & \longleftrightarrow & & & \text{all-to-all} \\
\hline
\end{array}
$$

Note that the final map "$i_0 i_1|i_2 i_3 i_4$" is a consecutive block map of naturally ordered X_r, $r = i_0 i_1 i_2 i_3 i_4$, and that the last communication step must redistribute the computed output so that $a[i_3 i_4 i_2 i_1 i_0] = X_{i_0 i_1 i_2 i_3 i_4}$ is relocated from $A[i_2 i_1 i_0]$ in processor $P_{i_3 i_4}$ to $A[i_2 i_3 i_4]$ in processor $P_{i_0 i_1}$.

Since bits i_1 and i_0 come from the Local M in $P_{i_3 i_4}$, the sequence $i_0 i_1$ takes on all values of 0, 1, 2, and 3. Therefore, when the sequence $i_0 i_1$ is used to identify the destination processors, all processors are named. That is, each $P_{i_3 i_4}$ must send data to all other processors. This is obviously the worst case.

When other CBM mappings are used, each processor may need to send data to only a subset of all other processors. Hence the name of "possibly all-to-all" is used to label this communication step.

It was proposed in [36] that each processor collect all data with the same destination and send them as one message. Since each processor will receive data from several or all other processors, it will need to arrange its final share of data in appropriate order in the local array.

The parallel algorithms resulting from using the other possible CBM mappings are given below for $N = 2^n = 32$ and $P = 2^d = 4$.

$$|i_3i_2|i_4i_1i_0 \quad |i_3i_2|i_4i_1i_0 \quad |\tau_4i_2|i_3i_1i_0 \quad |\tau_4\tau_3|i_2i_1i_0 \quad |\tau_4\tau_3|\tau_2i_1i_0 \quad |\tau_4\tau_3|\tau_2\tau_1i_0 \quad |\tau_1\tau_2|\tau_0\tau_3\tau_4$$

Initial $\qquad\qquad\longleftrightarrow \quad \longleftrightarrow$ possibly $\rightarrow \cdots \rightarrow$ all-to-all

$|\text{PID}|$Local M

$$|i_2i_1|i_4i_3i_0 \quad |i_2i_1|i_4i_3i_0 \quad |i_2i_1|\tau_4i_3i_0 \quad |\tau_4i_1|i_2\tau_3i_0 \quad |\tau_4\tau_2|i_1\tau_3i_0 \quad |\tau_4\tau_2|\tau_1\tau_3i_0 \quad |\tau_2\tau_3|\tau_0\tau_1\tau_4$$

Initial $\qquad\qquad\longleftrightarrow \quad \longleftrightarrow$ possibly $\rightarrow \cdots \rightarrow$ all-to-all

$|\text{PID}|$Local M

$$|i_1i_0|i_4i_3i_2 \quad |i_1i_0|i_4i_3i_2 \quad |i_1i_0|\tau_4i_3i_2 \quad |i_1i_0|\tau_4\tau_3i_2 \quad |\tau_4i_0|i_1\tau_3\tau_2 \quad |\tau_4\tau_1|i_0\tau_3\tau_2 \quad |\tau_3\tau_4|\tau_0\tau_1\tau_2$$

Initial $\qquad\qquad\longleftrightarrow \quad \longleftrightarrow$ possibly $\rightarrow \cdots \rightarrow$ all-to-all

$|\text{PID}|$Local M

Observe that in the second last case, each $P_{i_4i_2}$ is required to send data to destinations identified by $P_{i_2i_3}$. Since bit i_2 in the PID of the receiver is also part of the sender's PID, its value is fixed to be either 0 or 1 by each sender. Accordingly, for $i_2 = 0$, each P_{i_40} will need to send data to the two processors identified by P_{0i_3}, i.e., P_{00} and P_{01}. Similarly, for $i_2 = 1$, each P_{i_41} will need to send data to the two processors identified by P_{1i_3}, i.e., P_{10} and P_{11}. Therefore, the communication is not all-to-all in this case.

In general, the communication algorithm for output rearrangement varies from mapping to mapping as well as from problem to problem, as is its cost.

A new general algorithm will be presented in the next chapter, which requires only $1.5d$ more "concurrent exchanges" in the worst case. Thus the algorithm proposed in Chapter 22 appears to deal with the question in [36] concerning the communication requirement for solving the data rearrangement problems arising in FFT or other similar algorithms.

21.2.4 Maintaining specific maps for input and output

The work in [95, 104] concerns two specific CBM mappings; in each case the initial mapping for input $x_{i_4i_3i_2i_1i_0}$ is required to be maintained for output $X_{i_0i_1i_2i_3i_4}$. These two parallel algorithms are depicted in Table 21.1 using given N and P values. For the ordered parallel FFT in [95], the communication cost of $2d + 1 = 5$ exchanges for $N = 2^n = 32$ and $P = 2^d = 4$ are marked by five occurrences of $\longleftrightarrow$ in the first column. The algorithm differs depending on whether $d = n/2$, $d < n/2$, $d = n - 1$, or $n/2 < d < n - 1$. Consequently, the communication costs range from $1.5d + 2$ to $2d + 1$ concurrent exchanges as indicated by the following theorem.

Theorem [95] *An ordered FFT of length $N = 2^n$ can be implemented on a hypercube of dimension d with $n/2 + d + 1$ parallel transmissions if $n/2 < d$ and n is even. If n is odd and $(n + 1)/2 < d$, then $(n + 1)/2 + d + 1$ parallel transmissions are required. For the remaining cases the ordered FFT can be implemented with $2d + 1$ parallel transmissions.*

Note that a hypercube of dimension d has $P = 2^d$ processors, and that the condition $n/2 < d$ is equivalent to $\frac{N}{P} < P$, and $(n+1)/2 < d$ is equivalent to $\frac{N}{P} < \frac{P}{2}$, so they represent fine-grain cases. Consequently, for more common medium-grain and large-grain cases, $2d + 1$ parallel exchanges will be required.

The cyclic-order parallel FFT in [104] refers to an FFT with naturally ordered input x and output X both CBM mapped to the processors using one element per block. That is, the initial cyclic mapping of x is maintained for X. The case $\frac{N}{P} < P$ was considered in [104] for the massively parallel Connection Machine (CM). The communication cost was derived in proving the following lemma.

Lemma [104] *A cyclic-order FFT of length $N = 2^n$ can be implemented on a hypercube of dimension d (where $n/2 < d$) with $2d - n/2$ parallel transmissions if n is even and $2d - (n-1)/2$ parallel transmissions if n is odd.*

Since the condition $n/2 < d$ is equivalent to $\frac{N}{P} < P$, this lemma again applies to the fine-grain cases, and the number of parallel message exchanges is $2d - n/2 > d$ (if n is even) or $2d - (n-1)/2 > d$ (if n is odd). In the finest-grain case, $\frac{N}{P} = 1$ if n is even, i.e., $n = d$, and the number of parallel transmissions is given by $2d - n/2 = 1.5d$. In the example cited in Table 21.1, the condition $n/2 < d$ is satisfied, and there are $2d - n/2 = 6$ concurrent exchanges as identified by six occurrences of $\longleftrightarrow$ in column 2.

In both algorithms, the pivot is always the *current leftmost bit* in the Local M, but reordering of the address bits in Local M is also performed by exchanging the pivot with another bit in the Local M. This implies that other bits from Local M can effectively serve as the pivot bit, although they must first be permuted into the fixed pivot location.

In [95, 104], the permutation of any other bit with the fixed pivot bit is defined as a single i-cycle. The algorithms were developed by first decomposing the required permutation (or final mapping) into disjoint cycles, and each disjoint cycle can then be implemented by a sequence of i-cycles. Some of the i-cycles are followed by butterfly computations, and other i-cycles are used only for the purpose of rearranging local data or permuting data between processors.

The pseudo-code (similar to CM Fortran) FFT algorithm given in [104] uses an i-cycle subroutine assuming that $\frac{N}{\tilde{P}} = 2$, where $\tilde{P}$ could represent the number of virtual processors when $\frac{N}{P} > 2$. In the latter case, $\tilde{P} > P$, and the cost of $1.5\tilde{d}$ concurrent exchanges, where $\tilde{d} = \log_2 \tilde{P}$, includes the communication between virtual processors. In the next chapter alternative ways are proposed having the same or lower communication cost and without the restriction to fixed-pivot i-cycles. In addition, the "$\frac{N}{P}$" ratio is not restricted to a specific value.

Table 21.1 Some ordered parallel FFTs.

Swarztrauber [95] Example: $N = 2^n = 32$, $P = 2^d = 4 \left(\frac{N}{P} \geq P \right)$	Tong & Swarztrauber [104] Example: $N = 2^n = 256$, $P = 2^d = 32 \left(\frac{N}{P} < P \right)$
$\|i_4 i_3\|i_2 i_1 i_0$ (block map of x)	$\|i_4 i_3 i_2 i_1 i_0\|i_7 i_6 i_5$ (cyclic map of x)
$\|i_4 i_3\|i_1 i_2 i_0$	$\|i_4 i_3 i_2 i_1 i_0\|i_6 \tau_7 i_5$
$\|i_1 i_3\|i_4 i_2 i_0$ $\longleftarrow \longrightarrow$	$\|i_4 i_3 i_2 i_1 i_0\|i_5 \tau_7 \tau_6$
$\|i_1 i_3\|i_0 i_2 \tau_4$ $\longleftarrow \longrightarrow$	$\|\tau_5 i_3 i_2 i_1 i_0\|i_4 \tau_7 \tau_6$ $\longleftarrow \longrightarrow$
$\|i_1 i_0\|i_3 i_2 \tau_4$ $\longleftarrow \longrightarrow$	$\|\tau_5 \tau_4 i_2 i_1 i_0\|i_3 \tau_7 \tau_6$ $\longleftarrow \longrightarrow$
$\|i_1 i_0\|i_2 \tau_3 \tau_4$	$\|\tau_3 \tau_4 i_2 i_1 i_0\|\tau_5 \tau_7 \tau_6$ $\longleftarrow \longrightarrow$
$\|\tau_2 i_0\|i_1 \tau_3 \tau_4$ $\longleftarrow \longrightarrow$	$\|\tau_3 \tau_4 \tau_5 i_1 i_0\|i_2 \tau_7 \tau_6$ $\longleftarrow \longrightarrow$
$\|\tau_2 \tau_1\|i_0 \tau_3 \tau_4$ $\longleftarrow \longrightarrow$	$\|\tau_3 \tau_4 \tau_5 i_1 i_0\|\tau_6 \tau_7 \tau_2$
$\|\tau_0 \tau_1\|\tau_2 \tau_3 \tau_4$ $\longleftarrow \longrightarrow$ (block map of X)	$\|\tau_3 \tau_4 \tau_5 i_6 i_0\|i_1 \tau_7 \tau_2$ $\longleftarrow \longrightarrow$
	$\|\tau_3 \tau_4 \tau_5 \tau_6 i_0\|\tau_7 \tau_1 \tau_2$
	$\|\tau_3 \tau_4 \tau_5 \tau_6 \tau_7\|i_0 \tau_1 \tau_2$ $\longleftarrow \longrightarrow$ (cyclic map of X)

From E. Chu and A. George [28], *Linear Algebra and its Applications*, 284:95–124, 1998. With permission.

21.3 A Summary Table

The initial map, the final map, and the communication cost of all algorithms discussed in this chapter are summarized in Table 21.2 for $N = 32$ on $P = 2^d = 4$, 8, or 16 processors.

21.4 Notes and References

The idea of parallelizing FFTs with inter-processor permutations can be traced back to Singleton's work [83] in 1967. He developed a method for computing the FFT with auxiliary memory and limited high-speed storage. Instead of using data from the local memories of two processors, Singleton's method in its simplest form processes data from two input tapes. Thus it is not surprising that Singleton correctly predicted in [83] that

> "*This method of computing the fast Fourier transform offers interesting possibilities for parallel computation.*"

Singleton was not alone in his observation; he acknowledged in [83] that

> "*M. Pease of the Stanford Research Institute independently noted the parallel computation possibilities of this arrangement of the fast Fourier transform, and is exploring the idea further.*"

Readers are referred to [74, 75] for Pease's further contributions.

Table 21.2 Parallel FFTs in the literature.

Some One-Dimensional (1-D) Parallel FFTs

using $P = 2^d = 4$, 8, or 16 Processors

Input Data: $N = 2^n = 32$

References for the examples	Initial Mapping of $x_{i_4 i_3 i_2 i_1 i_0}$	Final Mapping of $X_{i_0 i_1 i_2 i_3 i_4}$	# Concurrent Exchanges	Message Length
Walton [107],1986	$\lvert i_4 i_3 \rvert i_2 i_1 i_0$ (block)	$\lvert \tau_4 \tau_3 \rvert \tau_2 \tau_1 \tau_0$ (unordered)	$2d$	$\frac{1}{2}\frac{N}{P}$
Jamieson, Mueller & Siegel [56], 1986	$\lvert i_3 i_2 i_1 i_0 \rvert i_4$ (cyclic)	$\lvert \tau_4 \tau_3 \tau_2 \tau_1 \rvert \tau_0$ (unordered)	d	$\frac{1}{2}\frac{N}{P}$
	$\lvert i_2 i_1 i_0 \rvert i_3 i_4$ (cyclic-variant)	$\lvert \tau_3 \tau_2 \tau_1 \rvert \tau_4 \tau_0$ (unordered)	d	$\frac{1}{2}\frac{N}{P}$
Swarztrauber [95], 1987	$\lvert i_4 i_3 \rvert i_2 i_1 i_0$ (block)	$\lvert \tau_0 \tau_1 \rvert \tau_2 \tau_3 \tau_4$ (block)	$2d + 1\ (\frac{N}{P} \ge P)$ $1.5d + 2\ (\frac{N}{P} = 1)$	$\frac{1}{2}\frac{N}{P}$
Chamberlain [23], 1988	$p[i_4 i_3 i_2]\lvert i_1 i_0$ (block w. PID in Gray code)	$p[\tau_4 \tau_3 \tau_2]\lvert \tau_1 \tau_0$ (unordered)	d	$\frac{N}{P}$
Tong & Swarztrauber [104], 1991	$\lvert i_2 i_1 i_0 \rvert i_4 i_3$ (cyclic)	$\lvert \tau_2 \tau_3 \tau_4 \rvert \tau_0 \tau_1$ (cyclic)	From d to $1.5d$ $(\frac{N}{P} < P)$	$\frac{1}{2}\frac{N}{P}$
Johnsson & Krawitz [59], 1992	$\lvert i_4 i_3 \rvert i_2 i_1 i_0$ (block)	$\lvert \tau_4 \tau_3 \rvert \tau_2 \tau_1 \tau_0$ (unordered)	$d - 1 + \frac{N}{P}$ (on all channels)	1
Dubey, Zubair & Grosch [36], 1994	$\lvert i_4 i_3 \rvert i_2 i_1 i_0$ (block)	$\lvert \tau_0 \tau_1 \rvert \tau_2 \tau_3 \tau_4$ (block)	d exchanges plus all to all in the worst case.	$\frac{1}{2}\frac{N}{P}$ for d & varied lengths for all to all.
	$i_4 \lvert i_3 i_2 \rvert i_1 i_0$ $i_4 i_3 \rvert i_2 i_1 \rvert i_0$ $i_4 i_3 i_2 \rvert i_1 i_0 \rvert$ (cyclic)	$\tau_0 \lvert \tau_1 \tau_2 \rvert \tau_3 \tau_4$ $\tau_0 \tau_1 \rvert \tau_2 \tau_3 \rvert \tau_4$ $\tau_0 \tau_1 \tau_2 \rvert \tau_3 \tau_4 \rvert$ (cyclic)		
Yang [110], 1996	$\lvert i_1 i_0 \rvert i_4 i_3 i_2$ (cyclic)	$\lvert \tau_1 \tau_0 \rvert \tau_2 \tau_3 \tau_4$ (block-equivalent)	d	$\frac{N}{P}$
Fabbretti et. al. [46] (radix-4 and local split-radix), 1996	$\lvert i_4 i_3 \rvert i_2 i_1 i_0$ (block)	$\lvert \tau_4 \tau_3 \rvert \tau_2 \tau_1 \tau_0$ (unordered)	d	$\frac{N}{P}$

From E. Chu and A. George [28], *Linear Algebra and its Applications*, 284:95–124, 1998. With permission.

Chapter 22

Further Improvement and a Generalization of Parallel FFTs

The algorithms developed in this chapter improve and/or generalize those presented in Chapters 20 and 21. Thus, readers should be familiar with the various data mapping techniques and algorithms discussed in the preceding two chapters.

22.1 Algorithms with Specific Mappings for Ordered Output

The two algorithms specified in Table 22.1 are developed in this section. They each require the ordered output to be mapped to processors according to specified mappings.

Comparing to the parallel algorithms which achieve the same or equivalent final mappings in Chapter 21, these two algorithms have lower communication cost either because the message length is halved or because the number of concurrent exchanges is reduced.

22.1.1 Algorithm I

The main idea of this algorithm is as follows. Consider a general case, in which the initial cyclic map of naturally ordered $x_{i_{n-1}\cdots i_1 i_0}$ is transformed to the final map of $X_{i_0 i_1 \cdots i_{n-1}}$ given by

$$| d \text{ PID bits } \overbrace{i_d i_{d+1} \cdots i_{n-2} i_{n-1}},$$

where PID $= i_0 i_{d-1} i_{d-2} \cdots i_2 i_1$, i.e., all bits in the initial PID are cyclic-shifted to the right by one position. Since the rightmost $n - d$ bits of X_r's subscript form the local M, the final mapping is equivalent to a consecutive block map of naturally ordered X.

Using what has been developed earlier in connection with ordered sequential FFTs and inter-processor permutations, an algorithm achieving these objectives can be constructed in a straightforward manner. The entire algorithm is depicted below for

Table 22.1 New distributed-memory parallel FFTs.

New Parallel FFTs using $P = 2^d = 8$ or 32 Processors

Input Data: $N = 2^n = 256$

New Algorithms	Initial Mapping of $x_{i_7 i_6 i_5 i_4 i_3 i_2 i_1 i_0}$	Final Mapping of $X_{i_0 i_1 i_2 i_3 i_4 i_5 i_6 i_7}$	# Concurrent Exchanges	Message Length
Algorithm I	$\lvert i_2 i_1 i_0\rvert i_7 i_6 i_5 i_4 i_3$ (cyclic)	$\lvert \tau_0 \tau_2 \tau_1\rvert \tau_3 \tau_4 \tau_5 \tau_6 \tau_7$ (block-equivalent)	$d+1$	$\frac{1}{2}\frac{N}{P}$
Algorithm II	$\lvert i_2 i_1 i_0\rvert i_7 i_6 i_5 i_4 i_3$ (cyclic)	$\lvert \tau_5 \tau_6 \tau_7\rvert \tau_0 \tau_1 \tau_2 \tau_3 \tau_4$ (cyclic if $\frac{N}{P} \geq P$)	d	$\frac{1}{2}\frac{N}{P}$
	$\lvert i_4 i_3 i_2 i_1 i_0\rvert i_7 i_6 i_5$ (cyclic)	$\lvert \tau_7 \tau_6 \tau_5 \tau_4 \tau_3\rvert \tau_0 \tau_1 \tau_2$ (cyclic-equivalent if $\frac{N}{P} < P$.)	d	$\frac{1}{2}\frac{N}{P}$

From E. Chu and A. George [28], *Linear Algebra and its Applications*, 284:95–124, 1998.   With permission.

$N = 32$ and $P = 2^d = 8$. As explained in Chapter 20, the communication cost is $d+1$ concurrent exchanges of $\frac{1}{2}\frac{N}{P}$ elements.

$$\lvert i_2 i_1 i_0\rvert i_4 i_3 \quad \lvert i_2 i_1 i_0\rvert i_3 i_4 \quad \lvert i_2 i_1 i_0\rvert i_3 \tau_4 \quad \lvert \tau_3 i_1 i_0\rvert i_2 \tau_4 \quad \lvert \tau_3 \tau_2 i_0\rvert i_1 \tau_4 \quad \lvert \tau_3 \tau_2 \tau_1\rvert i_0 \tau_4 \quad \lvert \tau_0 \tau_2 \tau_1\rvert \tau_3 \tau_4$$

Initial
|PID|Local M (ordered local FFT)   (block-
(cyclic)   equivalent)

Observe that Algorithm I combines a specific initial mapping (from Chapter 17) with inter-processor permutations (from Chapter 20) and an ordered local FFT (from Chapters 6 or 9). By symmetry, if so desired, the same algorithm can also transform block mapped x to cyclic-equivalent mapped X, and vice versa.

22.1.2   Algorithm II

In Table 22.1, the objectives achieved by Algorithm II are shown to be the following:

- If $\frac{N}{P} \geq P$, the initial cyclic map is maintained for naturally ordered output.

- If $\frac{N}{P} < P$, the final map is cyclic-equivalent because its PID bits are arranged in reverse order compared to those in a cyclic map.

Note that for $N = 2^n$ and $P = 2^d$, the condition $\frac{N}{P} \geq P$ implies $n - d \geq d$, or $n \geq 2d$; the condition $\frac{N}{P} < P$ implies $n - d < d$, or $n < 2d$.

To provide an example in each case, the initial map of input element x_m, $m = i_7 i_6 i_5 i_4 i_3 i_2 i_1 i_0$, and the final map of output element X_r, $r = i_0 i_1 i_2 i_3 i_4 i_5 i_6 i_7$, are given in Table 22.1. Observe that in both cases, the initial PID is formed by the rightmost d bits of m, and the final PID is formed by rearranging the leftmost d bits of m.

For simplicity, assume $n \geq 2d$ so that the rightmost d bits do not overlap the leftmost d bits; the case $n < d$ will be considered later.

The main ideas which guide the development of Algorithm II are identified by ◆ when they are introduced. The algorithm begins with the idea of combining unordered and ordered sequential FFT in local computation.

> ◆ Assuming $n \geq 2d$, the algorithm is designed to perform the first d stages of local computation without reordering the local data, followed by $n - 2d$ more stages of local computation using an ordered sequential FFT.

The initial cyclic map would thus be changed to

$$= \overbrace{|i_{d-1}i_{d-2}\cdots i_1 i_0}^{\text{| Initial rightmost } d \text{ bits}} \overbrace{|\tau_{n-1}\tau_{n-2}\cdots\tau_{n-d+1}\tau_{n-d}}^{\text{| Initial leftmost } d \text{ bits}} \overbrace{\tau_d\tau_{d+1}\cdots\tau_{n-d-1}}^{\text{Reversed } n - 2d \text{ bits}} .$$

To continue, inter-processor permutations must be used in performing the next d stages of butterfly computation so that a final cyclic map given by

$$|\tau_{n-d}\tau_{n-d+1}\cdots\tau_{n-2}\tau_{n-1}\overbrace{\tau_0\tau_1\cdots\tau_{d-2}\tau_{d-1}}^{}\overbrace{\tau_d\tau_{d+1}\cdots\tau_{n-d-1}}^{}$$

can be obtained. To minimize the communication cost, the following idea is employed.

> ◆ Instead of using a fixed pivot from Local M, a PID bit is switched with a different pivot before each stage of butterfly computation begins, and the pivots are chosen in the order they appear in the final PID.

Accordingly, when butterfly computation ends, i_{d-1} would have been switched with τ_{n-d}, i_{d-2} with τ_{n-d+1}, $\cdots$, i_1 with τ_{n-2}, and i_0 with τ_{n-1} as desired. Observe that all d potential pivot bits are τ_ρ instead of i_ρ, and that they are all permuted into their respective final positions in the PID after d concurrent exchanges of $\frac{1}{2}\frac{N}{P}$ elements.

For the case $n < 2d$, it turns out that one may achieve the following cyclic-equivalent final map of $X_{i_0 i_1 \cdots i_{n-1}}$ at the same communication cost.

$$|\tau_{n-1}\tau_{n-2}\cdots\tau_{d+1}\tau_d\tau_{d-1}\cdots\tau_{n-d}\overbrace{\tau_0\tau_1\cdots\tau_{n-d-1}}^{} .$$

Note that the final PID bit-reverses the rightmost d bits of X's subscript. In order to describe how Algorithm II is generalized to include this case, the condition $n - d < d$ is reflected in the initial map by expressing it as

$$= \overbrace{|i_{d-1}i_{d-2}\cdots i_{n-d}i_{n-d-1}\cdots i_1 i_0}^{\text{| Initial rightmost } d \text{ bits}} \overbrace{|i_{n-1}\cdots i_{d+1}i_d}^{\text{| } n-d < d \text{ bits}} .$$

After the first $n - d$ stages of local computations (without reordering of the local data), the mapping becomes

$$|i_{d-1}i_{d-2}\cdots i_{n-d}i_{n-d-1}\cdots i_1 i_0 \overbrace{|\tau_{n-1}\cdots\tau_{d+1}\tau_d}^{} .$$

> ◆ The next $d - (n - d) = 2d - n$ inter-processor permutations switch the leftmost $2d - n$ bits in PID with the bits in the Local M in the order from left to right; if it reaches the end of Local M, begin from the leftmost bit in Local M until all $2d - n$ bits in PID are processed.

The following map results:

$$\left|\underbrace{\tau_{n-1}\tau_{n-2}\cdots\tau_{k+1}\tau_k\ i_{n-d-1}}\quad\cdots\quad i_1 i_0\right|\overbrace{\tau_{m-1}\tau_{m-2}\cdots\tau_{n-d}}\ \overbrace{\tau_{k-1}\tau_{k-2}\cdots\tau_m}\ .$$

Now the local reordering can be applied to rearrange the bits in Local M to obtain

$$\left|\underbrace{\tau_{n-1}\tau_{n-2}\cdots\tau_{k+1}\tau_k\ i_{n-d-1}}\quad\cdots\quad i_1 i_0\right|\overbrace{\tau_{n-d}\cdots\tau_{m-2}\tau_{m-1}}\ \overbrace{\tau_m\cdots\tau_{k-2}\tau_{k-1}}\ .$$

(Note that the bits braced from τ_{m-1} to τ_{n-d} are reversed, and the bits braced from τ_{k-1} to τ_m are reversed. Since the bit-reversing operations can be incorporated into the computation, the internal reordering does not necessarily increase processing time in an efficient implementation.)

The final $n-d$ inter-processor permutations in Algorithm II switches the remaining $n-d$ bits in PID into the Local M, resulting in the final mapping

$$\left|\tau_{n-1}\tau_{n-2}\cdots\tau_{k+1}\tau_k\ \underbrace{\tau_{k-1}\tau_{k-2}\cdots\tau_m\tau_{m-1}\tau_{m-2}\cdots\tau_{n-d}}\right|\overbrace{\tau_0\tau_1}\quad\cdots\quad\tau_{n-d-1}\ .$$

This is a cyclic-equivalent mapping for $X_{i_0 i_1 i_2 \cdots i_{n-1}}$ because the Local M is formed by the leftmost $n-d$ bits of X_r's subscript. Note that the rightmost d bits of X_r's subscript can be obtained by reversing the PID bits, and vice versa.

22.2 A General Algorithm and Communication Complexity Results

This algorithm allows the choice of *any* initial CBM for naturally ordered input x and *any* final CBM for naturally ordered output X.

For $N = 128$ and $P = 32$, the four possible initial CBMs of x are shown in the first four entries in column 1 in Table 22.2; in each case, the initial CBM is required to be maintained for X as shown in column 2. In column 3, both actual and the maximum (worst-case) communication complexities are presented for $N = 2^n$ and $P = 2^d$, assuming only that $\frac{N}{P} \geq 2$. That is, the specific relationships between n and d in different cases are not exploited in the algorithm.

The last three entries in Table 22.2 contain same information for cases when the final map is different from the initial map.

The construction of this algorithm and the derivation of its communication cost is presented next.

22.2.1 Phase I of the general algorithm

For any given initial mapping, the corresponding parallel algorithm with inter-processor permutations from Chapter 20 is used in phase I of this general algorithm. This process is depicted below once again for a familiar example using $N = 32$ and $P = 2^d = 8$.

$\left\|i_2 i_1 i_0\right\|i_4 i_3$	$\left\|i_2 i_1 i_0\right\|i_4 i_3$	$\left\|i_2 i_1 i_0\right\|\tau_4 i_3$	$\left\|\tau_4 i_1 i_0\right\|i_2 \tau_3$	$\left\|\tau_4 \tau_2 i_0\right\|i_1 \tau_3$	$\left\|\tau_4 \tau_2 \tau_1\right\|i_0 \tau_3$	$\left\|\tau_0 \tau_2 \tau_1\right\|\tau_4 \tau_3$
Initial			$\longleftarrow\longrightarrow$	$\longleftarrow\longrightarrow$	$\longleftarrow\longrightarrow$	$\longleftarrow\longrightarrow$
$\left\|\text{PID}\right\|\text{Local } M$						
(cyclic)						(unordered)

Table 22.2 General communication results for CBM mappings.

A General Algorithm for

Arbitrary Initial and Final CBM Mappings

Assumption: $N = 2^n$, $P = 2^d$, and $\frac{N}{P} \geq 2$

Any initial CBM of $x_{i_6 i_5 i_4 i_3 i_2 i_1 i_0}$	Any final CBM of $X_{i_0 i_1 i_2 i_3 i_4 i_5 i_6}$	# Concurrent Exchanges	Message Length
$\|i_6 i_5 i_4 i_3\|i_2 i_1 i_0$ (block)	$\|\tau_0 \tau_1 \tau_2 \tau_3\|\tau_4 \tau_5 \tau_6$ (block)	$2d + 1 \leq 2.5d + 1$	$\frac{1}{2}\frac{N}{P}$
$i_6\|i_5 i_4 i_3 i_2\|i_1 i_0$	$\tau_0\|\tau_1 \tau_2 \tau_3 \tau_4\|\tau_5 \tau_6$	$2d + 2 \leq 2.5d + 1$	$\frac{1}{2}\frac{N}{P}$
$i_6 i_5\|i_4 i_3 i_2 i_1\|i_0$	$\tau_0 \tau_1\|\tau_2 \tau_3 \tau_4 \tau_5\|\tau_6$	$2d + 2 \leq 2.5d + 1$	$\frac{1}{2}\frac{N}{P}$
$i_6 i_5 i_4\|i_3 i_2 i_1 i_0\|$ (cyclic)	$\tau_0 \tau_1 \tau_2\|\tau_3 \tau_4 \tau_5 \tau_6\|$ (cyclic)	$2d + 1 \leq 2.5d + 1$	$\frac{1}{2}\frac{N}{P}$
$i_6 i_5 i_4\|i_3 i_2 i_1 i_0\|$ (cyclic)	$\|\tau_0 \tau_1 \tau_2 \tau_3\|\tau_4 \tau_5 \tau_6$ (block)	$2d \leq 2.5d + 1$	$\frac{1}{2}\frac{N}{P}$
$i_6 i_5 i_4\|i_3 i_2 i_1 i_0\|$ (cyclic)	$\tau_0 \tau_1\|\tau_2 \tau_3 \tau_4 \tau_5\|\tau_6$	$2d + 2 \leq 2.5d + 1$	$\frac{1}{2}\frac{N}{P}$
$\|i_6 i_5 i_4 i_3\|i_2 i_1 i_0$ (block)	$\tau_0 \tau_1 \tau_2\|\tau_3 \tau_4 \tau_5 \tau_6\|$ (cyclic)	$2d \leq 2.5d + 1$	$\frac{1}{2}\frac{N}{P}$

Note that Phase I is completed after $d+1$ concurrent exchanges of $\frac{1}{2}\frac{N}{P}$ elements. At this point, all arithmetic work has been done, the initial PID bits have been cyclic-shifted to the right by one position, and the mapping of the output is "unordered."

To achieve any specific CBM map for ordered X, the currently unordered output is to be re-distributed and reordered in Phase II of this algorithm.

22.2.2 Phase II of the general algorithm

To change the mapping of computed output, apparently more inter-processor permutations may be added in Phase II to switch the desired τ_ρ bits into the PID and arrange them in final order, and that can be followed by local reordering so that the bits in the Local M are also arranged as desired.

To construct the algorithm and determine the maximum communication cost, consider the two possibilities:

Case (i) If the initial PID does not overlap the final PID, then exactly d more inter-processor permutations are needed to switch in the final PID bits. In total, $2d+1$ concurrent exchanges are needed.

This simple reordering process is depicted below for an example which satisfies the criterion.

$$|\tau_0\,\tau_2\,\tau_1|\tau_7\,\tau_6\,\tau_5\,\tau_4\,\tau_3 \qquad |\tau_5\,\tau_2\,\tau_1|\tau_7\,\tau_6\,\tau_0\,\tau_4\,\tau_3 \qquad |\tau_5\,\tau_6\,\tau_1|\tau_7\,\tau_2\,\tau_0\,\tau_4\,\tau_3 \qquad |\tau_5\,\tau_6\,\tau_7|\tau_1\,\tau_2\,\tau_0\,\tau_4\,\tau_3$$

Map at the $\longleftarrow\!\longrightarrow$ $\longleftarrow\!\longrightarrow$ $\longleftarrow\!\longrightarrow$
end of Phase I
(unordered)

$$|\tau_5\,\tau_6\,\tau_7|\tau_0\,\tau_1\,\tau_2\,\tau_3\,\tau_4$$
Local reordering
to get desired
final map (cyclic)

Case (ii) If the initial PID overlaps the final PID, then one may first switch in the $\lambda\,\tau_\rho$ bits currently in Local M so that these λ PID bits are put in their final positions. Repeat this process until no desired PID bits exist in the local M. These steps are depicted below for an example which exhibits such characteristics.

In the example, bits τ_3 and τ_4 are first brought into their final positions in the PID. The process is then repeated for bit τ_2. Observe that even τ_2 was in the PID at the beginning of phase II, it was switched out because it occupies bit τ_3's final position in the PID.

For this example, all PID bits are now in their final positions. Therefore, only local reordering is needed to achieve the desired output distribution.

$$|\tau_0\,\tau_2\,\tau_1|\tau_5\,\tau_4\,\tau_3 \quad |\tau_0\,\tau_3\,\tau_1|\tau_5\,\tau_4\,\tau_2 \quad |\tau_0\,\tau_3\,\tau_4|\tau_5\,\tau_1\,\tau_2 \quad |\tau_2\,\tau_3\,\tau_4|\tau_5\,\tau_1\,\tau_0 \quad |\tau_2\,\tau_3\,\tau_4|\tau_0\,\tau_1\,\tau_5$$

Map at the $\longleftarrow\!\longrightarrow$ $\longleftarrow\!\longrightarrow$ $\longleftarrow\!\longrightarrow$ Local reordering to
end of Phase I   get desired final
(unordered)   CBM (blocksize=2)

Example 1:   Phase II

However, depending on the problem, it may not be true that all PID bits are in their final positions at this point. In order to show how this could happen as well as the steps possibly required, the initial map used in the next example is assumed to be

$$i_6|i_5 i_4 i_3 i_2|i_1 i_0 \quad \text{or} \quad |i_5 i_4 i_3 i_2|i_6 i_1 i_0$$

The initial map assumed
in Example 2.

And it is assumed that the same map is maintained for the output as specified below.

$$\tau_0|\tau_1 \tau_2 \tau_3 \tau_4|\tau_5 \tau_6 \quad \text{or} \quad |\tau_1 \tau_2 \tau_3 \tau_4|\tau_0 \tau_5 \tau_6$$

The final map assumed
in Example 2.

The steps described above are now performed on Example 2 as depicted below. Observe that bit τ_1 in the local M is permuted into its final position in the PID first. Next, bit τ_2, which occupies bit τ_1's final position and is thus switched out earlier, is permuted with τ_5. Now, all PID bits are in, but they are not in the right order!

$$|\tau_2\tau_5\tau_4\tau_3|\tau_6\tau_1\tau_0 \quad |\tau_1\tau_5\tau_4\tau_3|\tau_6\tau_2\tau_0 \quad |\tau_1\tau_2\tau_4\tau_3|\tau_6\tau_5\tau_0 \quad |\tau_1\tau_2\tau_4\tau_3|\tau_0\tau_5\tau_6$$

Map at the	$\longleftarrow \longrightarrow$	$\longleftarrow \longrightarrow$	Local
end of Phase I			reordering
(unordered)			(unordered PID)

Example 2.: Phase II in progress

To describe the remaining steps in Phase II, suppose k (in total) τ_ρ bits have been permuted into the PID so far. (Note that $\lambda = 2, k = 3$ in Example 1, and $\lambda = 1, k = 2$ in Example 2.) The remaining task now involves reordering the $v = d - k$ potentially out-of-order bits in the PID.

To rearrange the PID bits through inter-processor permutations, observe that one need only consider a fixed (arbitrary) pivot from Local M, because every bit in the current Local M is a non-PID bit in the final map.

How many inter-processor permutations are needed to reorder v PID bits via a single pivot? Observe that in the worst case, the non-PID pivot bit is permuted back in Local M after two PID bits happen to switch their positions, i.e., three permutations could be required for placing every two bits. This is indeed the case when τ_4 and τ_3 are switched in Example 2 as show below.

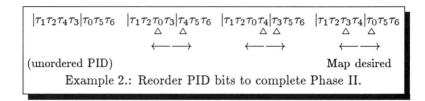

(unordered PID)   Map desired

Example 2.: Reorder PID bits to complete Phase II.

Therefore, at most, $1.5v$ inter-processor permutations are required to reorder the v PID bits. Since v can be as large as d, in total $2.5d + 1$ concurrent exchanges are needed.

Although the two cases above were considered separately to help explain the algorithm, note that case (ii) reduces to case (i) if $\lambda = d$.

Note that by performing only the permutation steps *symbolically*, the actual communication costs for given N and P can be predicted for the entire class of CBM mappings as shown in Table 22.2.

Since the reordering phase requires only $1.5d$ more concurrent exchanges in the worst case, this algorithm appears to have addressed the question in [36] concerning the communication requirement for solving the data rearrangement problems arising in FFT or other similar algorithms.

In addition, since the reordering phase is independent of the computation phase, the algorithm and the communication complexity results are applicable even if the initial and final CBM maps are different. The last three entries in Table 22.2 are such examples. Thus, the proposed algorithm deals with cases that do not appear to be handled by the algorithm proposed in [36].

Chapter 23

Parallelizing Two-dimensional FFTs

23.1   The Computation of Multiple 1D FFTs

The need to compute a set of 1D FFTs arises naturally in many applications. If the FFTs have the same length and properties (e.g., all real or complex), an appropriate sequential FFT algorithm may be applied to them one after the other, or it may be applied to them all at once stage by stage. Since the same set of twiddle factors are applied, it is inefficient in this context to compute them on the fly. Instead, the twiddle factors should be pre-computed (once), stored, and reused for each FFT in the set. Using the latter approach, the storage for the twiddle factors is the same as that required for a single FFT. It is thus straightforward to adapt FFT software to compute multiple 1D FFTs on a uniprocessor machine,

The simplest parallel algorithm to compute multiple 1D FFTs is "embarrassingly parallel"; the set of M 1D FFTs (of length N each) can simply be evenly divided among the p processors, and each processor simply applies an appropriate sequential algorithm to compute its share of $\lceil M/p \rceil$ 1D FFTs. In this case, there is no communication, but the twiddle factors need to be pre-computed and stored in each processor. This was referred to as the "independent processors" approach in [46].

If the computation associated with each single FFT is divided among several processors in some way, additional inter-processor communication is required. This "co-operative processors" approach was also explored in [46], and two algorithms using this approach were compared with the "independent processors" idea on an nCUBE 2 hypercube consisting of 128 processors. As expected, the timing results reported in [46] confirmed efficiency values of 99 to 99.9% for the "independent processors" approach, which was in contrast to efficiency values of 29 to 49% for the two implementations of the cooperating processors.

Although the computation of a two-dimensional FFT may be viewed as computing multiple 1D FFTs in each direction (as shown in the next section), the parallelization of a two-dimensional FFT presents another challenge because the highly efficient "independent processors" approach cannot be used on both directions without incurring inter-processor data communications, and it is no longer the clear winner. The 2D FFTs are the focus of the remainder of this chapter.

23.2   The Sequential 2D FFT Algorithm

In this section a fast serial algorithm for computing the DFT on a two-dimensional image consisting of $N_1 \times N_2$ signals is reviewed. The signals are stored in an $N_1 \times N_2$ matrix $\boldsymbol{x}$. An entry in the signal matrix is denoted by x_{ℓ_1, ℓ_2}. The 2D-DFT of $\boldsymbol{x}$ is defined by the following equation [70, 72]:

(23.1)
$$X_{r_1, r_2} = \sum_{\ell_1=0}^{N_1-1} \sum_{\ell_2=0}^{N_2-1} x_{\ell_1, \ell_2} \omega_{N_1}^{r_1 \ell_1} \omega_{N_2}^{r_2 \ell_2},$$

$$\text{for} \quad r_1 = 0, 1, \cdots, N_1 - 1, \quad \text{and} \quad r_2 = 0, 1, \cdots, N_2 - 1.$$

If the equation above is used in a straightforward (naive) way, $\Theta(N_1 N_2)$ arithmetic operations are required to compute each X_{r_1, r_2}, yielding a total cost of $\Theta(N_1^2 N_2^2)$, or $\Theta(N^4)$ if $N_1 = N_2 = N$. Fortunately, this may be reduced very significantly by separating the 2D-DFT into a series of 1D-DFTs, which can each be implemented using a fast 1D-FFT algorithm. This process is shown below.

(23.2)
$$X_{r_1, r_2} = \sum_{\ell_1=0}^{N_1-1} \sum_{\ell_2=0}^{N_2-1} x_{\ell_1, \ell_2} \omega_{N_1}^{r_1 \ell_1} \omega_{N_2}^{r_2 \ell_2}$$

$$= \sum_{\ell_1=0}^{N_1-1} \omega_{N_1}^{r_1 \ell_1} \left(\sum_{\ell_2=0}^{N_2-1} x_{\ell_1, \ell_2} \omega_{N_2}^{r_2 \ell_2} \right)$$

$$= \sum_{\ell_1=0}^{N_1-1} \omega_{N_1}^{r_1 \ell_1} \left(\tilde{X}_{\ell_1, r_2} \right)$$

$$= \sum_{\ell_1=0}^{N_1-1} \left(\tilde{X}_{\ell_1, r_2} \right) \omega_{N_1}^{r_1 \ell_1}, \quad r_1 = 0, 1, \cdots, N_1 - 1, \ r_2 = 0, 1, \cdots, N_2 - 1.$$

Thus, by effecting a series of (ordered) 1D-FFTs on the N_1 rows (of length N_2 each) of $\boldsymbol{x}$, the data in row ℓ_1 are transformed to $\tilde{X}_{\ell_1, r_2}$ for $0 \leq r_2 \leq N_2 - 1$. The total cost of this row-transform phase is $\Theta(N_1 N_2 \log_2 N_2)$. This is followed by a series of (ordered) 1D-FFTs on the N_2 columns (of length N_1 each) of $\tilde{X}$, and column r_2 of $\tilde{X}$ is transformed to X_{r_1, r_2} for $0 \leq r_1 \leq N_1 - 1$. The cost of the column-transform phase is $\Theta(N_2 N_1 \log_2 N_1)$. The efficiency of the discrete Fourier transform of a digital image consisting of $N_1 \times N_2$ signals is thus improved from $\Theta(N_1^2 N_2^2)$ to $\Theta(N_1 N_2 \log_2(N_1 N_2))$. When $N_1 = N_2 = N$, the computing cost is reduced from $\Theta(N^4)$ to $\Theta(N^2 \log_2 N)$. Therefore, the computational efficiency of the 2D-FFTs is even greater than that of the 1D-FFTs.

Note that if an in-place unordered FFT is used in both row-transform phase and column-transform phase, then the entry x_{ℓ_1, ℓ_2} would be finally overwritten by X_{m_1, m_2}, where the binary representation of m_1 bit-reverses that of ℓ_1, and m_2 is related to ℓ_2 in the same way.

For obvious reasons, the approach based on equation (23.2) is called the "row-column" method [41, 70] or the "separable" method [46] in the literature.

23.2.1 Programming considerations

As noted by Duhamel and Vetterli [41], the matrix which contains the data of a 2D transform grows quickly. For example, if $N_1 = N_2 = 1024$, there are more than one million complex numbers in the 1024×1024 matrix. Depending on the programming language used, this large matrix is stored either column-by-column or row-by-row in computer memory. In order to minimize the number of memory accesses, an efficient method must be used to access blocks of consecutive rows or blocks of consecutive columns in a manner compatible with the storage scheme. To achieve this, the row-column FFT is often performed by including a matrix transposition between the FFTs on the columns and the FFTs on the rows in order to allow access to the data by blocks [41]. A fast method for matrix transposing was proposed by Eklundh in [45]. The two possible implementations of the row-column method are depicted in Figure 23.1.

Figure 23.1 Sequential row-column 2D FFT algorithm—two implementations.

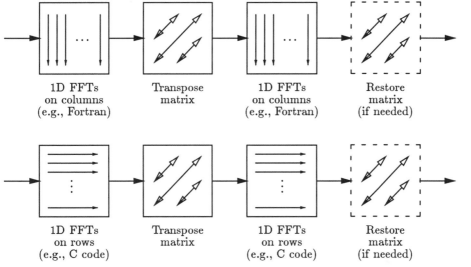

1D FFTs on columns (e.g., Fortran)	Transpose matrix	1D FFTs on columns (e.g., Fortran)	Restore matrix (if needed)
1D FFTs on rows (e.g., C code)	Transpose matrix	1D FFTs on rows (e.g., C code)	Restore matrix (if needed)

23.2.2 Computing a single 1D FFT stored in a 2D matrix

If the data of a single 1D FFT of length $N = 2^n$ is stored in a 2D matrix of dimension $N_1 \times N_2$, where $N = N_1 \times N_2$, $N_1 = 2^{n_1}$, and $N_2 = 2^{n_2}$, it can also be computed by independent 1D FFTs on the rows and columns of the matrix. However, note that in this context the twiddle factors used are derivatives of w_N^ℓ rather than derivatives of $w_{N_1}^\ell$ and $w_{N_2}^\ell$. For example, a 2D matrix was used in [51] to store a 1D FFT for distribution to processors that are connected by a hypercube or a 2D mesh network.

Consequently, except for using different twiddle factors, one can use the 2D FFT algorithm to compute a single 1D FFT if the latter is stored in a 2D matrix. It follows that parallel 2D FFT algorithms can be used to compute a single 1D FFT as well.

23.2.3   Sequential algorithms for matrix transposition

The standard method depicted in Figure 23.2 applies to any block partitioned matrix. Since each A_{ij} block could be of dimensions 1×1, $k \times k$, or $m \times \ell$, the matrix A may be square or rectangular of any size.

Figure 23.2 Transpose a matrix A by the standard method.

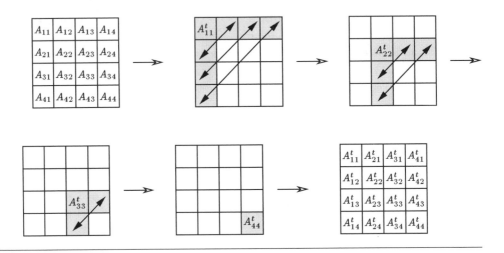

The divide-and-conquer method depicted in Figure 23.3 is recursive by nature. A square or rectangular matrix may be divided into four submatrices at each level.

Figure 23.3 Transpose a matrix A by the recursive method.

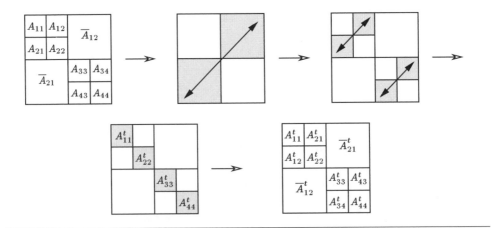

The method depicted in Figure 23.4 was originally proposed by Eklundh [45] to facilitate the out-of-core (when core memory was still in use and the memory size is very limited) matrix transposition. The idea is that two rows are read in each time, the appropriate elements are permuted, the modified two rows are then written out to the disk. The example in Figure 23.4 shows that the first permutation step involves row 1 and row 2, the second step involves row 3 and row 4, the third step involves row 1 and row 3, and the fourth step involves row 2 and row 4. (Each row can be a single row as originally proposed, or it can be a block row if each A_{ij} is a submatrix instead of a single element.)

Note that unlike the other two algorithms, some blocks will be moved multiple times in Eklundh's method. For example, block A_{14} was moved twice in Figure 23.4.

Note that this more complicated permutation scheme can be easily described using the binary representation of the row and column indices of block A_{ij}, which is denoted as $A[i, j]$ below. Suppose $i = b_2 b_1 b_0$, and $j = u_2 u_1 u_0$; then three pairwise exchanges may switch $A[b_2 b_1 b_0, u_2 u_1 u_0]$ with $A[b_2 b_1 u_0, u_2 u_1 b_0]$, followed by $A[b_2 u_1 u_0, u_2 b_1 b_0]$, and finally with $A[u_2 u_1 u_0, b_2 b_1 b_0]$.

Although this method requires that the number of (block) rows and (block) columns be the same, since i and j must have the same number of bits to effect pair-wise exchange throughout, there is no restriction on the dimension of each block itself. That is, $A[i, j]$ could again be a single element, a square, or a rectangular matrix.

Figure 23.4 Transpose a matrix A by Eklundh's method [45].

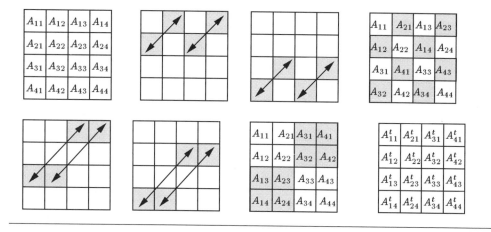

23.3   Three Parallel 2D FFT Algorithms for Hypercubes

Readers are assumed to be familiar with the hypercube multiprocessors introduced in Chapter 18 and the large number of parallel 1D FFTs described in previous chapters. As one would expect, different parallel algorithms are available to handle the column-oriented mapping scheme, row-oriented mapping scheme, and the 2D-block mapping scheme. Three sample algorithms are used to introduce some basic ideas in the following sections.

23.3.1   The transpose split (TS) method

The *transpose split* (TS) method used in [22, 24] parallelizes the row-column 2D FFT algorithm. In the example shown in Figure 23.5, the four processors P_0, P_1, P_2, and P_3, are each allocated a block of consecutive columns or rows. Clearly, only the matrix transposition phase(s) will incur inter-processor communication, and an efficient parallel algorithm for matrix transposition is all that is needed.

Figure 23.5 The TS (*transpose split*) method with column or row data allocation.

Each processor performs 1D FFTs on allocated columns:

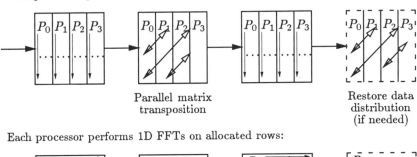

Parallel matrix
transposition

Restore data
distribution
(if needed)

Each processor performs 1D FFTs on allocated rows:

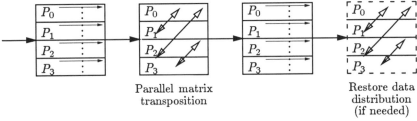

Parallel matrix
transposition

Restore data
distribution
(if needed)

23.3.2   The local distributed (LD) method

The *local distributed* (LD) method in [22, 24, 46] does not have a matrix transposition phase. For the example above, each processor first independently computes multiple 1D FFTs on allocated columns (or rows). In the next phase, since each individual row (or column) of the updated signal matrix is shared by all four processors as depicted in Figure 23.6, an appropriate "parallel" 1D FFT algorithm from previous chapters is used to transform the multiple rows (or columns) all at once. (Note that only one set of twiddle factors is needed.) An implementation proposed in [46] used the sequential "split-radix" algorithm to the rows, and the parallel "radix-4" to the columns.

Figure 23.6 The LD (*local distributed*) method with column or row wise data allocation.

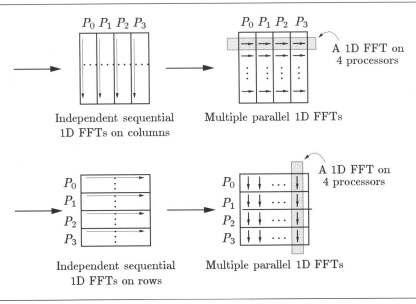

Since a processor may apply each stage of the FFT transformation to all 1D FFTs at once, the same communication algorithm for a single parallel 1D FFT may be easily modified to include the data needed for all 1D FFTs in each message, i.e., *the message size is increased, but the number of messages remains identical to that incurred by parallelizing a single 1D FFT*. Therefore, on machines with large communication bandwidth, the communication cost is expected to impact the performance of 2D FFT less significantly.

23.3.3   The 2D block distributed method

The basic idea of the 2D block distributed method [22] is depicted by a simple example on a 2-by-2 processor grid in Figure 23.7 below.

Figure 23.7 The 2D block distributed method on a 2×2 processor grid.

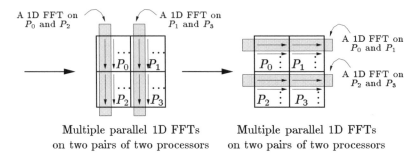

A 1D FFT on P_0 and P_2 A 1D FFT on P_1 and P_3 A 1D FFT on P_0 and P_1 A 1D FFT on P_2 and P_3

Multiple parallel 1D FFTs   Multiple parallel 1D FFTs
on two pairs of two processors   on two pairs of two processors

Even if one wishes to do so, it is not always possible to configure the available p processors as a $\sqrt{p} \times \sqrt{p}$ grid. For example, if $p = 2^d$ and d is an odd number, the p processors cannot be arranged as a square grid. In what follows, assume $p = 2^d = 2^{d_1 + d_2} = 2^{d_1} \times 2^{d_2}$. A desirable objective is that the 2D block distributed method work for all possible values of d_1 and d_2. To achieve this, it is apparent that the dimensions of the rectangular processor grid should be input parameters to the program.

When the p processors form a hypercube, the processors on each row and each column of the grid form a subcube, hence the name *subcube-grid* [26, 27, 29]. One may then choose any convenient dimensions, because the performance of the 2D block distributed method is not affected by the dimensions of the subcube-grid. The reasons for this are that the following observations hold, regardless of how the subcube-grid is configured.

- each processor has $(N_1 \times N_2)/p$ data elements,

- each message is either of length $(N_1 \times N_1)/p$ or one half of it (depending on the chosen parallel 1D FFT algorithm),

- all subcube-doubling message exchanges involve only neighboring processors,

- the total number of messages is $d_1 + d_2 = d$ always,

- the subcube-doubling communication algorithm does not cause traffic congestion.

Note that the generalized 2D block method includes the LD method as a special case corresponding to a $1 \times p$ or $p \times 1$ subcube-grid.

23.3.4   Transforming a rectangular signal matrix on hypercubes

Although a square signal matrix was shown in all examples in the previous sections, in reality the signal matrix may not be square. A little reflection leads to the conclusion that all three algorithms work without significant modification when the matrix is not square. The only proviso is that the TS (transpose split) method requires that its parallel matrix transposition algorithm handle rectangular matrices of any dimension.

23.4   The Generalized 2D Block Distributed (GBLK) Method for Subcube-grids and Meshes

Recall that the signal data for a 2D FFT are stored naturally in an $N_1 \times N_2$ matrix, and that by viewing the hypercube as various 2D subcube-grids, the generalized 2D block distributed (GBLK) method may be regarded as partitioning the matrix on a corresponding subcube-grid. Furthermore, the performance of the GBLK method is not affected by the aspect ratio of the subcube-grid for reasons discussed in the last section. In other words, given a hypercube consisting of $p = 2^d = 2^{d_1+d_2}$ processors, the $N_1 \times N_2$ data matrix may be mapped to any subcube-grid of dimensions $\gamma_1 \times \gamma_2$, where $\gamma_1 = 2^{d_1}$, and $\gamma_2 = 2^{d_2}$. Given below are the four possible subcube-grids for $p = 8$, together with the corresponding data mappings.

Figure 23.8 The four GBLK data mappings on 8-node subcube-grids.

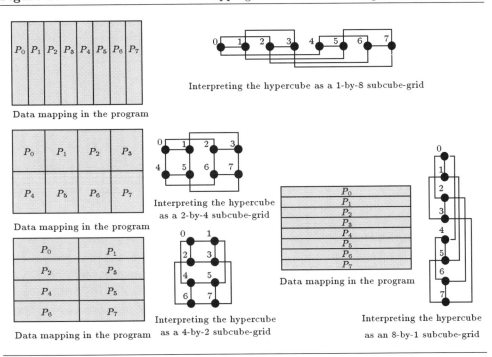

23.4.1   Running hypercube (subcube-grid) programs on meshes

The 512 computing nodes on the Intel Touchstone DELTA computer [44] are connected as a 16-by-32 two-dimensional mesh, and disjoint sub-meshes of dimensions (*row, col*), with *row* ≤ 16 and *col* ≤ 32, can be allocated to individual users [44]. For example, Figure 23.9 shows three 8-processor physical sub-meshes allocated from a 4-by-8 mesh.

Figure 23.9 Three 8-node physical sub-meshes allocated from a 4-by-8 mesh.

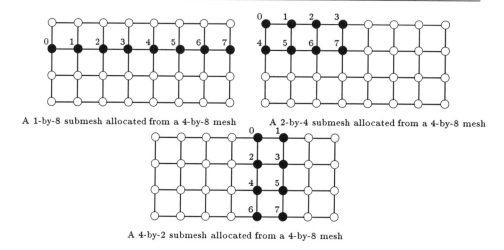

A 1-by-8 submesh allocated from a 4-by-8 mesh   A 2-by-4 submesh allocated from a 4-by-8 mesh

A 4-by-2 submesh allocated from a 4-by-8 mesh

From E. Chu [25], *The International Journal of High Performance Computing Applications*, 13(2):124–145, 1999. With permission.

A mesh has fewer communication channels than a hypercube, and it is not possible to have the allocated processors in Figure 23.9 form a subcube-grid. However, since a mesh is a connected network, there is a path between any two processors, a hypercube program implementing the subcube-doubling communication algorithm will run correctly on a mesh as shown by the 8-node examples in Figure 23.10. Regardless of whether a subcube-grid or a mesh is used, the matrix elements can always be distributed to processors using the same mapping scheme, and the communication algorithm can still be understood as passing the same sequences of messages between the same designated pairs of processors, and the length of each message remains unchanged.

Thus a different physical network topology will not affect the "correctness" of the algorithm or the software. However, a different physical network topology can

(i) increase the "physical distance" (measured by the number of hardware channels or hops) between communicating processors, and

(ii) cause "contention of communication channels" when logically-disjoint message paths overlap badly on the physical network

and hence compromise the effectiveness of the logical topology in achieving its objective.

If the *hop (distance) penalty* is low, the first problem will not affect the performance much. However, the contention of communication channels may be a serious problem because it can cause severe traffic congestion. In the next section, the extent of traffic congestion is directly related to the physical distance a message travels when using the subcube-doubling technique on a mesh, and the question of how to reduce traffic congestion by using an optimal aspect ratio to configure the physical mesh (at runtime) is addressed.

Figure 23.10 The four GBLK data mappings on four 8-node meshes.

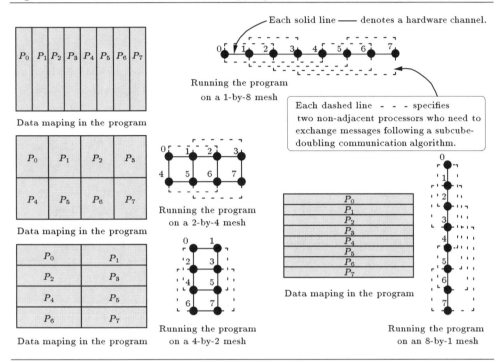

23.5 Configuring an Optimal Physical Mesh for Running Hypercube (Subcube-grid) Programs

The objective in configuring an optimal physical mesh is to minimize communication overhead due to the multi-hop (distance) penalty and traffic congestion. Since circuit-switching is used by the DELTA mesh and other currently available message-passing multiprocessors to manage the network, the contention of communication channels is resolved in a particular manner. It is, therefore, useful to show directly in Section 23.5.3 that the effect of channel contention on a circuit-switching physical mesh is also minimized by the optimal aspect ratio derived in Theorem 23.1 in Section 23.5.1.

23.5.1 Minimizing multi-hop penalty

To support the subcube-doubling communication algorithm on an arbitrary γ_1-by-γ_2 subcube-grid using a μ_1-by-μ_2 physical mesh, consider first how to minimize the total physical distance the messages travel. In the following analysis it is assumed that $\gamma_1 = 2^{d_1}$, $\gamma_2 = 2^{d_2}$, $\mu_1 = 2^{\delta_1}$, $\mu_2 = 2^{\delta_2}$, and $p = \gamma_1 \times \gamma_2 = \mu_1 \times \mu_2 = 2^d$, where $d_1, d_2,$

δ_1 and δ_2 are non-negative integers. It is also assumed without loss of generality that the p processors are numbered consecutively row by row in both the subcube-grid and the mesh: e.g., processors $P_0, \cdots, P_{\gamma_2-1}$ form the first row of the subcube-grid, and processors $P_0, \cdots, P_{\mu_2-1}$ form the first row of the mesh.

When the subcube-doubling algorithm is used for concurrent message exchanges among all pairs of processors, the communication requirement is the same for every processor (see Figures 23.8 and 23.10). It is thus sufficient to examine the requirement of processor P_0 in what follows.

Figures 23.8 and 23.10 demonstrate that regardless of the aspect ratio of the subcube-grid or the mesh, P_0 communicates with P_i, $i = 2^\ell$, $0 \leq \ell \leq d - 1$. However, the physical distance between P_0 and each such P_i varies with the physical meshes used to run the program. For example, according to Figure 23.10, using a 2-by-4 mesh, P_0 is one hop away from either P_1 or P_4, and two hops away from P_2; using a linear array, P_0 is one hop away from P_1, two hops away from P_2, and four hops away from P_4.

Theorem 23.1 is next stated and proved, which shows that the total physical distance between P_0 and all designated P_i's is a function of d and δ_1.

Theorem 23.1 Assume that the p processors denoted by $P_0, \cdots, P_{p-1}$ are arranged row by row on a $\mu_1 \times \mu_2$ physical mesh, where $\mu_1 = 2^{\delta_1}$, and $\mu_2 = 2^{\delta_2}$. If processor P_0 communicates with the $d = \delta_1 + \delta_2$ processors required by the subcube-doubling technique, then the total physical distance is given in hops by $H_{mesh}(d, \delta_1) = 2^{\delta_1} + 2^{d-\delta_1} - 2$, and is minimized when $\delta_1 = d/2$, i.e., $\mu_1 = \mu_2 = \sqrt{p}$, assuming that d is an even number.

Proof: Since each row of the physical mesh is a linear array, the physical distance between P_0 and P_j, $j = 2^{\ell_2}$, $0 \leq \ell_2 \leq \delta_2 - 1$, is exactly 2^{ℓ_2} hops. Since each column of the physical mesh is also a linear array, the physical distance between P_0 and P_i, $i = 2^{\delta_2+\ell_1}$, $0 \leq \ell_1 \leq \delta_1 - 1$, is exactly 2^{ℓ_1} hops. Therefore, the total distance between the d pairs of communicating processors can be computed by

(23.3)
$$H_{mesh}(d, \delta_1) = \sum_{\ell_2=0}^{\delta_2-1} 2^{\ell_2} + \sum_{\ell_1=0}^{\delta_1-1} 2^{\ell_1}$$
$$= 2^{d-\delta_1} + 2^{\delta_1} - 2 \,.$$

Minimizing $H_{mesh}(d, \delta_1)$ with respect to δ_1 yields $\delta_1 = d/2$. Hence $\mu_1 = \mu_2 = 2^{d/2} = \sqrt{p}$. ∎

23.5.2 Minimizing traffic congestion

To quantify the traffic congestion caused by the subcube-doubling communication on the physical mesh, a traffic weight $w_{i,j}^{(k)}$ is associated with each communication channel $C_{i,j}$ which physically connects processors P_i and P_j on the mesh, and $w_{i,j}^{(k)}$ is defined to be the number of overlapped communication paths on that channel during the kth communication step. Since the subcube-doubling communication is performed independently within each row and within each column of the mesh, it is sufficient to examine the extent of traffic congestion within one row and one column, which are linear arrays of sizes μ_2 and μ_1 on a μ_1-by-μ_2 mesh. Figure 23.11 shows the overlapped communication paths caused by each subcube-doubling communication step on a linear array consisting of processors P_0, P_1, $\cdots$, P_7. The values of $w_{i,i+1}^{(k)}$ defined for each channel connecting the neighboring processors P_i and P_{i+1} on the linear array are given in Table 23.1, where $0 \leq i \leq 6$ and $1 \leq k \leq 3$. The extent of traffic congestion can be quantified by the total weight $\sum_{k=1}^{3} \sum_{i=0}^{6} w_{i,i+1}^{(k)} = 28$ for this example.

Figure 23.11 The overlapped communication paths incurred by the subcube-doubling algorithm ($p = 8$).

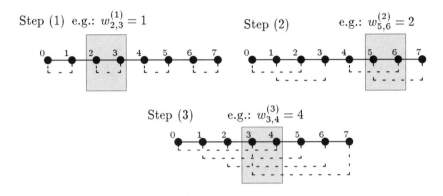

Step (1) e.g.: $w_{2,3}^{(1)} = 1$ Step (2)   e.g.: $w_{5,6}^{(2)} = 2$

Step (3)   e.g.: $w_{3,4}^{(3)} = 4$

From E. Chu [25], *The International Journal of High Performance Computing Applications*, 13(2):124–145, 1999. With permission.

Table 23.1 Traffic weights for communication channels $C_{i,i+1}$ in Figure 1.

Step (k)	$w_{0,1}^{(k)}$	$w_{1,2}^{(k)}$	$w_{2,3}^{(k)}$	$w_{3,4}^{(k)}$	$w_{4,5}^{(k)}$	$w_{5,6}^{(k)}$	$w_{6,7}^{(k)}$	$\sum_{i=0}^{6} w_{i,i+1}^{(k)}$
$k = 1$	1	0	1	0	1	0	1	4
$k = 2$	1	2	1	0	1	2	1	8
$k = 3$	1	2	3	4	3	2	1	16
$\sum_{k=1}^{3} \sum_{i=0}^{6} w_{i,i+1}^{(k)}$								28

From E. Chu [25], *The International Journal of High Performance Computing Applications*, 13(2):124–145, 1999. With permission.

Now consider the total traffic weight caused by the subcube-doubling algorithm on a linear array consisting of $\mu = 2^\delta$ processors in Lemma 23.2 below.

Lemma 23.2 The total traffic weight imposed by the subcube-doubling communication on a linear array of size $u = 2^\delta$ is given by $W(\delta) = \sum_{k=1}^{\delta} \sum_{i=0}^{\mu-2} w_{i,i+1}^{(k)} = 2^{2\delta-1} - 2^{\delta-1}$.

Proof: Observe that during the kth subcube-doubling communication step, each pair of communicating processors is $m = 2^{k-1}$ hops apart. Since the traffic weights on the m channels connecting processors P_0, P_1, $\cdots$, P_m is summed up by $\sum_{\ell=1}^{m} \ell$, and the traffic weights on the $m - 1$ channels connecting processors P_m, P_{m+1}, $\cdots$, P_{2m-1} is summed up by $\sum_{\ell=1}^{m-1} \ell$, the total weight on channels connecting each disjoint group of

$2m = 2^k$ processors, can be computed by

$$
(23.4) \qquad \sum_{\ell=1}^{m} \ell + \sum_{\ell=1}^{m-1} \ell = m^2 = 2^{2k-2} .
$$

Since there are $2^{\delta-k}$ groups of 2^k processors performing the kth subcube-doubling communication step independently, the total weight from all k communication steps on the entire linear array can be computed by

$$
(23.5) \qquad W(\delta) = \sum_{k=1}^{\delta} \sum_{i=0}^{\mu-2} w_{i,i+1}^{(k)} = \sum_{k=1}^{\delta} 2^{\delta-k} \left(\sum_{\ell=1}^{2^{k-1}} \ell + \sum_{\ell=1}^{2^{k-1}-1} \ell \right) = 2^{2\delta-1} - 2^{\delta-1} .
$$

■

Theorem 23.3 The total traffic weight imposed by the subcube-doubling communication on a μ_1-by-μ_2 mesh, where $\mu_1 = 2^{\delta_1}$, $\mu_2 = 2^{\delta_2}$, and $p = 2^{\delta_1+\delta_2} = 2^d$, is given by $W_{mesh}(d, \delta_1) = 2^{d-1}(2^{\delta_1} + 2^{d-\delta_1} - 2)$, and is minimized when $\delta_1 = d/2$, i.e., $\mu_1 = \mu_2 = \sqrt{p}$, assuming that d is an even number.

Proof: As noted earlier, the subcube-doubling communication is performed independently within each row and each column of the μ_1-by-μ_2 mesh. Lemma 23.2 implies

$$
(23.6) \qquad W_{row}(\delta_2) = 2^{2\delta_2-1} - 2^{\delta_2-1}
$$

and

$$
(23.7) \qquad W_{column}(\delta_1) = 2^{2\delta_1-1} - 2^{\delta_1-1} .
$$

Since there are $\mu_1 = 2^{\delta_1}$ rows and $\mu_2 = 2^{\delta_2}$ columns, the total traffic weight on the μ_1-by-μ_2 mesh is given by

$$
(23.8) \qquad
\begin{aligned}
W_{mesh}(d, \delta_1) &= 2^{\delta_2} \times W_{column}(\delta_1) + 2^{\delta_1} \times W_{row}(\delta_2) \\
&= 2^{d-1}(2^{\delta_1} + 2^{d-\delta_1} - 2) \\
&= 2^{d-1} \times H_{mesh}(d, \delta_1) .
\end{aligned}
$$

Therefore, the value $\delta_1 = d/2$ that minimizes $H_{mesh}(d, \delta_1)$ in Theorem 23.1 also minimizes $W_{mesh}(d, \delta_1)$.   ■

Corollary 23.4 follows immediately from Theorems 23.1 and 23.3.

Corollary 23.4 If the given physical mesh consists of $p = 2^d$ processors, where d is an odd number, then $H_{mesh}(d, \lfloor d/2 \rfloor) = H_{mesh}(d, \lceil d/2 \rceil)$, and $W_{mesh}(d, \lfloor d/2 \rfloor) = W_{mesh}(d, \lceil d/2 \rceil)$.

The results above are depicted in Figure 23.12 for the 8-processor example, i.e., either a 2-by-4 mesh or a 4-by-2 mesh should be used to run the hypercube program regardless of how the matrix is partitioned among the processors.

Figure 23.12 Optimal 8-node meshes for running hypercube programs.

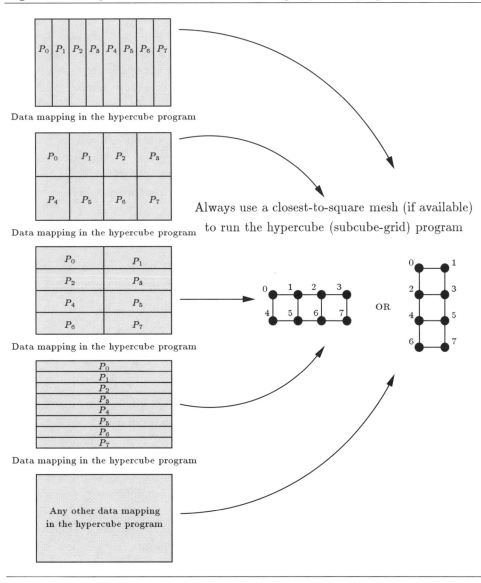

Always use a closest-to-square mesh (if available) to run the hypercube (subcube-grid) program

23.5.3   Minimizing channel contention on a circuit-switched network

When a message is to be sent from one processor to another on a circuit-switched network, a header packet is sent to reserve all of the channels required to build its path. After this "circuit" is established, the message is transmitted, and an end-of-message indicator releases the channels [44]. Therefore, when the paths of several concurrent messages overlap, the establishment of each corresponding circuit must wait for the shared channel(s) to be released from one previously built route. Such wait time can be eliminated if there is no overlapped concurrent communication paths. From the analysis of subcube-doubling communication on a mesh in the previous section, it is clear that there are exactly "m" m-hop paths overlapping each other when a processor sends a message to a destination m hops away within each row or each column of the physical mesh; i.e., the number of overlapped paths is the same as the physical distance a message travels (see Figure 23.11 for an example). Consequently, the physical mesh chosen to minimize the total physical distance a processor's messages travel in Theorem 23.1 also minimizes the total number of overlapped communication paths, and hence the effect of channel contention on a circuit-switched network.

Thus, Theorems 23.1, 23.3, and Corollary 23.4 imply that when a hypercube program is run (or emulated) on a mesh,

for best performance, a closest-to-square physical mesh should be used.

The user still has the flexibility of choosing a particular data mapping to facilitate memory access, and/or to simplify data structures, and/or for programming convenience.

23.6   Pipelining Subcube-doubling Communications on All Hypercube Channels

The idea of pipelining subcube-doubling communications on all hypercube channels was used by Calvin in [22] to overlap communication and computation in implementing parallel 2D FFT algorithms. To help explain this idea, recall from Chapter 18 the d communication steps in the basic subcube doubling algorithm—they are depicted again in Figure 23.13 ($d = 3$ in the example). Note that while there are $d\left(2^{d-1}\right)$ channels in a hypercube of dimension d, only 2^{d-1} channels are used per communication step.

To make use of *all* available channels simultaneously, each processor is required to *pipeline* its outgoing messages to all its neighbors. For example, P_0 is shown to pipeline its messages to P_1, P_2, and P_4 using a *non-blocking* send in each step as shown in Figure 23.14, so does P_1 as well as every other processor.

Since the pipelining technique typically involves sending multiple shorter messages instead of a single long message, the startup time caused by multiple sends must be "overlapped" (or "masked") by arithmetic work to a large extent if the pipelining method is to be effective. Instead of simply displaying a data mapping which can accomplish this objective, the computation of N_1 1D FFTs of length N_2 on $p = 2^d = 8$ processors is used as an example to "construct" and "demonstrate" such a mapping step by step.

Figure 23.13 The d synchronous exchanges in the subcube-doubling algorithm $(d = 3)$.

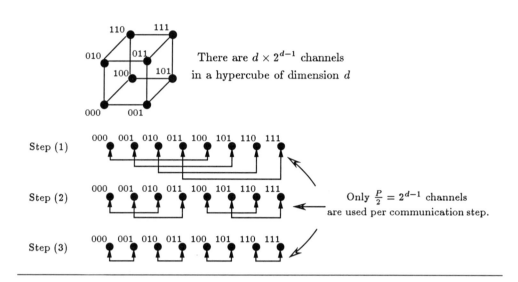

Figure 23.14 Pipelining subcube-doubling "send" on all channels $(d = 3)$.

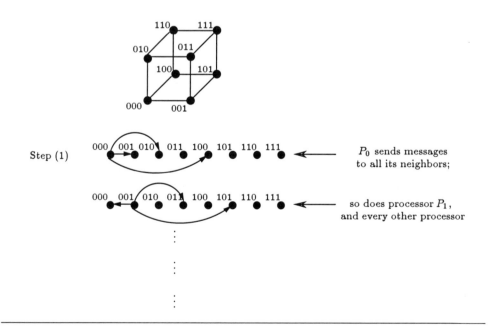

Referring to Figure 23.15, observe that P_0 has been assigned the first block of $\frac{N_2}{p}$ columns ($p = 8$ in the example). Instead of applying FFT steps to the entire block all at once and exchanging a single message with one neighbor, the data in P_0 are now partitioned into $d = \log_2 p$ portions ($d = 3$ in the example), and P_0 interleaves its local computation on each portion of data with message passing to each neighbor as described in Algorithm 23.6.

Algorithm 23.1 The actions by P_0 in step 1.

---

begin

    $d := \log_2 p$ d is the hypercube dimension

    for $k = 1$ **to** d iterate on d portions of data

        P_0 performs local computation on the kth portion;

        P_0 sends the updated data from this portion to its kth neighbor;

    end for

end

After P_0 completes the initial step, ideally the message P_0 expects from P_4 has already arrived, and P_0 can now use the incoming data to update the first portion of its data. (If the data is always ready when the processor needs it, the communication time is said to be fully masked.) Referring to Figure 23.15 again, observe that after P_0 updates the first portion of data, P_0 immediately sends the newly updated data from this portion to P_1. *Since the data in each block row are divided evenly among the p processors, P_0 must rotate its d neighbors each step.* By this time, ideally the data P_0 needs to update the second portion of data has arrived from P_2. After updating the second portion, P_0 immediately sends the necessary data from this portion to P_4, and so on. The actions by P_0 in the second step are described in Algorithm 582. With the understanding that the list of appropriate neighbors is rotated by one position (see Figure 23.15), the generic description of step 2 may be used to describe step 3. (For $d > 3$, the same description may be used for step 2, step 3, ..., and step d.)

Algorithm 23.2 A generic description of actions by P_0 in steps 2, 3,$\cdots$, d.

---

begin

    $d := \log_2 p$ d is the hypercube dimension

    for $k = 1$ **to** d iterate on d portions of data

        P_0 receives data from an appropriate   communication cost is masked if

            neighbor;   data have arrived when needed

        P_0 updates an appropriate portion of data;

        P_0 sends the updated data from this portion

            to an appropriate neighbor;

    end for

end

For a hypercube of dimension d, step $d + 1$ is the last step.

Algorithm 23.3 A generic description of actions by P_0 in step d+1 – the last step.

<u>**begin**</u>
    $d := \log_2 p$ d is the hypercube dimension
    <u>**for**</u> $k = 1$ <u>**to**</u> d iterate on d blocks of data
        P_0 *receives data from an appropriate*   communication cost is masked if
        neighbor;   data have arrived when needed
        P_0 *updates an appropriate portion of data*;
    <u>**end for**</u>
<u>**end**</u>

Figure 23.15 Pipelining subcube-doubling "send" from P_0 on all channels $(p = 8)$.

The hypercube channels used by P_0

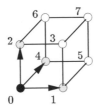

Example: To parallelize N_1 1D FFTs

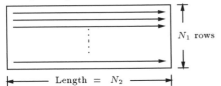

The 1st step of pipelined "SEND" from P_0

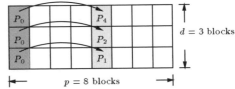

The 2nd step of pipelined "SEND" from P_0

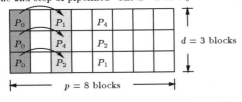

The 3rd step of pipelined "SEND" from P_0

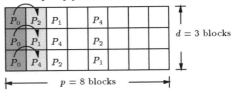

Observe from Figure 23.15 that the data mapping has been partially constructed in the process of developing the algorithm for P_0. For $p = 8$, one only needs to follow the actions of P_7, which mirrors that of P_0, and the data mapping is completed in Figure 23.16.

Figure 23.16 Pipelining subcube-doubling "send" from P_7 on all channels $(p = 8)$.

The hypercube channels used by P_7

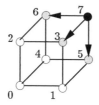

Example: To parallelize N_1 1D FFTs

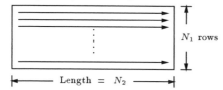

The 1st step of pipelined "SEND" from P_7

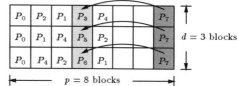

The 2nd step of pipelined "SEND" from P_7

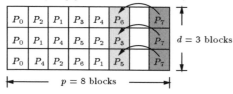

The 3rd step of pipelined "SEND" from P_7

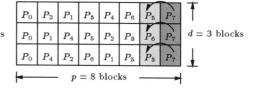

With the entire map constructed in Figure 23.16, one can now visualize the actions of any processor from the map. Keep in mind that all processors perform the same actions (pairing up with appropriate neighbors) "concurrently." As one more example, the actions by P_1 are shown in Figure 23.17.

Figure 23.17 Pipelining subcube-doubling "send" from P_1 on all channels ($p = 8$).

The hypercube channels used by P_1

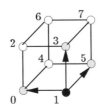

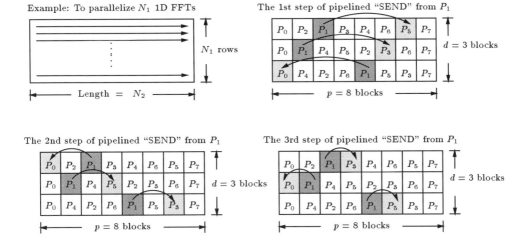

Example: To parallelize N_1 1D FFTs

N_1 rows

Length = N_2

The 1st step of pipelined "SEND" from P_1

$d = 3$ blocks

$p = 8$ blocks

The 2nd step of pipelined "SEND" from P_1

$d = 3$ blocks

$p = 8$ blocks

The 3rd step of pipelined "SEND" from P_1

$d = 3$ blocks

$p = 8$ blocks

It should be understood that the generic description of the $d + 1$ steps of the algorithm given in this section is not tailored to P_0's actions at all, but instead reflects the concurrent actions of *all* p processors. Note that in order to mask the communication cost, the matrix must be sufficiently large so that each processor can be kept busy computing before the message it waits for arrives. An analytical model was used in [22] to derive the minimum size of such a matrix, which, as expected, is a function of the number of processors and the hardware parameters of the machine being used.

The method described above for computing multiple 1D FFTs may be viewed as employing an "all processor-to-all neighbor" communication scheme depicted in Figure 23.18. Although the simplest problem of computing many 1D FFTs is used in this section to make various aspects of this communication scheme easily understood, the method is not designed and should not be used for this simple case—because the "independent processor" method incurs no communication at all. However, this method is useful for FFT of higher dimensions, which is revisited in the next section.

Figure 23.18 All processor–to–all neighbor communication scheme ($p = 8$).

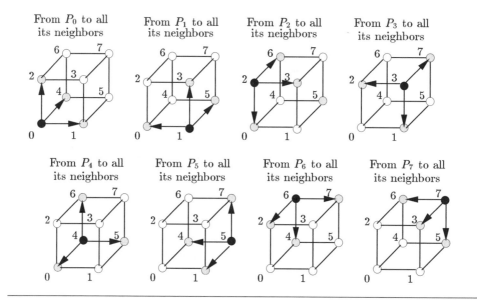

23.7 Changing Data Mappings During Parallel 2D FFT Computation

The data mappings required in implementing the following four methods for computing the 2D FFT are depicted in Figures 23.19–23.22.

- The TS (*transpose split*) method: Two different data mappings are required in phases I and II (see Figure 23.19). Note that by distributing consecutive rows to the processors in Phase II, one has effectively transposed the data matrix as desired. Accordingly, an efficient parallel algorithm for changing the data mapping is an efficient parallel algorithm for matrix transposition, and such an algorithm will be presented in the next section.

Figure 23.19 The TS method: Different data mappings used in phases I and II ($p = 8$).

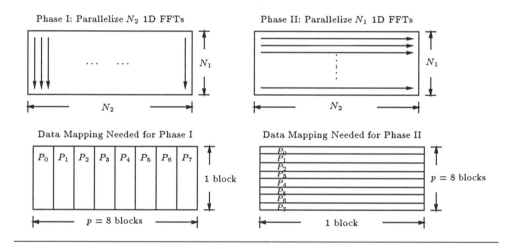

- The LD (*local distributed*) method: Identical data mappings are used in phases I and II (see Figure 23.20).

- The GBLK (*generalized block distributed*) method: Identical data mappings are used in phases I and II (see Figure 23.21).

- Calvin's method [22]: Two different data mappings are required in phases I and II (see Figure 23.22). This method appears to be called the LD method with overlap in [22]. It is not clear how the cost for changing the mapping can be masked from the very brief description in [22].

23.8 Parallel Matrix Transposition By Changing Data Mapping

As indicated in the previous section, an efficient parallel algorithm for changing the mappings from distributing the matrix columns to distributing the matrix rows is an

Figure 23.20 The LD method: Identical data mappings used in phases I and II $(p = 8)$.

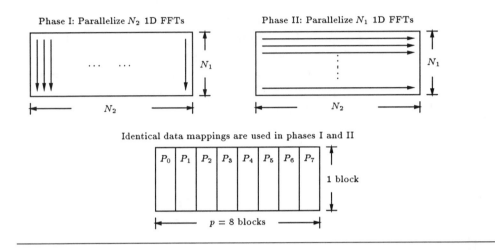

Figure 23.21 The GBLK method: Identical data mappings used in phases I and II $(p = 8)$.

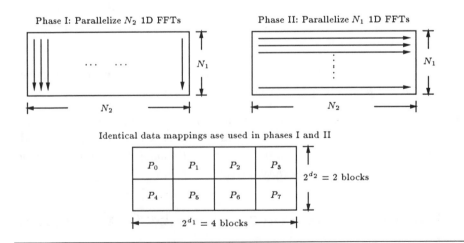

Figure 23.22 Calvin's method: Different data mappings used in phases I and II ($p = 8$).

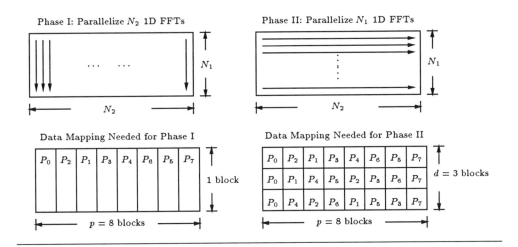

efficient parallel algorithm for matrix transposition. Observe from Figure 23.23 that a data mapping by columns may be viewed as distributing the $N_1 \times N_2$ matrix A on a $1 \times p$ subcube-grid, and a data mapping by rows may be viewed as distributing the same matrix on a $p \times 1$ subcube-grid, where $p = 2^d$. That is, each row of the matrix is initially shared by $p = 2^d$ processors, and is finally stored in its entirely in a single processor. This objective can be accomplished by halving the number of processors on each row of the subcube-grid, and doubling the number of processors on each column of the subcube-grid in d steps. Observe from Figure 23.23 that each time the subcube-grid changes its dimensions this way, all that is required is that every processor exchange one half of its data with a directly connected processor. Accordingly, the total communication cost for transposing an $N_1 \times N_2$ matrix is exactly d concurrent exchanges of $\frac{1}{2} \frac{N_1 \times N_2}{p}$ complex numbers among all pairs of processors. (If desired, the initial column mapping can be restored by reversing the steps with the same communication cost.)

Note that the communication cost in Phase II of the LD (local distributed) method for a 2D FFT (without inter-processor data permutation) requires d concurrent exchanges of $\frac{N_1 \times N_2}{p}$ complex numbers [98], and the communication cost of this matrix transposition algorithm is one half of that amount. Therefore, if the initial column mapping needs not to be restored, the TS (transpose split) method incurs half the communication cost of the LD method; if the initial column mapping must be restored, the communication cost of the TS method becomes the same as that of the LD method.

23.9 Notes and References

As noted in Section 23.3.3, the performance of the 2D block distributed method is not affected by the *aspect ratio* of the subcube-grid for reasons identified there. However, this is not the case for many parallel matrix algorithms, i.e., *the choice of the*

Figure 23.23 Parallel matrix transposition by changing data mapping ($p = 8$).

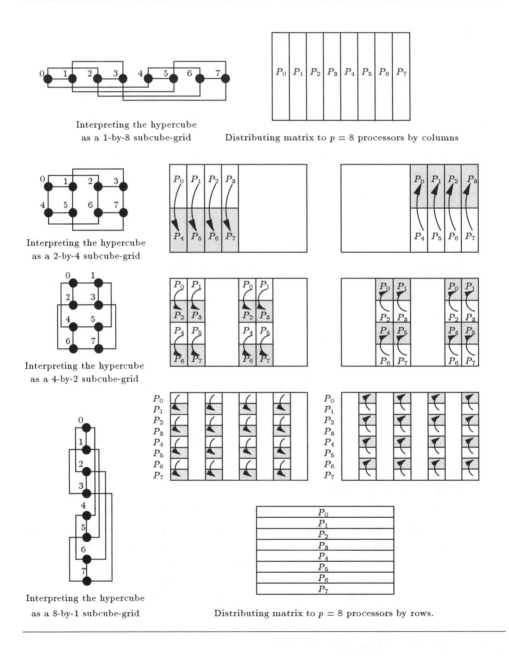

aspect ratio for the subcube-grid can severely impact the performance of parallel algo-rithms, and the subcube-grid is an important and versatile physical network topology. For example, Chu and George show in [26, 27, 29] that *an optimal aspect ratio* can be determined at run time for a class of fundamental numerical algorithms including Gaussian elimination with partial pivoting, QR factorization (with column pivoting [27]), Gauss-Jordan inversion, and multiple least squares updating algorithms. The significant net saving in execution time and storage usage gained from using an optimal subcube-grid was demonstrated by numerical experiments on iPSC/2 and iPSC/860 hypercubes in [26, 27, 29].

Furthermore, the authors reported in [27] the iPSC/2 and iPSC/860 execution times to demonstrate an efficient data relocation algorithm which dynamically changes the data mapping between the subcube-grids, and the same algorithm was used in the last section for changing the aspect ratio from $1 \times p$ to $p \times 1$, which effectively transposes the distributed matrix among the p processors as desired.

The interplay of optimal physical and logical network topologies in the design and implementation of parallel matrix algorithms was investigated further by Chu in [25].

Other interesting algorithms for computing the 2D FFTs include the class of vector-radix algorithms as well as the class of polynomial transform algorithms. The basic principles underlying these two classes of sequential algorithms were reviewed in [41], and their parallel implementation on hypercube and mesh machines was recently examined by Angelopoulos and Pitas in [2]. Readers are referred to [65, 66, 79, 80] for more details on the vector-radix algorithms, and [69, 70] on the polynomial transforms originally proposed by Nussbaumer for the computation of 2D cyclic convolutions.

Chapter 24

Computing and Distributing Twiddle Factors in the Parallel FFTs

It was assumed in the previous chapters that all $\omega_N^{r\ell}$ values, which are commonly referred to as *twiddle factors*, are pre-computed and available to the computer programs implementing the various transform algorithms. One could argue that this is a reasonable assumption to make, since FFT codes are usually applied sequentially to large numbers of vectors, and thus pre-computation of the twiddle factors is an efficient strategy. The argument appears to be valid for single processor machines, and for shared-memory multiprocessors, since access to the twiddle factors is straightforward. As pointed out by Chamberlain [23], this is also an efficient strategy for distributed-memory machines if multiple transforms are performed by each processor all at once, which is certainly the case in computing multiple 1D FFTs or 2D FFTs as discussed in Chapter 23.

However, for transforming a *single* large vector on a distributed-memory machine, the best strategy is not at all obvious. If every processor has a copy of all the twiddle factors, then they may consume more memory than the data being transformed—a somewhat incongruous circumstance. If this represents an unacceptable amount of storage, then the FFT algorithm must arrange that the (pre-computed) twiddle factors are conveyed among the processors so that they are available when needed. A final option is to compute the twiddle factors "on the fly" as required, which relieves the storage and communication burdens at the expense of additional computation.

The choice of strategy depends on a number of factors: the relative speeds of communication and computation, the amount of memory available compared to the size of the problem being solved, and the algorithm itself. Some specific strategies were considered and compared under special circumstances in [23, 51, 98, 104], but they do not seem to generalize because the distribution of twiddle factors can be drastically different for different algorithms, as discussed in the following two sections.

271

24.1   Twiddle Factors for Parallel FFT Without Inter-Processor Permutations

Since the twiddle factors are completely dictated by the data elements involved in each butterfly computation, the twiddle factors required by each processor are easily identified by the data it owns, which are different for different mappings. Assuming that naturally ordered input data are transformed by DIF FFT without inter-processor permutations, an example using a consecutive block mapping is given in Figure 24.1, and another example using a cyclic mapping is given in Figure 24.2.

Observe that in either case, one processor may need to use more twiddle factors than the other. For comparison, the distribution of the twiddle factors are tabulated in Table 24.1 for the consecutive block map in Figure 24.1, and in Table 24.2 for the cyclic map in Figure 24.2. Apparently, in the former case, the twiddle factors are not evenly distributed among the processors, whereas in the latter case, a more balanced (but still not fully balanced) distribution results from using the cyclic data map in parallelizing the DIF FFT algorithm.

Figure 24.1 $\mathrm{DIF_{NR}}$ FFT twiddle factors required if a consecutive block map is used.

$i_4i_3i_2i_1i_0$	$P_{i_4i_3}$ $\vert i_4i_3\vert i_2i_1i_0$	$\vert\tau_4i_3\vert i_2i_1i_0$	$\vert\tau_4\tau_3\vert i_2i_1i_0$	$\vert\tau_4\tau_3\vert\tau_2i_1i_0$	$\vert\tau_4\tau_3\vert\tau_2\tau_1i_0$
00000	$P_0\;\;\omega_N^{i_3i_2i_1i_0}$	$P_0\;\;\omega_N^{\tau_2i_1i_00}$	$P_0\;\;\omega_N^{i_1i_000}$	$P_0\;\;\omega_N^{i_0000}$	$P_0\;\;\omega_N^0=1$
00001	P_0 (Stage 1)	P_0 (Stage 2)	P_0 (Stage 3)	P_0 (Stage 4)	$P_0\,\omega_N^0$ (Stage 5)
00010	P_0	P_0	P_0	$P_0\,\omega_N^0$	P_0
00011	P_0	P_0	P_0	$P_0\,\omega_N^8$	$P_0\,\omega_N^0$
00100	P_0	P_0	$P_0\,\omega_N^0$	P_0	P_0
00101	P_0	P_0	$P_0\,\omega_N^4$	P_0	$P_0\,\omega_N^0$
00110	P_0	P_0	$P_0\,\omega_N^8$	$P_1\,\omega_N^0$	P_0
00111	P_0	P_0	$P_0\,\omega_N^{12}$	$P_1\,\omega_N^8$	$P_0\,\omega_N^0$
01000	P_1	$P_1\,\omega_N^0$	P_1	P_1	P_1
01001	P_1	$P_1\,\omega_N^2$	P_1	P_1	$P_1\,\omega_N^0$
01010	P_1	$P_1\,\omega_N^4$	P_1	$P_1\,\omega_N^0$	P_1
01011	P_1	$P_1\,\omega_N^6$	P_1	$P_1\,\omega_N^8$	$P_1\,\omega_N^0$
01100	P_1	$P_1\,\omega_N^8$	$P_1\,\omega_N^0$	P_1	P_1
01101	P_1	$P_1\,\omega_N^{10}$	$P_1\,\omega_N^4$	P_1	$P_1\,\omega_N^0$
01110	P_1	$P_1\,\omega_N^{12}$	$P_1\,\omega_N^8$	$P_1\,\omega_N^0$	P_1
01111	P_1	$P_1\,\omega_N^{14}$	$P_1\,\omega_N^{12}$	$P_1\,\omega_N^8$	$P_1\,\omega_N^0$
10000	$P_2\,\omega_N^0$	P_2	P_2	P_2	P_2
10001	$P_2\,\omega_N^1$	P_2	P_2	P_2	$P_2\,\omega_N^0$
10010	$P_2\,\omega_N^2$	P_2	P_2	$P_2\,\omega_N^0$	P_2
10011	$P_2\,\omega_N^3$	P_2	P_2	$P_2\,\omega_N^8$	$P_2\,\omega_N^0$
10100	$P_2\,\omega_N^4$	P_2	$P_2\,\omega_N^0$	P_2	P_2
10101	$P_2\,\omega_N^5$	P_2	$P_2\,\omega_N^4$	P_2	$P_2\,\omega_N^0$
10110	$P_2\,\omega_N^6$	P_2	$P_2\,\omega_N^8$	$P_2\,\omega_N^0$	P_2
10111	$P_2\,\omega_N^7$	P_2	$P_2\,\omega_N^{12}$	$P_2\,\omega_N^8$	$P_2\,\omega_N^0$
11000	$P_3\,\omega_N^8$	$P_3\,\omega_N^0$	P_3	P_3	P_3
11001	$P_3\,\omega_N^9$	$P_3\,\omega_N^2$	P_3	P_3	$P_3\,\omega_N^0$
11010	$P_3\,\omega_N^{10}$	$P_3\,\omega_N^4$	P_3	$P_3\,\omega_N^0$	P_3
11011	$P_3\,\omega_N^{11}$	$P_3\,\omega_N^6$	P_3	$P_3\,\omega_N^8$	$P_3\,\omega_N^0$
11100	$P_3\,\omega_N^{12}$	$P_3\,\omega_N^8$	$P_3\,\omega_N^0$	P_3	P_3
11101	$P_3\,\omega_N^{13}$	$P_3\,\omega_N^{10}$	$P_3\,\omega_N^4$	P_3	$P_3\,\omega_N^0$
11110	$P_3\,\omega_N^{14}$	$P_3\,\omega_N^{12}$	$P_3\,\omega_N^8$	$P_3\,\omega_N^0$	P_3
11111	$P_3\,\omega_N^{15}$	$P_3\,\omega_N^{14}$	$P_3\,\omega_N^{12}$	$P_3\,\omega_N^8$	$P_3\,\omega_N^0$

Table 24.1 DIF$_{\text{NR}}$ FFT twiddle factors required by each processor in Figure 24.1.

Processors	$\omega_N^{i_3 i_2 i_1 i_0}$ (Stage 1)	$\omega_N^{i_2 i_1 i_0 0}$ (Stage 2)	$\omega_N^{i_1 i_0 0 0}$ (Stage 3)	$\omega_N^{i_0 0 0 0}$ (Stage 4)	$\omega_N^0 = 1$ (Stage 5)
P_0			ω_N^0 ω_N^4 ω_N^8 ω_N^{12}	ω_N^0 ω_N^8	$\omega_N^0 = 1$
P_1		ω_N^0 ω_N^2 ω_N^4 ω_N^6 ω_N^8 ω_N^{10} ω_N^{12} ω_N^{14}	ω_N^0 ω_N^4 ω_N^8 ω_N^{12}	ω_N^0 ω_N^8	$\omega_N^0 = 1$
P_2	ω_N^0 ω_N^1 ω_N^2 ω_N^3 ω_N^4 ω_N^5 ω_N^6 ω_N^7		ω_N^0 ω_N^4 ω_N^8 ω_N^{12}	ω_N^0 ω_N^8	$\omega_N^0 = 1$
P_3	ω_N^8 ω_N^9 ω_N^{10} ω_N^{11} ω_N^{12} ω_N^{13} ω_N^{14} ω_N^{15}	ω_N^0 ω_N^2 ω_N^4 ω_N^6 ω_N^8 ω_N^{10} ω_N^{12} ω_N^{14}	ω_N^0 ω_N^4 ω_N^8 ω_N^{12}	ω_N^0 ω_N^8	$\omega_N^0 = 1$

Figure 24.2 DIF$_{\mathrm{NR}}$ FFT twiddle factors required if a cyclic map is used.

$i_4i_3i_2i_1i_0$	$P_{i_1i_0}$	$\vert i_1i_0\vert i_4i_3i_2$		$\vert i_1i_0\vert\tau_4i_3i_2$		$\vert i_1i_0\vert\tau_4\tau_3i_2$		$\vert i_1i_0\vert\tau_4\tau_3\tau_2$		$\vert\tau_1i_0\vert\tau_4\tau_3\tau_2$	
00000	P_0	P_0	$\omega_N^{i_3i_2i_1i_0}$	P_0	$\omega_N^{i_2i_1i_00}$	P_0	$\omega_N^{i_1i_000}$	P_0	$\omega_N^{i_0000}$	P_0	$\omega_N^0=1$
00001	P_1	P_1	(Stage 1)	P_1	(Stage 2)	P_1	(Stage 3)	P_1	(Stage 4)	P_1	ω_N^0 (Stage 5)
00010	P_2	P_2		P_2		P_2		P_2	ω_N^0	P_2	
00011	P_3	P_3		P_3		P_3		P_3	ω_N^8	P_3	ω_N^0
00100	P_0	P_0		P_0		P_0	ω_N^0	P_0		P_0	
00101	P_1	P_1		P_1		P_1	ω_N^4	P_1		P_1	ω_N^0
00110	P_2	P_2		P_2		P_2	ω_N^8	P_2	ω_N^0	P_2	
00111	P_3	P_3		P_3		P_3	ω_N^{12}	P_3	ω_N^8	P_3	ω_N^0
01000	P_0	P_0		P_0	ω_N^0	P_0		P_0		P_0	
01001	P_1	P_1		P_1	ω_N^2	P_1		P_1		P_1	ω_N^0
01010	P_2	P_2		P_2	ω_N^4	P_2		P_2	ω_N^0	P_2	
01011	P_3	P_3		P_3	ω_N^6	P_3		P_3	ω_N^8	P_3	ω_N^0
01100	P_0	P_0		P_0	ω_N^8	P_0	ω_N^0	P_0		P_0	
01101	P_1	P_1		P_1	ω_N^{10}	P_1	ω_N^4	P_1		P_1	ω_N^0
01110	P_2	P_2		P_2	ω_N^{12}	P_2	ω_N^8	P_2	ω_N^0	P_2	
01111	P_3	P_3		P_3	ω_N^{14}	P_3	ω_N^{12}	P_3	ω_N^8	P_3	ω_N^0
10000	P_0	P_0	ω_N^0	P_0		P_0		P_0		P_0	
10001	P_1	P_1	ω_N^1	P_1		P_1		P_1		P_1	ω_N^0
10010	P_2	P_2	ω_N^2	P_2		P_2		P_2	ω_N^0	P_2	
10011	P_3	P_3	ω_N^3	P_3		P_3		P_3	ω_N^8	P_3	ω_N^0
10100	P_0	P_0	ω_N^4	P_0		P_0	ω_N^0	P_0		P_0	
10101	P_1	P_1	ω_N^5	P_1		P_1	ω_N^4	P_1		P_1	ω_N^0
10110	P_2	P_2	ω_N^6	P_2		P_2	ω_N^8	P_2	ω_N^0	P_2	
10111	P_3	P_3	ω_N^7	P_3		P_3	ω_N^{12}	P_3	ω_N^8	P_3	ω_N^0
11000	P_0	P_0	ω_N^8	P_0	ω_N^0	P_0		P_0		P_0	
11001	P_1	P_1	ω_N^9	P_1	ω_N^2	P_1		P_1		P_1	ω_N^0
11010	P_2	P_2	ω_N^{10}	P_2	ω_N^4	P_2		P_2	ω_N^0	P_2	
11011	P_3	P_3	ω_N^{11}	P_3	ω_N^6	P_3		P_3	ω_N^8	P_3	ω_N^0
11100	P_0	P_0	ω_N^{12}	P_0	ω_N^8	P_0	ω_N^0	P_0		P_0	
11101	P_1	P_1	ω_N^{13}	P_1	ω_N^{10}	P_1	ω_N^4	P_1		P_1	ω_N^0
11110	P_2	P_2	ω_N^{14}	P_2	ω_N^{12}	P_2	ω_N^8	P_2	ω_N^0	P_2	
11111	P_3	P_3	ω_N^{15}	P_3	ω_N^{14}	P_3	ω_N^{12}	P_3	ω_N^8	P_3	ω_N^0

Table 24.2 DIF$_{\text{NR}}$ FFT twiddle factors required by each processor in Figure 24.2.

Processors	$\omega_N^{i_3 i_2 i_1 i_0}$ (Stage 1)	$\omega_N^{i_2 i_1 i_0 0}$ (Stage 2)	$\omega_N^{i_1 i_0 00}$ (Stage 3)	$\omega_N^{i_0 000}$ (Stage 4)	$\omega_N^0 = 1$ (Stage 5)
P_0	ω_N^0 ω_N^4 ω_N^8 ω_N^{12}	ω_N^0 ω_N^8	ω_N^0		
P_1	ω_N^1 ω_N^5 ω_N^9 ω_N^{13}	ω_N^2 ω_N^{10}	ω_N^4		$\omega_N^0 = 1$
P_2	ω_N^2 ω_N^6 ω_N^{10} ω_N^{14}	ω_N^4 ω_N^{12}	ω_N^8	ω_N^0	
P_3	ω_N^3 ω_N^7 ω_N^{11} ω_N^{15}	ω_N^6 ω_N^{14}	ω_N^{12}	ω_N^8	$\omega_N^0 = 1$

24.2 Twiddle Factors for Parallel FFT With Inter-Processor Permutations

Referring to Figures 20.1, 20.2, 20.3, 20.4, 20.5, and 20.6 in Chapter 20 on parallel FFTs with inter-processor permutations, one can tabulate the twiddle factors $\omega_N^{i_3 i_2 i_1 i_0}$, $\omega_N^{i_2 i_1 i_0 0}$, $\omega_N^{i_1 i_0 0 0}$, $\omega_N^{i_0 0 0 0}$, and ω_N^0 (inferred from global $m = i_4 i_3 i_2 i_1 i_0$) required by each processor as shown in Table 24.3 below. (Note that $p = 4$ and $N = 32$ in the example.) Again, it is assumed that a DIF$_{NR}$ FFT is used. Observe that in this case, each processor needs to compute almost all $N/2$ twiddle factors either in advance or on the fly (to save storage).

Table 24.3 DIF$_{NR}$ FFT twiddle factors required by each processor in Figures 20.1, 20.2, 20.3, 20.4, 20.5, and 20.6. ($p = 4$ and $N = 32$)

Processors	$\omega_N^{i_3 i_2 i_1 i_0}$ (Stage 1)	$\omega_N^{i_2 i_1 i_0 0}$ (Stage 2)	$\omega_N^{i_1 i_0 0 0}$ (Stage 3)	$\omega_N^{i_0 0 0 0}$ (Stage 4)	$\omega_N^0 = 1$ (Stage 5)
P_0	ω_N^0	ω_N^0	ω_N^0	ω_N^0	$\omega_N^0 = 1$
	ω_N^1	ω_N^2	ω_N^4	ω_N^8	
	ω_N^2	ω_N^4	ω_N^8		
	ω_N^3	ω_N^6	ω_N^{12}		
P_1	ω_N^8	ω_N^0	ω_N^0	ω_N^0	$\omega_N^0 = 1$
	ω_N^9	ω_N^2	ω_N^4	ω_N^8	
	ω_N^{10}	ω_N^4	ω_N^8		
	ω_N^{11}	ω_N^6	ω_N^{12}		
P_2	ω_N^4	ω_N^8	ω_N^0	ω_N^0	$\omega_N^0 = 1$
	ω_N^5	ω_N^{10}	ω_N^4	ω_N^8	
	ω_N^6	ω_N^{12}	ω_N^8		
	ω_N^7	ω_N^{14}	ω_N^{12}		
P_3	ω_N^{12}	ω_N^8	ω_N^0	ω_N^0	$\omega_N^0 = 1$
	ω_N^{13}	ω_N^{10}	ω_N^4	ω_N^8	
	ω_N^{14}	ω_N^{12}	ω_N^8		
	ω_N^{15}	ω_N^{14}	ω_N^{12}		

Part IV

Appendices

Appendix A

Fundamental Concepts of Efficient Scientific Computation

In this appendix some fundamental concepts of efficient scientific computation are introduced and demonstrated through examples relevant to the FFT computation.

A.1 Time and Space Consumed by the DFT and FFT Algorithms

In Chapters 2 and 3 the arithmetic cost of the DFT and FFT algorithms were shown to be proportional to N^2 (via the naive matrix times vector computation) and $N \log_2 N$, respectively. The constant of proportionality is not specified there because its value depends on what kinds of operations are counted in determining the arithmetic cost. For example, one could count the number of complex multiplications and additions; alternatively, the number of real multiplications and additions might be counted. Thus, in comparing algorithms, it is important that the reported costs refer to the same kinds of operations. Obviously, the constant of proportionality becomes important when one compares the efficiency of two algorithms having the same order of complexity. For example, an FFT variant requiring $4N \log_2 N$ operations is more efficient than an FFT variant requiring $5N \log_2 N$ operations.

A.1.1 Relating operation counts to execution times

The operation counts of the DFT algorithm and several FFT variants are listed for sample N values in Table A.1. The reported operation counts represent the total number of real additions and real multiplications, assuming that the twiddle factors are pre-computed. The cost of evaluating the twiddle factors is thus excluded in the operation count. The FFT variants referred to in the table were developed in early chapters of this book.

Table A.1 Operation counts of the DFT and FFT variants.

Problem Length $N = 2^s$	DFT $8N^2$	Some FFT variants				
		Radix-2 $(N = 2^s)$ $5N \log_2 N$	Radix-4 $(N = 2^{2m})$ $4\frac{1}{4}N \log_2 N$	Radix-8 $(N = 2^{3m})$ $4\frac{1}{12}N \log_2 N$	Radix-16 $(N = 2^{4m})$ $4\frac{1}{32}N \log_2 N$	Split-Radix $(N = 2^{2m})$ $4N \log_2 N$
64	32768	1920	1632	1568	–	1536
128	131072	4480	–	–	–	–
256	524288	10240	8704	–	8256	8192
512	2097152	23040	–	18816	–	–
1024	8388608	51200	43520	–	–	40960
2048	33554432	112640	–	–	–	–
4096	134217728	245760	208896	200704	198144	196608
8192	536870912	532480	–	–	–	–
16384	2147483648	1146880	974848	–	–	917504

To help put things in perspective, it is instructive to see what the differences in operation counts might imply in terms of expected execution times. In Table A.2, the expected times are calculated assuming that the additions and multiplications each require one 1 μsec (10^{-6} seconds).

Table A.2 Expected execution times of the DFT and FFT variants.

Problem Length $N = 2^s$	DFT $8N^2$	Some FFT Variants				
		Radix-2 $(N = 2^s)$ $5N \log_2 N$	Radix-4 $(N = 2^{2m})$ $4\frac{1}{4}N \log_2 N$	Radix-8 $(N = 2^{3m})$ $4\frac{1}{12}N \log_2 N$	Radix-16 $(N = 2^{4m})$ $4\frac{1}{32}N \log_2 N$	Split-Radix $(N = 2^{2m})$ $4N \log_2 N$
64	0.033 sec	0.0019 sec	0.0016 sec	0.0016 sec	–	0.0015 sec
128	0.131 sec	0.0045 sec	–	–	–	–
256	0.524 sec	0.0102 sec	0.0087 sec	–	0.0083 sec	0.0082 sec
512	2.097 sec	0.0230 sec	–	0.0188 sec	–	–
1024	8.389 sec	0.0512 sec	0.0435 sec	–	–	0.0410 sec
2048	33.554 sec	0.1126 sec	–	–	–	–
4096	2.233 min	0.2458 sec	0.2089 sec	0.2007 sec	0.1981 sec	0.1966 sec
8192	8.933 min	0.5325 sec	–	–	–	–
16384	35.783 min	1.1469 sec	0.9748 sec	–	–	0.9175 sec

A.1.2　Relating MFLOPS to execution times and operation counts

A common measure of performance is the "megaflop rate"; that is, the rate of execution measured in MFLOPS (millions of floating-point operations per second). It is calculated from the total flop (floating-point operation) count and the actual execution time (in seconds), i.e.,

$$\text{MFLOPS Rate} \equiv \frac{\text{Total Flop Count}}{\text{Actual Execution Time in Seconds}} \times 10^{-6}.$$

Of course the use of the MFLOPS number is sensitive to context. It is useful in determining how well a particular algorithm has been implemented on a particular machine. Of course one assumes implicitly that only those operations that are necessary are being performed! It is also useful in comparing the performance of different machines; implementations of the same algorithm on different machines may provide useful information about their relative desirability for certain applications.

A.2   Comparing Algorithms by Orders of Complexity

In this section techniques and notations for determining, expressing, and comparing complexity results are presented.

A.2.1   An informal introduction via motivating examples

Recall that the naive matrix times vector computation in the DFT algorithm requires $8N^2$ flops, whereas various FFT algorithms require from $4N \log_2 N$ to $5N \log_2 N$ flops to compute the same result. For N ranging from $N = 64$ to $N = 16384$, the corresponding flop counts and the estimated execution times are displayed in Tables A.1 and A.2. As N grows, the flop count of the DFT algorithm grows much faster than that of any FFT algorithm, while the ratio of the flop counts of any two FFT variants is a constant. The ratio of $8N^2$ and $4N \log_2 N$ is $2N/\log_2 N$, which goes to infinity with N. For completeness, L'Hôpital's rule from the calculus is provided below.

Theorem A.1 (L'Hôpital's Rule) If $f(x)$ and $g(x)$ are both differentiable with derivatives $f'(x)$ and $g'(x)$ respectively, and if

$$\lim_{x \to \infty} f(x) = \lim_{x \to \infty} g(x) = \infty \, ,$$

then

$$\lim_{x \to \infty} \frac{f(x)}{g(x)} = \lim_{x \to \infty} \frac{f'(x)}{g'(x)}.$$

Note that L'Hôpital's rule can be applied again to the resulting limit, provided that the derivative functions satisfy the same conditions. This process is repeated until the limit can be determined. It is a simple exercise to use L'Hôpital's rule to show that

$$\lim_{N \to \infty} \frac{f(N)}{g(N)} = \lim_{N \to \infty} \frac{8N^2}{4N \log_2 N} = \infty.$$

In general, there are three possible outcomes in applying L'Hôpital's rule to compare the complexity of two algorithms.

Case (i)

$$\lim_{N\to\infty} \frac{f(N)}{g(N)} = c > 0,$$

where c is a constant. The two algorithms are said to have the same order of complexity.

Case (ii)

$$\lim_{N\to\infty} \frac{f(N)}{g(N)} = \infty.$$

This implies that $f(N)$ grows asymptotically faster than $g(N)$, and the algorithm costing $g(N)$ is said to have lower complexity. Informally, one denotes $g(N) < f(N)$.

Case (iii)

$$\lim_{N\to\infty} \frac{f(N)}{g(N)} = 0,$$

which is equivalent to

$$\lim_{N\to\infty} \frac{g(N)}{f(N)} = \infty.$$

This means that the algorithm costing $f(N)$ has lower complexity. Informally, one denotes $f(N) < g(N)$.

Complexity functions commonly have more than one term. When there are multiple terms, the fastest growing term is the *dominant term*. A few examples are displayed below.

1. $g_1(N) = N + \log_2 N + 5 = N + \text{lower order terms}$.

2. $g_2(N) = 5N \log_2 N + 3N + \log_2 N + 3 = 5N \log_2 N + \text{lower order terms}$.

3. $g_3(N) = (\log_2 N)^5 + \sqrt{N} + 18 = \sqrt{N} + \text{lower order terms}$.

4. $g_4(N) = 2^N + N^{99} = 2^N + \text{lower order terms}$.

5. $g_5(N) = \log_3 N + N^{0.1} = N^{0.1} + \text{lower order terms}$.

To compare the orders of complexity of any two functions listed above, one only needs to compare their leading terms. For example, $g_5(N)$ is more efficient than $g_1(N)$ because $N^{0.1}/N \to 0$ as $N \to \infty$.

A.2.2   Formal notations and terminologies

By applying the techniques introduced in the last section, it is a simple exercise to verify that $f(N)$ and $g(N)$ given below have the same order, as do $g(N)$ and $h(N)$. It seems natural to conclude that $f(N)$, $g(N)$, and $h(N)$ have the same order of complexity.

1. $f(N) = 3N \log_3 N$.

2. $g(N) = 7N \log_5 N + N - 17$.

3. $h(N) = N \log_2 N + 10N - 8$.

This explains why each order of complexity is *formally* defined as a "set" containing an infinite number of member functions.

Definition A.2 Define $\Theta(f(N))$, which is called the *order* of $f(N)$, to be the set of complexity functions $g(N)$ for which there exists some positive real constants c_1 and c_2 and some nonnegative integer N_o such that, for all $N \geq N_o$,

$$c_1 \times f(N) \leq g(N) \leq c_2 \times f(N) .$$

To distinguish two different orders of complexity from each other, the following definition is used.

Definition A.3 For a given complexity function $f(N)$, $o(f(N))$, pronounced *small-oh* or *little-oh* of $f(N)$, is the set of all complexity functions $g(N)$ satisfying the following:
For *every* positive real constant c there exits a nonnegative integer N_o such that, for all $N \geq N_o$,

$$g(N) \leq c \times f(N)$$

The next theorem from [68] allows the definitions of $\Theta(f(N))$ and $o(f(N))$ to be tested by limits.

Theorem A.4

$$\lim_{N \to \infty} \frac{f(N)}{g(N)} = \begin{cases} c \text{ implies } f(N) \in \Theta(g(N)) \text{ if } c > 0, \\ 0 \text{ implies } f(N) \in o(g(N)), \\ \infty \text{ implies } g(N) \in o(f(N)). \end{cases}$$

A.2.3   The big-Oh and big-Omega notations

For completeness, two notations which are commonly used in the literature and text books are defined and related to the Θ-notation below.

Definition A.5 For a given complexity function $f(N)$, $O(f(N))$, pronounced *big-Oh* of $f(N)$, is the set of complexity functions $g(N)$ for which there exists some positive real constant c_2 and some nonnegative integer N_o such that for all $N \geq N_o$

$$g(N) \leq c_2 \times f(N) .$$

Definition A.6 For a given complexity function $f(N)$, $\Omega(f(N))$, pronounced *big-Omega* of $f(N)$, is the set of complexity functions $g(N)$ for which there exists some positive real constant c_1 and some nonnegative integer N_o such that for all $N \geq N_o$

$$g(N) \geq c_1 \times f(N) .$$

The order $\Theta(f(N))$ in Definition A.2 can then be expressed as the intersection of the set $O(f(N))$ and the set $\Omega(f(N))$. This leads to the following alternative definition of $\Theta(f(N))$.

Definition A.7 For a given complexity function $f(N)$,

$$\Theta(f(N)) = O(f(N)) \cap \Omega(f(N)) .$$

Note that one cannot know the order of $g(N)$ from a claim that $g(N) \in O(N^5)$, because this claim is true for $g(N) = \log_2 N$, $g(N) = N$, $g(N) = N^2$, ..., $g(N) = N^5$. On the other hand, $g(N) \in \Theta(f(N))$ guarantees that $g(N)$ and $f(N)$ have the same order. Thus, the Θ-notation provides more specificity about the order of complexity, which explains why its use is standard in text books on algorithmic analysis.[1]

A.2.4 Some common uses of the Θ-notation

The examples given at the end of Section A.2.1 are used again to demonstrate some common uses of the "order" notation.

1. $g_1(N) = N + \log_2 N + 5 \in \Theta(N)$.

2. $g_2(N) = 5N \log_2 N + 3N + \log_2 N + 3 \in \Theta(N \log_2 N)$.

3. $g_3(N) = (\log_2 N)^5 + \sqrt{N} + 18 \in \Theta\left(\sqrt{N}\right)$.

4. $g_4(N) = 2^N + N^{99} \in \Theta\left(2^N\right)$.

5. $g_5(N) = \log_3 N + N^{0.1} \in \Theta\left(N^{0.1}\right)$.

It is also common to preserve the leading term, and use the "order" notation to represent the remaining terms.

1. $g_1(N) = N + \log_2 N + 5 = N + \Theta\left(\log_2 N\right)$.

2. $g_2(N) = 5N \log_2 N + 3N + \log_2 N + 3 = 5N \log_2 N + \Theta(N)$.

3. $g_3(N) = (\log_2 N)^5 + \sqrt{N} + 18 = \sqrt{N} + \Theta\left(\log_2^5 N\right)$.

4. $g_4(N) = 2^N + N^{99} = 2^N + \Theta\left(N^{99}\right)$.

5. $g_5(N) = \log_3 N + N^{0.1} = N^{0.1} + \Theta\left(\log_3 N\right)$.

---

[1] It is often true that authors use the notion $g(N) \in O(f(N))$ when they really mean $g(N) \in \Theta(f(N))$. It most contexts it is trivially obvious that $g(N) \in \Omega(f(N))$, so no real ambiguity results.

Appendix B

Solving Recurrence Equations by Substitution

This appendix is devoted to the solution of recurrence equations. This topic is covered in depth here because it is a useful analytical tool in the study of various fast Fourier transform (FFT) algorithms in this book.

B.1 Deriving Recurrences From a Known Function

Consider a function which defines the factorial of a positive integer N. By definition,

$$(B.1) \qquad N! \equiv \prod_{K=1}^{N} K = \left(\prod_{K=1}^{N-1} K \right) \times N = ((N-1)!) \times N .$$

Defining $T(N) = N!$, which implies $T(N-1) = (N-1)!$, yields

$$(B.2) \qquad \boxed{T(N) = \begin{cases} T(N-1) \times N & \text{if } N \geq 2, \\ 1 & \text{if } N = 1. \end{cases}}$$

Equations (B.2) relate the function values of $T(N)$ and $T(N-1)$; together with the boundary condition $T(1) = 1$, it defines $N!$. It is called a recurrence equation because the function T appears on both sides of (B.2). Solving a recurrence equation to determine the analytical form of an unknown function is the topic of the next section.

B.2 Solving Recurrences to Determine an Unknown Function

Example B.1 Consider again the example in (B.2), and adopt the view that its solution is not known. The solution method is first demonstrated for $N = 5$. Observe that in the following solution process, the right-hand side of the identity $T(N) = T(N-1) \times N$ is substituted for $T(N)$ when $N = 5, 4, 3, 2$, and the boundary condition

287

$T(1) = 1$ is last substituted to terminate the process when $N = 1$.

$$
\begin{aligned}
T(5) &= (T(4) \times 5) \\
&= (T(3) \times 4) \times 5 \\
&= (T(2) \times 3) \times 4 \times 5 \\
&= (T(1) \times 2) \times 3 \times 4 \times 5 \\
&= 1 \times 2 \times 3 \times 4 \times 5. \\
&= 5!\,.
\end{aligned}
$$

The analytical solution expressed in terms of N is obtained in exactly the same manner as shown below. The solution is reached when T disappears from the right-hand side.

$$
\begin{aligned}
T(N) &= (T(N-1) \times N) \\
&= (T(N-2) \times (N-1)) \times N \\
&= (T(N-3) \times (N-2)) \times (N-1) \times N \\
&\;\;\vdots \\
&= (T(1) \times 2) \times 3 \times \cdots \times (N-2) \times (N-1) \times N \\
&= 1 \times 2 \times 3 \times \cdots \times (N-2) \times (N-1) \times N \\
&= N!\,.
\end{aligned}
$$

This method is called *the substitution method*.

Example B.2 Given below is another recurrence equation which can be solved by the same technique to determine the unknown function $T(N)$.

(B.3)

> Solving
>
> $$
> T(N) = \begin{cases} T\left(\frac{N}{2}\right) + 1 & \text{if } N = 2^n \geq 2, \\ 1 & \text{if } N = 1. \end{cases}
> $$

The assumption $N = 2^n$ ensures that $\frac{N}{2}, \frac{N}{4}, \frac{N}{8}, \cdots, \frac{N}{2^k}$ ($k < n$) are always a power of 2, and it further ensures that the occurrence of $T\left(\frac{N}{2^k}\right)$ can always be replaced by $T\left(\frac{N}{2^k} \times \frac{1}{2}\right)$ using (B.3) until $T(1) = 1$ is finally substituted to make T disappear from the right-hand side. This process is shown below.

$$T(N) = \left(T\left(\frac{N}{2}\right) + 1\right)$$
$$= \left(T\left(\frac{N}{2^2}\right) + 1\right) + 1$$
$$= \left(T\left(\frac{N}{2^3}\right) + 1\right) + 1 + 1$$
$$\vdots$$
$$= \left(T\left(\frac{N}{2^n}\right) + 1\right) + (n - 1)$$
$$= T(1) + n$$
$$= 1 + n$$
$$= \log_2 N + 1 .$$

Note that the assumption $N = 2^n$ implies $n = \log_2 N$.

Example B.3 Now consider another example before this solution method is generalized further.

(B.4)

> Solving
> $$T(N) = \begin{cases} 2T\left(\frac{N}{2}\right) + N & \text{if } N = 2^n \geq 2, \\ 0 & \text{if } N = 1. \end{cases}$$

Observe that substituting N by $\frac{N}{2}$ in (B.4) yields

(∗)
$$T\left(\frac{N}{2}\right) = 2T\left(\frac{N}{4}\right) + \frac{N}{2} ,$$

substituting N by $\frac{N}{4}$ in (B.4) yields

(∗∗)
$$T\left(\frac{N}{4}\right) = 2T\left(\frac{N}{8}\right) + \frac{N}{4} ,$$

and so on. These are the identities used in solving (B.4). The process of repetitive substitutions is shown below. Observe that $T\left(\frac{N}{2}\right)$ is replaced by $T\left(\frac{N}{2^2}\right)$ using identity (∗), which is then replaced by $T\left(\frac{N}{2^3}\right)$ using identity (∗∗), $\cdots$, and eventually $T(2)$ is replaced by $T(1) = 0$. When T disappears from the right-hand side, the remaining terms, expressed in a closed-form, represent the analytical form of function $T(N)$.

$$T(N) = \left(2T\left(\frac{N}{2}\right) + N\right)$$

$$= 2\left(2T\left(\frac{N}{2^2}\right) + \frac{N}{2}\right) + N$$

$$= 2^2 T\left(\frac{N}{2^2}\right) + N + N$$

$$= 2^2 \left(2T\left(\frac{N}{2^3}\right) + \frac{N}{2^2}\right) + N + N$$

$$= 2^3 T\left(\frac{N}{2^3}\right) + N + N + N$$

$$= 2^3 T\left(\frac{N}{2^3}\right) + N \times 3$$

$$\vdots$$

$$= 2^n T\left(\frac{N}{2^n}\right) + N \times n$$

$$= 2^n T(1) + N \times n$$

$$= 0 + N \log_2 N$$

$$= N \log_2 N .$$

Thus $T(N) = N \log_2 N$ is the solution of (B.4).

Comparing the three examples above, it is evident that while the principle of substitution is straightforward, some skill is needed to detect the pattern in which the terms in the right-hand side emerge during the substitution process, and to collect and simplify these terms into a closed-form expression at the end. The closed-form expressions of some mathematical summations encountered in solving generalized recurrence equations are presented in the next section.

B.3   Mathematical Summation Formulas

The formulas reviewed in this section are useful in finding closed-form expressions of some commonly encountered summations.

Arithmetic series

(B.5) $$\sum_{K=1}^{N} K = 1 + 2 + \cdots + N = \frac{N(N+1)}{2} .$$

(B.6) $$\sum_{K=1}^{N} K^2 = 1 + 2^2 + \cdots + N^2 = \frac{N(N+1)(2N+1)}{6} .$$

Geometric series

$$(B.7) \qquad \sum_{k=0}^{n-1} x^k = 1 + x + x^2 + \cdots + x^{n-1}$$

$$= \frac{x^n - 1}{x - 1} , \text{ if } x \neq 1 \text{ is a real number.}$$

$$(B.8) \qquad \sum_{k=0}^{\infty} x^k = 1 + x + x^2 + \cdots$$

$$= \frac{1}{1 - x} , \text{ if } -1 < x < 1 .$$

Harmonic series

$$(B.9) \qquad \sum_{K=1}^{N} \frac{1}{K} = 1 + \frac{1}{2} + \frac{1}{3} + \frac{1}{4} + \cdots + \frac{1}{N}$$

$$= \ln N + \Theta(1) .$$

A telescoping sum

$$(B.10) \qquad \sum_{K=1}^{N-1} \frac{1}{K(K+1)} = \sum_{K=1}^{N-1} \left(\frac{1}{K} - \frac{1}{K+1} \right)$$

$$= \left(1 - \frac{1}{2} \right) + \left(\frac{1}{2} - \frac{1}{3} \right) + \cdots + \left(\frac{1}{N-1} - \frac{1}{N} \right)$$

$$= 1 - \frac{1}{N} .$$

Bounding a finite sum by integrals

$$(B.11) \qquad \int_{A-1}^{B} F(x)dx \leq \sum_{K=A}^{B} F(K) \leq \int_{A}^{B+1} F(x)dx ,$$

if $F(K)$ is a monotonically increasing function.

$$(B.12) \qquad \int_{A}^{B+1} F(x)dx \leq \sum_{K=A}^{B} F(K) \leq \int_{A-1}^{B} F(x)dx ,$$

if $F(K)$ is a monotonically decreasing function.

B.4   Solving Generalized Recurrence Equations

Theorem B.1 If the execution time of an algorithm satisfies

(B.13)
$$T(N) = \begin{cases} aT\left(\frac{N}{c}\right) + bN & \text{if } N = c^k > 1, \\ b & \text{if } N = 1. \end{cases}$$

where $a \geq 1$, $c \geq 2$, $b > 0$, and $c > 0$ are given constants, then the order of complexity of this algorithm is given by

(B.14)
$$T(N) = \begin{cases} \dfrac{bc}{c-a}N & \text{if } a < c, \\ bN \log_c N + \Theta(N) & \text{if } a = c, \\ \dfrac{ab}{a-c}N^{\log_c a} & \text{if } a > c. \end{cases}$$

Proof: Assuming $N = c^k > 1$ for $c \geq 2$,

$$
\begin{aligned}
T(N) &= aT\left(\frac{N}{c}\right) + bN \\
&= a\left(aT\left(\frac{N}{c^2}\right) + b\left(\frac{N}{c}\right)\right) + bN \\
&= a^2 T\left(\frac{N}{c^2}\right) + ab\left(\frac{N}{c}\right) + bN \\
&= a^2 \left(aT\left(\frac{N}{c^3}\right) + b\left(\frac{N}{c^2}\right)\right) + ab\left(\frac{N}{c}\right) + bN \\
&= a^3 T\left(\frac{N}{c^3}\right) + a^2 b\left(\frac{N}{c^2}\right) + ab\left(\frac{N}{c}\right) + bN
\end{aligned}
$$

(B.15)
$$\vdots$$

$$
\begin{aligned}
&= a^k T\left(\frac{N}{c^k}\right) + a^{k-1}b\left(\frac{N}{c^{k-1}}\right) + a^{k-2}b\left(\frac{N}{c^{k-2}}\right) + \cdots + ab\left(\frac{N}{c}\right) + bN \\
&= a^k T(1) + a^{k-1}bc + a^{k-2}bc^2 + \cdots + abc^{k-1} + bc^k \\
&= a^k b + a^k b\left(\frac{c}{a}\right) + a^k b\left(\frac{c}{a}\right)^2 + \cdots + a^k b\left(\frac{c}{a}\right)^{k-1} + a^k b\left(\frac{c}{a}\right)^k \\
&= a^k b \sum_{i=0}^{k}\left(\frac{c}{a}\right)^i.
\end{aligned}
$$

If $a = c$, then $\frac{c}{a} = 1$, and $N = c^k = a^k$, which implies $k = \log_c N$. Equation (B.15) becomes

(B.16)
$$T(N) = a^k b \sum_{i=0}^{k}\left(\frac{c}{a}\right)^i = Nb \sum_{i=0}^{k} 1 = Nb(k+1) = bN \log_c N + bN.$$

If $a > c$, then $\frac{c}{a} < 1$ and $\lim_{i \to \infty} \left(\frac{c}{a}\right)^i = 0$. As before, $N = c^k$ implies $k = \log_c N$. Equation (B.15) becomes

$$T(N) = a^k b \sum_{i=0}^{k} \left(\frac{c}{a}\right)^i = a^k b \left(\frac{1 - (c/a)^{k+1}}{1 - (c/a)}\right)$$

(B.17)
$$\approx a^k b \left(\frac{a}{a-c}\right) = a^{\log_c N} \left(\frac{ab}{a-c}\right)$$

$$= \frac{ab}{a-c} N^{\log_c a} \ .$$

If $a < c$, then reformulate equation (B.15) to obtain a geometric series in terms of $\frac{a}{c} < 1$. From equation (B.15),

$$T(N) = a^k b \sum_{i=0}^{k} \left(\frac{c}{a}\right)^i$$

$$= a^k b + a^{k-1} bc + a^{k-2} bc^2 + \cdots + abc^{k-1} + bc^k$$

$$= c^k b \left(\frac{a}{c}\right)^k + c^k b \left(\frac{a}{c}\right)^{k-1} + \cdots + c^k b \left(\frac{a}{c}\right) + c^k b$$

(B.18)
$$= c^k b \sum_{i=0}^{k} \left(\frac{a}{c}\right)^i$$

$$= Nb \left(\frac{1 - (a/c)^{k+1}}{1 - (a/c)}\right)$$

$$\approx bN \left(\frac{c}{c-a}\right)$$

$$= \frac{bc}{c-a} N \ .$$

∎

Theorem B.2 If the execution time of an algorithm satisfies

(B.19)
$$T(N) = \begin{cases} aT\left(\frac{N}{c}\right) + bN^r & \text{if } N = c^k > 1, \\ d & \text{if } N = 1. \end{cases}$$

where $a \geq 1$, $c \geq 2$, $r \geq 1$, $b > 0$, and $d > 0$ are given constants, then the order of complexity of this algorithm is given by

(B.20)
$$T(N) = \begin{cases} \frac{bc^r}{c^r - a} N^r + \Theta\left(N^{\log_c a}\right) & \text{if } a < c^r, \\ bN^r \log_c N + \Theta\left(N^r\right) & \text{if } a = c^r, \\ \left(\frac{ab}{a - c^r} + (d - b)\right) N^{\log_c a} & \text{if } a > c^r. \end{cases}$$

Proof: Assuming $N = c^k > 1$ for $c \geq 2$,

(B.21)

$$T(N) = aT\left(\frac{N}{c}\right) + bN^r$$

$$= a\left(aT\left(\frac{N}{c^2}\right) + b\left(\frac{N}{c}\right)^r\right) + bN^r$$

$$= a^2 T\left(\frac{N}{c^2}\right) + ab\left(\frac{N}{c}\right)^r + bN^r$$

$$= a^2\left(aT\left(\frac{N}{c^3}\right) + b\left(\frac{N}{c^2}\right)^r\right) + ab\left(\frac{N}{c}\right)^r + bN^r$$

$$= a^3 T\left(\frac{N}{c^3}\right) + a^2 b\left(\frac{N}{c^2}\right)^r + ab\left(\frac{N}{c}\right)^r + bN^r$$

$$\vdots$$

$$= a^k T\left(\frac{N}{c^k}\right) + a^{k-1}b\left(\frac{N}{c^{k-1}}\right)^r + a^{k-2}b\left(\frac{N}{c^{k-2}}\right)^r + \cdots + ab\left(\frac{N}{c}\right)^r + bN^r$$

$$= a^k T(1) + bN^r\left(\frac{a}{c^r}\right)^{k-1} + bN^r\left(\frac{a}{c^r}\right)^{k-2} + \cdots + bN^r\left(\frac{a}{c^r}\right) + bN^r$$

$$= a^k d + bN^r \sum_{i=0}^{k-1}\left(\frac{a}{c^r}\right)^i .$$

If $a = c^r$, then $\frac{a}{c^r} = 1$, and $a^k = c^{rk} = \left(c^k\right)^r = N^r$. Equation (B.21) becomes

(B.22)

$$T(N) = a^k d + bN^r \sum_{i=0}^{k-1}\left(\frac{a}{c^r}\right)^i = dc^{rk} + bN^r \sum_{i=0}^{k-1} 1$$

$$= dN^r + bN^r(k - 1 + 1) = bN^r \log_c N + dN^r .$$

If $a < c^r$, then $\frac{a}{c^r} < 1$ and $\lim_{i \to \infty}\left(\frac{a}{c^r}\right)^i = 0$. As before, $N = c^k$ implies $k = \log_c N$. Equation (B.21) becomes

$$T(N) = a^k d + bN^r \sum_{i=0}^{k-1}\left(\frac{a}{c^r}\right)^i$$

$$\approx a^k d + bN^r\left(\frac{c^r}{c^r - a}\right)$$

(B.23)

$$= da^{\log_c N} + bN^r\left(\frac{c^r}{c^r - a}\right)$$

$$= \frac{bc^r}{c^r - a}N^r + dN^{\log_c a}$$

$$= \frac{bc^r}{c^r - a}N^r + \Theta\left(N^{\log_c a}\right) .$$

Note that $a < c^r$ implies $\log_c a < \log_c c^r$; i.e., $\log_c a < r$, and the term $N^{\log_c a}$ is a lower order term compared to N^r.

Finally, consider the case $a > c^r$. Because $\frac{c^r}{a} < 1$, it is desirable to reformulate equation (B.21) to obtain a geometric series in terms of $\frac{c^r}{a}$ as shown below.

$$T(N) = a^k d + b N^r \sum_{i=0}^{k-1} \left(\frac{a}{c^r}\right)^i$$

$$= a^k d + a^{k-1} b \left(\frac{N}{c^{k-1}}\right)^r + a^{k-2} b \left(\frac{N}{c^{k-2}}\right)^r + \cdots + ab \left(\frac{N}{c}\right)^r + b N^r$$

$$= a^k d + a^{k-1} b \left(\frac{c^k}{c^{k-1}}\right)^r + a^{k-2} b \left(\frac{c^k}{c^{k-2}}\right)^r + \cdots + ab \left(\frac{c^k}{c}\right)^r + b (c^k)^r$$

$$= a^k d + a^{k-1} b c^r + a^{k-2} b (c^r)^2 + \cdots + ab (c^r)^{k-1} + b (c^r)^k$$

$$= a^k d + a^k b \left(\frac{c^r}{a}\right) + a^k b \left(\frac{c^r}{a}\right)^2 + \cdots + a^k b \left(\frac{c^r}{a}\right)^{k-1} + a^k b \left(\frac{c^r}{a}\right)^k$$

(B.24)
$$= a^k d + a^k b \sum_{i=1}^{k} \left(\frac{c^r}{a}\right)^i$$

$$= a^k d + a^k b \left(-1 + \sum_{i=0}^{k} \left(\frac{c^r}{a}\right)^i\right)$$

$$\approx a^k d - a^k b + a^k b \left(\frac{a}{a - c^r}\right)$$

$$= a^k \left((d - b) + \frac{ab}{a - c^r}\right) k$$

$$= \left(\frac{ab}{a - c^r} + (d - b)\right) a^{\log_c N}$$

$$= \left(\frac{ab}{a - c^r} + (d - b)\right) N^{\log_c a} . \qquad \blacksquare$$

Theorem B.3

Solving

(B.25)
$$T(N) = \begin{cases} \sum_{K=1}^{N} \frac{1}{N}(T(K-1) + T(N-K) + N - 1) & \text{if } N > 1, \\ 0 & \text{if } N = 1. \end{cases}$$

Solution:
(B.26)
$$T(N) = \sum_{K=1}^{N} \frac{1}{N}\left(T(K-1) + T(N-K) + N - 1\right)$$

$$= (N - 1) + \frac{1}{N}\left(T(0) + T(N-1) + T(1) + T(N-2) + \cdots + T(N-1) + T(0)\right)$$

$$= (N - 1) + \frac{1}{N}\left(2T(0) + 2T(1) + \cdots + 2T(N-2) + 2T(N-1)\right)$$

$$= (N - 1) + \frac{2}{N} \sum_{K=1}^{N} T(K - 1) .$$

To solve for $T(N)$, it is desirable to simplify (B.26) so that only $T(N-1)$ appears in the right-hand side. Before that can be done, the expression of $T(N-1)$ is first determined by substituting N by $N-1$ on both sides of (B.26), and the result is shown in identity (B.27).

$$\text{(B.27)} \qquad T(N-1) = (N-2) + \frac{2}{N-1} \sum_{K=1}^{N-1} T(K-1) \,.$$

Because $T(0)$, $T(1)$, $\cdots$, $T(N-2)$ appear in the right-hand sides of both identities (B.26) and (B.27), they can be cancelled out by subtracting one from the other if their respective coefficients can be made the same. To accomplish this, multiply both sides of (B.26) by $N/(N-1)$, and obtain

$$\text{(B.28)} \qquad \begin{aligned} \left(\frac{N}{N-1}\right) T(N) &= \left(\frac{N}{N-1}\right)\left((N-1) + \frac{2}{N}\sum_{K=1}^{N} T(K-1)\right) \\ &= N + \frac{2}{N-1}\sum_{K=1}^{N} T(K-1) \,. \end{aligned}$$

Subtracting (B.27) from (B.28), one obtains

$$\text{(B.29)} \qquad \begin{aligned} \left(\frac{N}{N-1}\right) T(N) - T(N-1) &= N - (N-2) + \frac{2}{N-1}T(N-1) \\ &= 2 + \frac{2}{N-1}T(N-1) \,. \end{aligned}$$

By moving $T(N-1)$ on the left-hand side of (B.29) to the right-hand side, $T(N)$ can be expressed as

$$\text{(B.30)} \qquad \begin{aligned} T(N) &= \left(\frac{N-1}{N}\right)\left(T(N-1) + 2 + \frac{2}{N-1}T(N-1)\right) \\ &= \left(\frac{N-1}{N}\right)\left(2 + \frac{N+1}{N-1}T(N-1)\right) \\ &= \left(\frac{N+1}{N}\right) T(N-1) + 2\left(\frac{N-1}{N}\right) \\ &\le \left(\frac{N+1}{N}\right) T(N-1) + 2 \\ &= \mathcal{T}(N) \,. \end{aligned}$$

Since $T(N) = \mathcal{T}(N)$ for all practical purposes, one may now solve the following recurrence involving $\mathcal{T}(N)$, which is much simpler than (B.25).

$$\text{(B.31)} \qquad \boxed{\begin{aligned} &\text{Solving} \\ &\mathcal{T}(N) = \begin{cases} 2 + \left(\frac{N+1}{N}\right)\mathcal{T}(N-1) & \text{if } N > 1, \\ 0 & \text{if } N = 1. \end{cases} \end{aligned}}$$

Observe that substituting N by $N-1$ on both sides of identity (B.31) yields

$$\text{(*)} \qquad \mathcal{T}(N-1) = 2 + \left(\frac{N}{N-1}\right)\mathcal{T}(N-2) \,,$$

substituting N by $N-2$ in (B.31) yields

(✿✿) $$T(N-2) = 2 + \left(\frac{N-1}{N-2}\right) T(N-3),$$

and so on. These are the identities to be used in solving (B.31). The process of repetitive substitutions is shown below. Observe that $T(N-1)$ is replaced by $T(N-2)$ using identity (✿), which is then replaced by $T(N-3)$ using identity (✿✿), $\cdots$, and eventually $T(2)$ is replaced by $T(1) = 0$.

$$
\begin{aligned}
T(N) &= 2 + \frac{N+1}{N} T(N-1) \\
&= 2 + \frac{N+1}{N}\left(2 + \frac{N}{N-1} T(N-2)\right) \\
&= 2 + 2\left(\frac{N+1}{N}\right) + \frac{N+1}{N-1} T(N-2) \\
&= 2 + 2\left(\frac{N+1}{N}\right) + \frac{N+1}{N-1}\left(2 + \frac{N-1}{N-2} T(N-3)\right) \\
&= 2 + 2(N+1)\left(\frac{1}{N} + \frac{1}{N-1}\right) + \frac{N+1}{N-2} T(N-3) \\
&\ \ \vdots \\
&= 2 + 2(N+1)\left(\frac{1}{N} + \frac{1}{N-1} + \cdots + \frac{1}{4}\right) + \frac{N+1}{3} T(2) \\
&= 2 + 2(N+1)\left(\frac{1}{N} + \frac{1}{N-1} + \cdots + \frac{1}{4}\right) + \frac{N+1}{3}\left(2 + \frac{3}{2} T(1)\right) \\
&= 2 + 2(N+1)\left(\frac{1}{N} + \frac{1}{N-1} + \cdots + \frac{1}{4} + \frac{1}{3}\right) \\
&= 2 + 2(N+1)\left(-1 - \frac{1}{2} + \sum_{K=1}^{N} \frac{1}{K}\right) \\
&= 2 + 2(N+1)\left(H_n - 1.5\right) \\
&= 2 + 2(N+1)\left(\ln N + \Theta(1)\right) \\
&= 2N \ln N + \Theta(N) \\
&= 1.386 N \log_2 N + \Theta(N)
\end{aligned}
$$

(B.32)

The closed-form expression $T(N) = 1.386 \log_2 N + \Theta(N)$ is the solution of the recurrence (B.31).

B.5   Recurrences and the Fast Fourier Transforms

Since FFT algorithms are recursive, it is natural that their complexity emerges as a set of recurrence equations. Thus, determining the complexity of an FFT algorithm involves solving these recurrence equations. Four examples are given below. Their solutions are left as exercises.

Example B.4 Arithmetic Cost of the Radix-2 FFT Algorithm

$$
\text{Solving } T(N) = \begin{cases} 2T\left(\frac{N}{2}\right) + 5N & \text{if } N = 2^n \geq 2, \\ 0 & \text{if } N = 1. \end{cases}
$$

Answer: $T(N) = 5N \log_2 N$.

Example B.5 Arithmetic Cost of the Radix-4 FFT Algorithm

$$
\text{Solving } T(N) = \begin{cases} 4T\left(\frac{N}{4}\right) + 8\frac{1}{2}N - 32 & \text{if } N = 4^n \geq 16, \\ 16 & \text{if } N = 4. \end{cases}
$$

Answer: $T(N) = 8\dfrac{1}{2}N \log_4 N - \dfrac{43}{6}N + \dfrac{32}{3}$

$\qquad\qquad = 4\dfrac{1}{4}N \log_2 N - \dfrac{43}{6}N + \dfrac{32}{3}$.

Example B.6 Arithmetic Cost of the Split-Radix FFT Algorithm

$$
\text{Solving } T(N) = \begin{cases} T\left(\frac{N}{2}\right) + 2T\left(\frac{N}{4}\right) + 6N - 16 & \text{if } N = 4^n \geq 16, \\ 16 & \text{if } N = 4, \\ 4 & \text{if } N = 2. \end{cases}
$$

Answer: $T(N) = 4N \log_2 N - 6N + 8$.

Example B.7 Arithmetic Cost of the Radix-8 FFT Algorithm

$$
\text{Solving } T(N) = \begin{cases} 8T\left(\frac{N}{8}\right) + 12\frac{1}{4}N - c & \text{if } N = 8^n \geq 64, \text{ and } c > 0, \\ d & \text{if } N = 8, \text{ and } d > 0. \end{cases}
$$

Answer: $T(N) = 12\dfrac{1}{4}N \log_8 N - \Theta(N)$

$\qquad\qquad = 4\dfrac{1}{12}N \log_2 N - \Theta(N)$.

Example B.8 Arithmetic Cost of the Radix-16 FFT Algorithm

$$\text{Solving } T(N) = \begin{cases} 16T\left(\frac{N}{16}\right) + 16\frac{1}{8}N - c & \text{if } N = 16^n \geq 256, \text{ and } c > 0, \\ d & \text{if } N = 16, \text{ and } d > 0. \end{cases}$$

$$\text{Answer: } T(N) = 16\frac{1}{8}N\log_{16}N - \Theta(N)$$

$$= 4\frac{1}{32}N\log_2 N - \Theta(N).$$

Bibliography

[1] A. V. Aho, J. E. Hopcroft, and J. D. Ullman. *The Design and Analysis of Computer Algorithms*. Addison-Wesley Publishing Company, Reading, MA, 1974.

[2] G. Angelopoulos and I. Pitas. Two-dimensional FFT algorithms on hypercube and mesh machines. *Signal Processing*, 30:355–371, 1993.

[3] G. D. Bergland. The fast Fourier transform recursive equations for arbitrary length records. *Math. Comp.*, 21(98):236–238, 1967.

[4] G. D. Bergland. A fast Fourier transform algorithm for real-valued series. *Comm. Assoc. Comput. Mach.*, 11(10):703–710, 1968.

[5] G. D. Bergland. A fast Fourier transform algorithm using base 8 iterations. *Math. Comp.*, 22:275–279, 1968.

[6] G. D. Bergland. Fast Fourier transform hardware implementations - A survey. *IEEE Transactions on Audio and Electroacoustics*, AU-17(2):109–119, 1969.

[7] G. D. Bergland. Fast Fourier transform hardware implementations - An overview. *IEEE Transactions on Audio and Electroacoustics*, AU-17(2):104–108, 1969.

[8] G. D. Bergland. A radix-eight fast Fourier transform subroutine for real-valued series. *IEEE Transactions on Audio and Electroacoustics*, AU-17(2):138–144, 1969.

[9] Å. Björck. *Numerical Methods for Least Squares Problems*. The Society for Industrial and Applied Mathematics, Philadelphia, PA, 1996.

[10] L. L. Bluestein. A linear filtering approach to the computation of discrete Fourier transform. In *1968 NEREM Record*, pages 218–219, Boston, MA, Nov. 6–8 1968. Reprinted in *Papers on Digital Signal Processing*, Ed. A. V. Oppenheim, pp. 171–172, Cambridge, MA: The M.I.T. Press, 1969.

[11] L. L. Bluestein. A linear filtering approach to the computation of discrete Fourier transform. *IEEE Transactions on Audio and Electroacoustics*, AU-18:451–455, 1970. Reprinted in *Digital Signal Processing*, Eds. L. R. Rabimer and C. M. Rader, pp. 317–321, New York: IEEE Press, 1972.

[12] L. Bomans and D. Roose. Communication benchmarks for the iPSC/2. In F. Andre and J. P. Verjus, editors, *Hypercube and distributed computers*, pp. 93–103. Elsevier Science Publishers B.B. (North-Holland), 1989.

[13] R. Bracewell. *The Hartley Transform*. Oxford University Press, New York, NY, 1986.

[14] R. N. Bracewell. Assessing the Hartley transform. *IEEE Transactions on Acoustics, Speech, and Signal Processing*, ASSP-38:2174–2176, 1990.

[15] W. L. Briggs and V. E. Hensen. *The DFT: An Owners Manual for the Discrete Fourier Transform*. The Society for Industrial and Applied Mathematics, Philadelphia, PA, 1995.

[16] W. L. Briggs and T. Turnbull. Fast Poisson solvers for MIMD computers. *Parallel Computing*, 6:265–274, 1988.

[17] D. O. Brigham. *The Fast Fourier Transform*. Prentice-Hall, Inc., Englewood Cliffs, NJ, 1974.

[18] O. Buneman. A compact non-iterative Poisson solver. Technical Report 294, Stanford University Institute for Plasma Research, Stanford, CA, 1969.

[19] O. Buneman. Conversion of FFT's to fast Hartley transforms. *SIAM J. Sci. Stat. Comput.*, 7:624–638, 1986.

[20] C. S. Burrus and P. W. Eschenbacher. An in-place, in-order prime factor FFT algorithm. *IEEE Transactions on Acoustics, Speech, and Signal Processing*, ASSP-29:806–817, 1981.

[21] B. L. Buzbee, G. H. Golub, and C. W. Nielson. On direct methods for solving Poisson's equations. *SIAM J. Numer. Anal.*, 7:627–656, 1970.

[22] C. Calvin. Implementation of parallel FFT algorithms on distributed memory machines with a minimum overhead of communication. *Parallel Computing*, 22:1255–1279, 1996.

[23] R. M. Chamberlain. Gray codes, fast Fourier transforms and hypercubes. *Parallel Computing*, 6:225–233, 1988.

[24] C. Y. Chu. Comparison of two-dimensional FFT methods on the hypercube. In G. Fox, editor, *The Third Conference on Hypercube Concurrent Computers and Applications*, 1988.

[25] E. Chu. Impact of physical/logical network topology on parallel matrix computation. *The International Journal of High Performance Computing Applications*, 13(2):124–145, 1999.

[26] E. Chu and A. George. QR factorization of a dense matrix on a hypercube multiprocessor. *SIAM J. Sci. Comput.*, 11:990–1028, 1990.

[27] E. Chu and A. George. Parallel algorithms and subcube embedding on a hypercube. *SIAM J. Sci. Comput.*, 14:81–94, 1993.

[28] E. Chu and A. George. FFT algorithms and their adaptation to parallel processing. *Linear Algebra and its Appl.*, 284:95–124, 1998.

[29] E. Chu, A. George, and D. Quesnel. Parallel submatrix inversion on a subcube-grid. *Parallel Computing*, 19:243–256, 1993.

[30] W. T. Cochran, J. W. Cooley, D. L. Favin, H. D. Helms, R. A. Kaenel, W. W. Lang, G. C. Maling, D. E. Nelson, C. M. Rader, and P. D. Welch. What is the fast Fourier transform. *IEEE Transactions on Audio and Electroacoustics*, AU-15(2):45–55, 1967.

[31] J. W. Cooley, P. A. W. Lewis, and P. D. Welch. The fast Fourier transform algorithm and its applications. Technical Report RC-1743, IBM, February 1967.

[32] J. W. Cooley, P. A. W. Lewis, and P. D. Welch. Historical notes on the fast Fourier transform. *IEEE Transactions on Audio and Electroacoustics*, 15:76–79, 1967.

[33] J. W. Cooley and J. W. Tukey. An algorithm for the machine calculation of complex Fourier series. *Math. Comp.*, 19:297–301, 1965.

[34] C. de Boor. FFT as nested multiplications, with a twist. *SIAM J. Sci. Stat. Comput.*, 1:173–178, 1980.

[35] F. W. Dorr. The direct solution of the discrete Poisson's equation on a rectangle. *SIAN Review*, 12:248–263, 1970.

[36] A. Dubey, M. Zubair, and C. E. Grosch. A general purpose subroutine for fast Fourier transform on a distributed memory parallel machine. *Parallel Computing*, 20:1697–1710, 1994.

[37] P. Dubois and A. N. Venetsanopoulos. A new algorithm for the radix-3 FFT. *IEEE Transactions on Acoustics, Speech, and Signal Processing*, ASSP-26:222–225, 1978.

[38] P. Duhamel. Implementation of "split-radix" FFT algorithms for complex, real and real-symmetric data. *IEEE Transactions on Acoustics, Speech, and Signal Processing*, ASSP-34:285–295, 1986.

[39] P. Duhamel and H. Hollmann. Split-radix FFT algorithms. *Electron. Lett.*, 20:14–16, 1984.

[40] P. Duhamel and M. Vetterli. Improved Fourier and Hartley transform algorithms: Application to cyclic convolution of real data. *IEEE Transactions on Acoustics, Speech, and Signal Processing*, ASSP-35:818–824, 1987.

[41] P. Duhamel and M. Vetterli. Fast Fourier transforms: A tutorial review and a state of the art. *Signal Processing*, 19:259–299, 1990.

[42] T. H. Dunigan. Performance of the Intel iPSC/860 Hypercube. Technical Report ORNL/TM-11491, Oak Ridge National Laboratory, Oak Ridge, TN, 1990.

[43] T. H. Dunigan. Performance of the Intel iPSC/860 and NCUBE 6400 Hypercube. Technical Report ORNL/TM-11790, Oak Ridge National Laboratory, Oak Ridge, TN, 1991.

[44] T. H. Dunigan. Communication performances of the Intel Touchstone DELTA Mesh. Technical Report ORNL/TM-11983, Oak Ridge National Laboratory, Oak Ridge, TN, 1992.

[45] J. O. Eklundh. A fast computer method for matrix transposition. *IEEE Transactions on Computers*, 21:801–803, 1972.

[46] G. Fabbretti, A. Farina, D. Laforenza, and F. Vinelli. Mapping the synthetic aperture radar signal processor on a distributed-memory MIMD architecture. *Parallel Computing*, 22:761–784, 1996.

[47] W. M. Gentleman and G. Sande. Fast Fourier transforms – for fun and profit. In *1966 Fall Joint Computer Conf., AFIPS Proc.*, Vol. 29, pp. 563–578. Washington, D.C., Spartan, 1966.

[48] S. Goedecker. Fast radix 2, 3, 4, and 5 kernels for fast Fourier transformations on computers with overlapping multiply-add instructions. *SIAM J. Sci. Comput.*, 18:1605–1611, 1997.

[49] G. H. Golub and C. F. Van Loan. *Matrix Computations*. The Johns Hopkins University Press, Baltimore, MD, 1989.

[50] I. J. Good. The interaction algorithm and practical Fourier analysis. *J. Roy. Statist. Soc.*, Ser. B, 20:361–372, 1958. Addendum, 22:372–375, 1960.

[51] A. Gupta and V. Kumar. The scalability of FFT on parallel computers. *IEEE Transactions on Parallel and Distributed Systems*, 4:922–932, 1993.

[52] M. T. Heideman, D. H. Johnson, and C. S. Burrus. Gauss and the history of the FFT. *IEEE Transactions on Acoustics, Speech, and Signal Processing*, Magazine, Vol. 1, No. 4:14–21, October 1984.

[53] R. W. Hockney. A fast direct solution of Poisson's equation using Fourier analysis. *J. Assoc. Comput. Mach.*, 12:95–113, 1965.

[54] M. B. Allen III and E. L. Isaacson. *Numerical Analysis for Applied Science*. John Wiley & Sons, Inc., NY 10158-0012, 1998.

[55] J. JáJá. *An Introduction to Parallel Algorithms*. Addison-Wesley Publishing Co., Reading, MA, 1992.

[56] L. H. Jamieson, P. T. Mueller, Jr., and H. J. Siegel. FFT algorithms for SIMD parallel processing systems. *J. Parallel Distrib. Comput.*, 3:48–71, 1986.

[57] H. W. Johnson and C. S. Burrus. Large DFT modules: 11, 13, 17, 19 and 25. Technical Report 8105, Dept. of Electrical Engineering, Rice University, Houston, TX, 1981.

[58] H. W. Johnson and C. S. Burrus. An in-place, in-order radix-2 FFT. In *Proc. IEEE International Conference on Acoustics, Speech and Signal Processing*, pp. 28A.2.1–4, San Diego, CA, March 19–21, 1984.

[59] S. L. Johnsson and R. L. Krawitz. Cooley-Tukey FFT on the Connection machine. *Parallel Computing*, 18:1201–1221, 1992.

[60] A. H. Karp. Bit reversal on uniprocessors. *SIAM Review*, 38:1–26, 1996.

[61] A. Kolawa and S. W. Otto. Performance of the Mark II and Intel Hypercubes. In M. Heath, Ed., *Hypercube Multiprocessors 1986*, pp. 272–275. SIAM, 1986.

[62] D. P. Kolba and T. W. Parks. A prime factor algorithm using high-speed convolution. *IEEE Trans. Acoust. Speech Signal Process*, ASSP-25:281–294, 1977.

[63] J. D. Lipson. The fast Fourier transform: Its role as an algebraic algorithm. In *Proc. ACM Annual Conference*, pp. 436–441. ACM, 1976.

[64] C. F. Van Loan. *Computational Frameworks for the Fast Fourier Transform*. The Society for Industrial and Applied Mathematics, Philadelphia, PA, 1992.

[65] R. M. Mersereau and T. C. Speake. A unified treatment of Cooley-Tukey algorithms for the evaluation of the multidimensional DFT. *IEEE Transactions on Acoustics, Speech, and Signal Processing*, 22(5):320–325, 1981.

[66] Z. J. Mou and P. Duhamel. In-place butterfly-style FFT of 2D real sequences. *IEEE Transactions on Acoustics, Speech, and Signal Processing*, 36(10):1642–1650, October 1988.

[67] J. G. Nagy. *Toeplitz least squares computations*. Ph.D. thesis, North Carolina State University, Raleigh, NC, 1991.

[68] R. Neapolitan and K. Naimipour. *Foundations of Algorithms*. D. C. Heath and Company, Lexington, MA, 1996.

[69] H. J. Nussbaumer. Digital filtering using polynomial transforms. *Electronics Letters*, 13(13):386–387, June 1977.

[70] H. J. Nussbaumer. *Fast Fourier Transform and Convolution Algorithms*. Springer Series in Information Sciences. Springer-Verlag, Berlin, Germany, 1981.

[71] D. P. O'leary and J. A. Simmons. A bidiagonalization-regularization procedure for large scale discretizations of ill-posed problems. *SIAM J. Sci. Stat. Comput.*, 2:474–489, 1981.

[72] A. V. Oppenheim and R. W. Schafer. *Digital Signal Processing*. Prentice-Hall, Englewood Cliffs, NJ, 1975.

[73] C. C. Paige and M. A. Saunders. Least squares estimation of discrete linear dynamic systems using orthogonal transformations. *SIAM J. Numer. Anal.*, 14:180–193, 1977.

[74] M. C. Pease. An adaptation of the fast Fourier transform for parallel processing. *J. Assoc. Comput. Mach.*, 15:253–264, 1968.

[75] M. C. Pease. The indirect binary n-cube microprocessor array. *IEEE Transactions on Computers*, c-26:458–473, 1977.

[76] M. Pickering. *Introduction to Fast Fourier Transform Methods for Partial Differential Equations with Applications.* Research Studies Press, Letchworth, U.K., 1986.

[77] W. H. Press, S. A. Teukolsky, W. T. Vetterling, and B. P. Flannery. *Numerical Recipes in C: The Art of Scientific Computing.* Cambridge University Press, Cambridge, U.K., 1992.

[78] K. R. Rao and P. Yip. *Discrete Cosine Transform: Algorithms, Advantages, Applications.* Academic Press, Inc., NY, 1990.

[79] G. E. Rivard. Algorithm for direct fast Fourier transform of bivariant functions. In *1975 Annual Meeting of the Optical Society of America,* Boston, MA, October 1975.

[80] G. E. Rivard. Direct fast Fourier transform of bivariant functions. *IEEE Transactions on Acoustics, Speech, and Signal Processing,* 25(3):250–252, June 1977.

[81] J. H. Rothweiler. Implementation of the in-order prime factor transform for variable sizes. *IEEE Transactions on Acoustics, Speech, and Signal Processing,* ASSP-30:105–107, 1982.

[82] R. Saatcilar, S. Ergintav, and N. Canitez. The use of the Hartley transform in geophysical applications. *Geophysics,* 55(11):1488–1495, November 1990.

[83] R. C. Singleton. A method for computing the fast Fourier transform with auxiliary memory and limited high-speed storage. *IEEE Transactions on Audio and Electroacoustics,* 15:91–98, 1967.

[84] R. C. Singleton. An algorithm for computing the mixed radix Fourier transform. *IEEE Transactions on Audio and Electroacoustics,* AU-17:93–103, 1969.

[85] H. V. Sorensen, C. S. Burrus, and M. T. Heideman. *Fast Fourier Transform Database.* PWS Publishing Co., Boston, MA 02116-4324, 1995.

[86] H. V. Sorensen, M. T. Heideman, and C. S. Burrus. On computing the split-radix FFT. *IEEE Transactions on Acoustics, Speech, and Signal Processing,* ASSP-34:152–156, 1986.

[87] H. V. Sorensen, D. L. Jones, C. S. Burrus, and M. T. Heideman. On computing the discrete Hartley transform. *IEEE Transactions on Acoustics, Speech, and Signal Processing,* ASSP-33:1231–1238, 1985.

[88] H. V. Sorensen, D. L. Jones, M. T. Heideman, and C. S. Burrus. Real-valued fast Fourier transform algorithms. *IEEE Transactions on Acoustics, Speech, and Signal Processing,* ASSP-35:849–863, 1987.

[89] T. G. Stockham. High speed convolution and correlation. In *1966 Fall Joint Computer Conf., AFIPS Proc.,* Vol. 28, pp. 229–233. Washington, D.C., Spartan, 1966.

[90] H. S. Stone. Parallel processing with the perfect shuffle. *IEEE Transactions on Computers,* c-20:153–161, 1971.

[91] G. Strang. The discrete cosine transform. *SIAM Review*, 41:135–147, 1999.

[92] Y. Suzuki, T. Sone, and K. Kido. A new FFT algorithm of radix-3, 6, and 12. *IEEE Transactions on Acoustics, Speech, and Signal Processing*, ASSP-34:380–383, 1986.

[93] P. N. Swarztrauber. The methods of cyclic reduction, Fourier analysis, and the FACR algorithm for the discrete solution of Poisson's equation on a rectangle. *SIAM Review*, 19:490–500, 1977.

[94] P. N. Swarztrauber. FFT algorithms for vector computers. *Parallel Computing*, 1:45–63, 1984.

[95] P. N. Swarztrauber. Multiprocessor FFTs. *Parallel Computing*, 5:197–210, 1987.

[96] P. N. Swarztrauber, R. A. Sweet, W. L. Briggs, V. E. Henson, and J. Otto. Bluestein's FFT for arbirary N on the hypercube. *Parallel Computing*, 17:607–617, 1991.

[97] R. A. Sweet. A cyclic reduction algorithm for solving block tridiagonal systems of arbitrary dimension. *SIAM J. Numer. Anal.*, 14:706–720, 1977.

[98] R. A. Sweet, W. L. Briggs, S. Oliveira, J. L. Porsche, and T. Turnbull. FFTs and three-dimensional Poisson solvers for hypercubes. *Parallel Computing*, 17:121–131, 1991.

[99] C. Temperton. Implementation of a self-sorting in-place prime factor FFT algorithm. *J. of Computational Physics*, 58:283–299, 1985.

[100] C. Temperton. Implementation of a prime factor FFT algorithm on Cray-1. *Parallel Computing*, 6:99–108, 1988.

[101] C. Temperton. Self-sorting in-place fast Fourier transforms. *SIAM J. Sci. Comput.*, 12:808–823, 1991.

[102] C. Temperton. A generalized prime factor FFT algorithm for any $n = 2^p 3^q 5^r$. *SIAM J. Sci. Comput.*, 13:676–686, 1992.

[103] T. J. Terrell. *Introduction to Digital Filters*. Macmillan Education Ltd., 2nd ed., 1988.

[104] C. Tong and P. N. Swarztrauber. Ordered fast Fourier transforms on a massively parallel hypercube multiprocessor. *J. Parallel Distrib. Comput.*, 12:50–59, 1991.

[105] M. Vetterli and P. Duhamel. Split-radix algorithms for length p^m DFTs. *IEEE Transactions on Acoustics, Speech, and Signal Processing*, ASSP-37:1(1):57–64, 1989.

[106] J. S. Walker. *Fast Fourier Transforms*. Studies in Advanced Mathematics. CRC Press, Boca Raton, FL, 1991.

[107] S. R. Walton. Fast Fourier transforms on the hypercube. Technical report, Ametek Computer Research Division, 610 N. Santa Anita Ave. Arcadia, CA 91006, September 1986.

[108] Z. Wang and B. Hunt. The discrete w-transform. *Appl. Math. Comput.*, 16:19–48, 1985.

[109] S. Winograd. On computing the DFT. *Math. Comp.*, 32:175–199, 1978.

[110] W. Yang. *Parallel Ordered FFT Algorithms on Distributed-Memory Multipro-cessors*. M.Sc. thesis, Department of Mathematics and Statistics, University of Guelph, Ontario, Canada, 1996.

Index